Molecular Electro-Optics

PART I

ELECTRO-OPTICS SERIES

Series Editor: Dr. Herbert Elion

Managing Director, Electro-Optics
Arthur D. Little, Inc.
Cambridge, Massachusetts

Volume 1.   Molecular Electro-Optics: Part 1—Theory and Methods; Part 2—
Applications to Biopolymers, *edited by Chester T. O'Konski*

Other volumes in preparation

Molecular Electro-Optics

PART I

Theory and Methods

Edited by Chester T. O'Konski

Department of Chemistry
University of California
Berkeley, California

MARCEL DEKKER, INC., New York and Basel

219314

MARCEL DEKKER, INC.

270 Madison Avenue, New York, New York 10016

LIBRARY OF CONGRESS CATALOG CARD NUMBER: 75-43047

ISBN: 0-8247-6395-5

Current printing (last digit):
10 9 8 7 6 5 4 3 2 1

PRINTED IN THE UNITED STATES OF AMERICA

PREFACE

The molecules of a fluid respond to an applied electric field by orientation or deformation, and this produces optical aniso-tropies on a macroscopic scale. The magnitudes and speeds of these responses are related to the molecular charge distribution, directional polarizabilities, and reorientational rates. Thus, electro-optic measurements provide a general and powerful approach for the determination of structural characteristics of molecules.

The Kerr electro-optic effect (electric double refraction) was discovered in 1875, but the characterization of macromolecules by electro-optic effects began only in the late 1940s with the introduction of pulsed electric fields and oscillographic photore-cording. This approach enabled direct measurement of relaxation times which relate to molecular rotational diffusion. Starting in 1950, many experimental contributions by this method have appeared. With the aid of interpretations based upon classical and quantum theory, these experiments have led to a better under-standing of the electric orientation phenomena. In this mono-graph we focus on biopolymers, but small molecules and colloidal dispersions are also discussed. Methods of measurement, the theories of various effects, interpretation of data in terms of macromolecular electric properties and size and shape, and a variety of applications are included.

The interdependence of experiment and theory often can be seen in reviewing the past two decades of progress in this field. An example is the use of high fields for orienting macromolecules to the saturation limit to facilitate analysis of the Kerr

constant. This development in our Berkeley laboratory was
accompanied by the extension of the electric birefringence theory
to high fields, and this is relevant to understanding the propagation of intense laser beams (see Chap. 12, "Nonlinear Electro-
optics"). The relaxation and saturation techniques have led to
new information about macromolecules, as discussed in Chap. 3 and
in Part 2. Highly monochromatic laser sources have facilitated
development of the electrophoretic doppler shift technique (Chap.
9). High intensity lasers will probably produce further extensions of technique and new discoveries.

The literature in this field is scattered in journals of
physics, chemistry, and biology of many lands. The fact that no
reference book encompassing both theory and applications of
electro-optic relaxation methods is available first prompted me to
consider a book on electro-optics in the early 1960s. In 1964
Marcel Dekker suggested that I write a book on the electric properties of macromolecules. I decided not to undertake that broad
a project but in 1970, after the stimulation of a professorship in
Uppsala which provided the incentive to organize some of the knowledge in this field, I proposed instead the present monograph.
Because there have been many new developments in the last decade,
it seemed best to arrange a collaboration. The authors of each
chapter were especially selected on the basis of their original
contributions and their command of the material in the respective
subtopics.

As can be seen from the chapter titles, the subject has been
divided according to the special interests of individually invited
contributors. In Part 1, "Theory and Methods," I introduce the
subject with a chapter outlining some history. The authors of the
remaining chapters present the theory of various effects, the
apparatus and methods for their measurement, and review illustrative applications. It was decided to exclude from this volume
crystal electro-optics in order to keep the monograph to a reasonable size. An exception occurs in Chap. 12, where some of the

nonlinear effects in crystals are discussed to present the theory
comprehensively. Part 2 consists of eight chapters which deal
primarily with the data obtained on various types of biological
systems, and with their interpretation. It is our hope that with
this foundation many workers not yet fully aware of the great
potential of electro-optic relaxation methods as an alternative or
supplement to ultracentrifugation, light scattering, transport
measurements, optical spectroscopy, and dielectric dispersion will
be persuaded to consider this relatively recent but powerful and
general approach to studies of the properties of molecules of all
kinds.

   Electro-optic researches in my laboratory have been pursued
by several students and visiting scholars, most of whom are among
the authors. The first Ph.D. student, Arthur Haltner, deserves
special mention not only for his interesting choice of subjects,
but also for excellence of experimental execution. I am grateful
for the interest of Sydney Fleming, who saw at an early date the
potential of the electric birefringence relaxation technique and
fostered early studies in industry.

   I am happy to acknowledge the interest of persons who
attended the small seminar and the small class at Berkeley in 1965
and 1973, respectively, and the kind interest and stimulating discussions of Prof. Stig Claesson at Uppsala University in the
spring of 1970. I thank my recent associates in electro-optics
research, John W. Jost, Michael C. Kwan, and Lloyd S. Shepard, for
helpful suggestions on a number of chapters in the book.

   My greatest appreciation goes to all the authors who have
joined to create this new treatise. Many of their names appear in
Chap. 1, in which the early (and sometimes also quite recent)
discoveries are outlined. Because of the way that their contributions have fitted together, the whole of the enterprise has become
greater than the sum of its outstanding parts.

                                                   Chester T. O'Konski

CONTRIBUTORS TO PART 1

Buckingham, A. D., Department of Theoretical Chemistry, University of Cambridge, Cambridge, England

Charney, Elliot, Laboratory of Chemical Physics, National Institute of Arthritis, Metabolism, and Digestive Diseases, National Institutes of Health, Bethesda, Maryland

Flygare, W. H., Department of Chemistry, University of Illinois, and Noyes Chemical Laboratory, Urbana, Illinois

Ha, Tae-Kyu, Laboratory of Physical Chemistry, Swiss Federal Institute of Technology, Zurich, Switzerland

Hartford, Steven L.,* School of Chemical Sciences, University of Illinois, Urbana, Illinois

Jennings, Barry R., Department of Physics, Brunel University, Uxbridge, Middlesex, England

Jernigan, Robert L.,** Physical Sciences Laboratory, Division of Computer Research and Technology, National Institutes of Health, Bethesda, Maryland

Kielich, Stanislaw, Nonlinear Optics Department, Institute of Physics, A. Mickiewicz University at Poznań, Poznań, Poland

Krause, Sonja, Department of Chemistry, Rensselaer Polytechnic Institute, Troy, New York

Liptay, Wolfgang, Institute of Physical Chemistry, University of Mainz, Mainz, Germany

---

*Current affiliation: Ames Research Center, Moffett Field, California.

**Current affiliation: Laboratory of Theoretical Biology, National Cancer Institute, National Institutes of Health, Bethesda, Maryland.

O'Konski, Chester T., Department of Chemistry, University of
    California at Berkeley, Berkeley, California

Paulson, Charles M., Jr., Engineering Physics Laboratory, E. I.
    du Pont de Nemours and Co., Inc., Wilmington, Delaware

Ridgeway, Don, Departments of Statistics and Physics, North
    Carolina State University, Raleigh, North Carolina

Thompson, Douglas S.,[*] Engineering Research and Development
    Division, Engineering Physics Laboratory, E. I. du Pont de
    Nemours and Co., Inc., Wilmington, Delaware

Tinoco, Ignacio, Jr., Department of Chemistry, University of
    California at Berkeley, Berkeley, California

Ware, B. R., Department of Chemistry, Harvard University,
    Cambridge, Massachusetts

[*] Current affiliation: Department of Chemistry, Hampden-Sydney College,
Hampden-Sydney, Virginia

CONTENTS

Cumulative author and subject indexes appear in Part 2.

CONTENTS OF PART 2

Applications to Biopolymers

Molecular Electro-Optics

PART I

Chapter 1

DISCOVERY AND EARLY
EXPLORATIONS OF ELECTRO-OPTIC EFFECTS

Chester T. O'Konski

Department of Chemistry
University of California
Berkeley, California

I. PHENOMENOLOGICAL DESCRIPTION AND
EQUATIONS OF THE KERR ELECTRO-OPTIC EFFECT

Before describing Kerr's discovery of the first electro-optic
effect to be detected, let us consider the physical principles of
the phenomenon. Figure 1 illustrates the electric double refraction, or birefringence, and introduces conventional terminology.
A plane-polarized light beam enters a sample placed between plane-
parallel electrodes connected to an external voltage source. The
direction of polarization of the light beam is specified by the
direction of its electric vector, $\underline{E}_i$, which is normally oriented

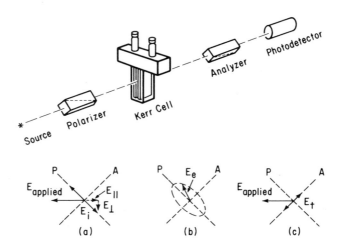

FIG. 1.   Apparatus for Kerr electro-optic effect (electric
birefringence) experiments. Below it are shown the states of
polarization of light: (a) after passing through the polarizer,
(b) after passing through the cell, and (c) after passing through
the analyzer. E_i is the electric vector of the plane-polarized
light incident upon the sample, and it can be decomposed into two
inphase components, $E_{\parallel}$ and $E_{\perp}$, parallel and perpendicular,
respectively, to the applied electric field. E_e is the vector of
the elliptically polarized light emerging from the birefringent
solution in the cell, and E_t is the component of E_e transmitted by the analyzer.

at 45° with respect to the direction of the applied electric field,
as illustrated in the view along the direction of the light beam in
Fig. 1a. The vector of the incident light, $\underline{E}_i$, may be decomposed
into two components, one parallel to the direction of the applied
electric field, designated $E_{\parallel}$, and the other, $E_{\perp}$, perpendicular
to the direction of the applied electric field. These two vectors
are in phase as the electromagnetic radiation enters the medium.
Under the influence of the external field the refractive indexes of
the medium change and become unequal for the parallel and the
perpendicular components; the sample behaves like a uniaxial crystal with its unique optic axis in the direction of the applied
field. The two component rays traverse the medium at different

velocities, and emerge out of phase, as illustrated in Fig. 1b. The resultant, E_e, produces a vector which describes an ellipse.

In the absence of an electric field, the light coming out of the Kerr cell is plane-polarized and is extinguished by the crossed analyzer, but in the presence of the field, the emergent light has an electric field component perpendicular to that of the incident polarized light, so that a portion of the light is transmitted by the analyzer, as shown by the vector E_t in Fig. 1c. The intensity of the transmitted light, relative to that of the incident light, is directly related to the degree of optical anisotropy produced by the applied electric field.

The optical pathlength of the cell, for the parallel component of the light, may be expressed by

$$N_{\parallel} = \frac{Ln_{\parallel}}{\lambda_0} ,$$

(1)

where N is the number of wavelengths of light in the cell of length L, n is the refractive index for the parallel light component, and λ_0 is the wavelength of light in vacuum. A similar equation applies to the perpendicular component. It follows that the phase retardation, or the optical pathlength difference in radians, δ, is given by

$$\delta = 2\pi(N_{\parallel} - N_{\perp}) = \frac{2\pi L\ \Delta n}{\lambda_0}$$

(2)

Here $\Delta n = n_{\parallel} - n_{\perp}$, and it is called the double refraction or the birefringence of the medium. The value of δ, and thus Δn, can be obtained from measurements of the increase in the intensity of light produced by application of the electric field to the specimen between crossed polarizer and analyzer oriented at 45° with respect to the applied field (see Chap. 3).

In most substances at low electric field intensities, the birefringence at a specified wavelength varies with the square of the electric field, i.e.,

$$\Delta n = n_{\parallel} - n_{\perp} = KnE^2 \tag{3}$$

where E is the applied electric field intensity, and n is the
mean refractive index. This is known as the Kerr law, and K is
called the Kerr constant.

The specific Kerr constant of a substance in solution or suspension is a measure of the electric and optical properties of the
molecular kinetic units; it is given by

$$K_{sp} = \frac{K}{C_v} \tag{4}$$

where C_v is the volume fraction of the solute. In using Eq. (3)
for a solute, the contribution of other components to Δn is
first subtracted out.

II.   KERR'S WORK

The discovery of the first electro-optic effect was reported
in 1875 by the Reverend John Kerr, Mathematical Lecturer in the
Free-Church Training College of Glasgow, in a treatise entitled
"A New Relation between Electricity and Light." He mentions
Maxwell's famous "Treatise on Electricity and Magnetism," first
published in 1873 [1], which marked the achievement of the
mathematical theory of electromagnetic radiation. But Kerr's
chief inspiration apparently was Faraday's persistent idea of, and
unsuccessful searches for, electro-optic phenomena. In the opening
paragraph of his paper [2], Kerr writes:

> The thought which led me to the following inquiry
> was briefly this: --that if a transparent and optically
> isotropic insulator were subjected properly to intense
> electrostatic force, it should act no longer as an iso-
> tropic body upon light sent through it. Faraday was
> often occupied with expectations of this kind; and he
> has mentioned in his memoir on the Magnetization of
> Light, and elsewhere in his "Researches," how he
> experimented in this very direction, upon electrolytes

as well as dielectrics, at different times and in many
ways, but always without success. As far as I remember, I have not read or heard of an attempt in this
field by any other naturalist. I proceed to offer a
few notes of some recent experiments of my own. The
investigation is not so complete as I should wish it
to be; but it has been carried forward as far as my
limited time and means would allow. At present I confine myself to solid dielectrics, reserving the case
of liquids for a second paper.

Kerr then goes on to describe his experiments on plate glass
"electrified" with the aid of a Ruhmkorff induction apparatus
which gave a spark of 20 to 25 cm in air. He observed the glass
plate between two crossed Nicol prisms. He noticed that when the
electric field was turned on, with its direction 45° from that of
the electric vector of the polarized beam, light began to appear
through the glass in about 2 sec, it brightened continuously for
about 30 sec, and then disappeared gradually to zero after the
electric field was removed. Several experiments were described.
Finding variables which affected the phenomenon was of interest in
seeking to understand its origins. Observations were reported on
clear amber resin and on quartz as well as on glass. He found that
the "electrified" quartz and glass acted upon transmitted light as
if they were compressed along the electric lines of force (positive
birefringence), while the electrified resin acted as if it were
stretched along the lines of force (negative birefringence). Previous experiments on strain birefringence played an important part
in the interpretation of these qualitative studies.

It is interesting to read Kerr's application of Faraday's
views as to the constitution and function of dielectrics in his
theoretical interpretation of the results:

> When the induction terminals were charged, the
> particles of the dielectric throughout the field are
> electrically polarized, and tend accordingly to arrange
> themselves end to end, and to cohere in files along the
> lines of force, just as iron filings do in a magnetic
> field. As far as this tendency of the polarized

particles toward a file arrangement along the lines of
force takes effect, there is a new molecular structure
induced in the dielectric ... and ... we may assume
... that the effect of electric force is to superinduce
a uniaxial structure upon the primitive structure.

Kerr considered some alternative possibilities. His last
paragraph in that first article is as follows:

I have made some experiments, and have had a good
many reflections, bearing on other explanations (other
than the one mentioned above) of the phenomena; and I
think it not unlikely that strains due to the mutual
actions of intensely charged shells of the dielectric,
or strains due to the changes of temperature, may have
something to do with the facts. But in the meantime I
offer the preceding remarks as a sketch of what appears
to me to be the only probable theory.

In his second paper [3] Kerr reported electric birefringence
observations in various liquids, including carbon disulfide, benzene, paraffin oil, kerosene, turpentine (two samples of opposite
optical rotations), olive oil, and castor oil. He noticed that the
olive oil gave a negative birefringence, like the glass in his
earlier experiments, whereas the other liquids displayed a positive
birefringence, like resin. Optical compensators were already known
at this time, but Kerr did not have one; therefore, he improvised
one with a slip of glass held with the hands and subjected to tension which he knew produced positive birefringence, with the slow
optic axis along the direction of stretch. He observed
semiquantitatively that the birefringence varied many fold, being
strongest in carbon disulfide and weakest in paraffin oil and
kerosene. He also noted that the birefringence appeared and disappeared in all liquids essentially instantaneously with the
application of the field.

He reiterated the theory that the particles of dielectrified
bodies tend to arrange themselves in files along the lines of
force, and suggested that the lines of electric force are lines of
compression in one class of dielectrics and lines of extension in
another class. He also stated that "the facts ... in whatever way

interpreted ... give promise of some new insight into that
interesting subject, the molecular mechanism of electric action."
The last paragraph of his second paper is:

> I cannot conclude without expressing a hope,
> amounting almost to a belief, that the plate-cell
> charged with carbon disulfide will develop from the
> present crude beginning into a valuable physical in-
> strument, a very delicate optical electrometer.

Thus, although his molecular explanation of the effect was partly
erroneous, by postulating a "new molecular structure" and attribut-
ing the birefringence to alignment of the molecules into "files,"
in addition to an (implied) orientation phenomenon, his hope that
his cell would become a valuable instrument was indeed prophetic.

He also noticed and reported the effect of the electric field
upon the dust particles commonly seen in liquid samples. When the
particles in benzene and carbon disulfide were numerous enough,
they formed a chain between the electrodes. At the instant of discharge, the chains were sometimes seen to break up violently, and
when the particles were so few as not to give chains, they appeared
as a set of "sparkling points ... which dart hither and thither
through the central parts of the electric field." Thus, both
pearl-chain formation and the peculiar kinetic effects observed
with small particles in high electric fields in liquids were observed in his pioneering experiments.

In two papers published in 1879 [4,5] Kerr extended his work
by making observations on various liquids. With the aid of a
"Government Fund" he improved his apparatus and began quantitative
measurements of retardation by measuring the weights on the compensator plate. A Thomson electrometer was introduced to measure
the applied electrostatic potentials. His optical observations
were entirely visual, of course; the light source was a flickering
paraffin candle! (Edison developed the first practical electric
lamp in 1880.) Most liquids Kerr studied displayed a positive
birefringence, and most of those were aromatic substances. Oils of
vegetable and animal origin generally showed a negative

birefringence. Kerr also observed and reported in detail quite a
number of curious effects due to conductivity, spark discharges,
inhomogeneities, and electrokinetic phenomena. Apparently these
have never been followed up systematically and could be interesting
to look at even today with the better apparatus and advanced
theoretical understanding we now possess.

In 1880, unsatisfied with the quantitative nature of the previous measurements, Kerr set out to determine the relationship
between birefringence and the applied electric field intensity:
"In the leisure of my last summer holidays I resumed the inquiry
with better means, carried it forward for some weeks with great
care." He now had a Jamin compensator [6], an "admirable instrument," and used sunlight reflected from a bright cloud as well as
the light from a paraffin candle as sources. After a number of
measurements on carbon disulfide, he concluded that the optical
retardation was proportional to the square of the applied electric
field intensity. One can infer that he had made quantitative
observations on other liquids and was confident that this was a
general law, although data on other liquids were not reported
systematically.

In 1882 Kerr published two more papers on qualitative observations of electric birefringence, calling attention to the
interesting repeats and variations of his own experiments by
Röntgen [7], who had added new observations on water, sulfuric
ether, and glycerine. Kerr added more than a hundred new liquids
to the previous 27 liquids on which he had reported earlier.
Among the liquids measured were bromine, molten phosphorus and
sulfur, some hydrocarbons, molten paraffin, various alcohols and
carboxylic acids, allyl ethers, iodides, bromides, and fluorides
[8]. The second paper in that year [9] reported on various
mercaptans, sulfides, esters of various carboxylic acids, nitrates
and nitrites, acetone and some aldehydes, some nitriles and amines,
additional chlorides and bromides, and miscellaneous oils and
inorganic and organic compounds. These observations were purely

qualitative, although the relative strengths of the effects were
sometimes mentioned. He noted the resemblance of most liquids to
positive uniaxial crystals, and that they were more numerous than
the negatively refracting liquids. He remarked that the electro-
optic and chemical characters of compounds are closely connected
and that regular progressions sometimes taking the form even of a
change of sign, occur in homologous series. He did not think in
terms of the orientation of an optically anisotropic molecule in
the electric field; rather his explanations were in more macroscopic terms as follows:

> Electro-optic double refraction is brought about by
> a special state of the dielectric, the state of essen-
> tially directional strain, which is concomitant and a
> condition of the maintenance of electric stress. ...
> There is, first, the directional transmission of elec-
> tric force from limit to limit of the field; there is,
> secondly, the special dioptric action of the force-
> transmitting medium, an action precisely similar to that
> of glass directionally stressed by tensions or pressures.
> ... I insist upon the essentially directional character
> of electric strain, in opposition to a theory of electro-
> optic action which has been advanced by Prof. Quincke.

Other electro-optic papers of Kerr include "Experiments on a
Fundamental Question in Electro-Optics" and "Reduction of Relative
Retardations to Absolute," both published in 1894 [10,11]. Kerr,
who published his first electro-optic work at the age of 51, died
at 83 in 1907, shortly before a mathematical theory of the Kerr
effect based upon orientation of anisotropic molecules was published by Langevin [12,13].

III.   DISPUTATIONS AND VERIFICATION OF KERR

Apparently the first person to attempt to repeat Kerr's
experiments with plate glass was Gordon in 1876 [14]. He got a
negative result. In the same year, Lodge [15] mentioned in a
brief footnote that he repeated Kerr's experiments with glass,

"but not without failures sufficient to excite my admiration for
the skill and patience involved in the discovery." Mackenzie [16]
failed to reproduce Kerr's phenomenon and thought it was produced
by heat only. Röntgen [7] got negative results for glass too, and
he decided that "accidental influences" were at work. Quincke [17]
confirmed Kerr's results for glass, using flint glass, and also for
carbon disulfide. He believed that the birefringence was a
secondary effect arising from mechanical stress produced by the
field; thus, he disagreed with Kerr's interpretation.

In 1882, following Kerr's two papers that year, Brongersma
[18] also succeeded in repeating Kerr's experiments on glass, although he remarked about difficulties with strain double refraction
produced by drilling the glass. Others also had reported problems
with this in attempting to reproduce Kerr's work. Brongersma did
not accept Kerr's assertion that the optical effects were due to a
primary action of the electric field upon molecular alignment, re-
marking: "In my opinion, it is not sufficiently proved that these
phenomena cannot be of secondary order. The motions of the molecules on their arrangement in a limited portion of the plate
(cell), may have for their consequence a development of heat; and
this ... may be the cause of the observed double refraction. "

In 1880 Röntgen [7] confirmed the existence of the Kerr effect
in liquids. This was further substantiated by Brongersma [18], who
remarked:

> Kerr saw in these phenomena a confirmation of
> Faraday's theory respecting dielectrics. That great
> physicist already regarded it as probable that under the
> influence of electricity an isotropic body passes into
> the anisotropic state, so that it behaves like a double
> refracting crystal. He did not, however, succeed in
> confirming this by experiment.

Another "Kerr effect" is the Kerr magneto-optic effect, reported in 1877 [19]. In this work, Kerr measured the rotation of
a beam of plane-polarized light reflected from the surface of
magnetized iron. He had expected to find the plane of

polarization rotated in the process of reflection, and confirmed
that this is so. This discovery was apparently considered more
important than his earlier one by the physicists of that day.

In 1898, Kerr was given the Royal Medal of the Royal Society
for his magneto-optic and electro-optic discoveries and studies.
The citation for his contributions is reproduced in this volume,
Appendix A, Sec. II. His most complete scientific biography gives
an interesting account of his discoveries, emphasizing the
magneto-optic discovery, and it is reproduced in Appendix A, Sec.
III.

IV.   DEVELOPMENT OF THE THEORY OF THE KERR EFFECT

The first clear proposal that the Kerr effect may be due to
molecular orientation appears to be that of Larmor [19a], who
stated in 1897: "The double refraction induced in dielectrics in
a strong electric field is possibly mainly due to molecular
orientation, as also that arising from mechanical strain." A
theory of the Kerr effect based upon a classical harmonic oscillator model was developed by Voigt [20] shortly after the discovery
of the electron. He showed that in an electric field, which produces a quasi-elastic force, the oscillator will no longer be harmonic and that its frequency will be split into two characteristic
values. The two absorptions will be parallel and perpendicular to
the applied field and can produce an optical anisotropy in the
medium. This proposal was critically examined in 1910 by Langevin
[12,13] who was the first to set up equations for the energy of an
electrically anisotropic molecule as a function of orientation, and
to use the Boltzmann equation to calculate the probability of
orientation as a function of angle (orientation distribution function). He followed through to obtain an equation for the Kerr
constant, but he did not introduce a permanent dipole moment, and
treated only the case of axial symmetry. By comparing the theories

with experimental data for CS_2 he showed that the Voigt effect was much too small to explain the Kerr constant. Later, Landenburg and Kopfermann [21] were able to observe the Voigt effect in sodium vapor at low pressures. They detected both a shift of the absorption line and the double refraction of the vapor produced by an electric field.

The wavelength dependence of the Kerr constant was discussed by Havelock [22] on the basis of ideas first suggested by Larmor [19a]. Havelock proposed a deformable cavity idea for the Kerr effect and also considered the possibility of orientation of anisotropic molecules as giving rise to the Kerr effect.

Enderle [23] and Voigt [23a] extended Langevin's theory to the more general case, but did not include a permanent dipole moment or optical activity. Pockels [24] considered the theory of simultaneously applied electric and magnetic fields, as well as separately applied fields, and developed equations for suspensions. This study related to experimental observations by Cotton and Mouton [25]. In 1912, Debye interpreted the temperature dependence of the dielectric constant of polar substances by deriving the expression for the mean electric moment from Boltzmann statistics [26], and near the same time, Thomson [27] gave the same result. These treatments resemble the calculation first carried out by Langevin [28] to find the mean magnetic moment of gas molecules having a permanent magnetic moment. A theory of the Kerr effect which included permanent dipole moment was given by Born [29]. These developments were later reviewed by Szivessy [30] and by Briegleb and Wolf [31]. Mallemann [32] extended the theory to optically active molecules. Debye [33] developed a quantitative treatment of the theory of the Kerr effect for generally anisotropic dipolar molecules in a gas.

Brief early reviews of magneto- and electro-optics were given by Wood [33a]. A comprehensive review of electric and magnetic double refraction was presented by Beams in 1932 [34]. While the differences of interpretation among the early workers are not

discussed, a full account of experiments and theory is given.
This includes an outline of the Langevin-Born theory for gases, its
modification for dense fluids by Raman and Krishnan [35], and a
discussion of the relationship of the Kerr effect to light scattering and molecular optical anisotropy. There is a discussion of the
theory in Born's [36] book on optics; the experimental verification of the Kerr law for liquids and the temperature and wavelength dependences of the Kerr effect also are reviewed. The use
of Kerr cells was discussed by Beams in 1930 [36a].

Several investigators have treated the quantum theory of the
Kerr effect [36,37-40]. Van Vleck [41] outlined how the Kerr
effect can be calculated from perturbation theory. Serber [40]
discussed the theory of depolarization and the Kerr effect. In
this volume, the current status of theories of electric birefringence in gases and liquids is summarized in Chap. 2; the quantum
theory and calculation of relevant molecular properties are given
in Chap. 14, and the related quantum theory of molecular absorption in electric fields is discussed in Chap. 6.

V. APPLICATIONS TO MOLECULAR STRUCTURE STUDIES

The relationship between the Kerr constant and the extent of
the depolarization of scattered light, which depends upon the optical anisotropy of a molecule, was first recognized by Gans [42].
The theoretical relationships between molecular optical properties
and light scattering were further developed by Debye [33] and Debye
and Sack [43]. During the 1930s a great deal of work was done on
the theory of optical anisotropy and molecular structure by Stuart
and others, and this has been summarized in the comprehensive review articles of Stuart [44,45]. More recent reviews were given by
Partington [46] and by Le Fèvre and Le Fèvre [47]. In these reviews, experimental methods of measurement of the Kerr effect in
gases and insulating liquids are outlined, and many results and
their interpretations are presented. A concise summary, including

representative results on gases and liquids, is given in Chap. 2
of this volume.

VI.   OTHER ELECTRO-OPTIC EFFECTS

The magnetic analog of the Kerr effect, double refraction
produced by a magnetic field transverse to the direction of propagation of a light beam, also is primarily due to orientation of the
molecules. This is known as the Cotton-Mouton effect, discovered
in 1905 [48], and it is not a subject of this book. Another related phenomenon, the Maxwell effect, or double refraction produced by flow, was first observed in Canada balsam in 1874 [49], a
date so close to Kerr's discovery that one wonders whether or not
this finding, not mentioned by Kerr, was in fact an important
encouragement for Kerr to renew Faraday's quest (see Sec. II).

There are several optical effects besides birefringence which
are related to the orientation produced by an electric field.
Since optical absorptions of molecules are generally anisotropic,
an electric field which causes a preferred orientation of the
molecules will cause the absorption to be different for plane-
polarized light vibrating along the direction of the orienting
electric field, as compared with the absorption perpendicular to
the direction of the orienting field. This is known as linear
electric dichroism (or, simply, electric dichroism) and it was
observed in microcrystalline and colloidal suspensions in the
early days of electro-optic studies [25,50-52]. Björnstahl [52a]
described the effects of an electric field on the optical transmission of nematic liquid crystals (see Chap. 22, Part 2 of
this treatise).

For ordinary-sized molecules electric dichroism is a weak
effect, but in the last 10 to 15 years there have been advances in
experimental methods and in theory which make it important. When
it was shown that macromolecules can be oriented very completely

in strong electric fields (around 10^3 to 10^4 V/cm) as the electric birefringence method was extended into the saturation region [53], it became evident that measurements of dichroism as a function of wavelength would be useful to determine the orientation of chromophores in ordered macromolecular structures. Studies were begun here on dyes and on a dye-polyelectrolyte complex by Bergmann in 1959, and shortly thereafter by Paulson on the triiodide amylose complex. These are discussed in Chap. 7, where it is shown that electric dichroism has become an important method for macromolecules.

Changes of the absorption of polarized light may occur purely by orientation when an electric field is applied, or there may also be changes in intrinsic molecular absorption coefficients due to oscillator strength or frequency changes or both. The intrinsic molecular optical changes may be referred to as electrochromism. Very high fields ordinarily are required to produce a measurable electrochromism in dyes [54], but a much stronger effect was found in a dye-polyelectrolyte complex [55]. The latter was shown, through examination of the time and wavelength dependences, to consist of a combination of orientation-produced electric dichroism and of electrochromism. The electrochromism was strong and is believed to be due to configurational and spectral changes involving dye-dye interactions. It was named "electrometachromism" [55].

The absorption changes produced in electric fields have been extensively studied by Lippert [56], Czekalla and coworkers [57], Liptay and Czekalla [58], and Labhart [59]. They have been reviewed by Liptay [60] and are developed in Chap. 6.

Observations of the state of polarization of the fluorescence of light emitted from molecules in a fluid which are irradiated with a polarized source can give information about the rotational relaxation times relative to the fluorescence emission time, as outlined in Chap. 16. Czekalla and coworkers [57,60a] showed that an electric field produces an additional polarization. Weber [61] has calculated this effect, using a model of rotationally

diffusing rigid spheres, for weak electric fields. Further
developments may make this effect very interesting for chemical
and biological studies, particularly if the relaxation effects can
be observed. See Chap. 21 for references to fluorescence intensity
modulation by electric fields applied to "labeled" membranes.

Electric field-induced changes in light scattering were observed in the early days of electro-optic work [62-64]. Interest
in this effect has been heightened by the advent of modern light-
scattering studies of macromolecules [65-67a] and by increased
interest in large biopolymers and certain inorganic polyelectrolytes, some of which can be oriented quite completely. The electric light-scattering method, discussed in Chap. 8, provides some
structural parameters for macromolecules not available by other
methods; thus, in spite of sensitivity limitations, it is quite
useful for certain problems.

Electric fields produce significant changes of optical rotation for certain interesting macromolecules, e.g., helices [68,69],
and useful new molecular parameters can be obtained, as shown in
Chap. 10. A closely related phenomenon is electric circular
dichroism, which has been investigated recently [70].

When polarized intense optical radiation is passed through a
substance the optical electric field will tend to align the molecules when they are optically anisotropic. Buckingham [71] calculated the magnitude of this effect and suggested the use of intense light flashes to observe it. The same phenomenon contributes
to self-focusing of intense laser pulses [72,72a]. It has been
discussed recently in relation to stimulated scattering as well as
the laser field Kerr effect [73].

The Stark effect is a well-known electro-optic phenomenon. It
has been used extensively for modulation in microwave spectrosocpy
[74]. Recently, the effect of electric fields on high-resolution
spectra has been shown to be a useful modulation method in high-
resolution optical spectroscopy, and the subject is being
reinvestigated [75-79] in relation to both birefringence and

absorption changes. See Chaps. 2 and 6 for reviews.

A relatively new method for studying macromolecules in solution is Doppler broadening of scattered laser radiation. This reveals the dynamics of macromolecular motions [80]. When combined
with electrophoresis, a powerful method of electrophoretic analysis
results. This is one of the most recent developments in electro-
optics and is discussed in Chap. 9.

An electric field will change the Gibbs free energy of liquid
phases, and thus should affect the critical mixing temperature of
partially miscible liquids. Since a strong opalescence occurs near
the critical mixing temperature, there should be an observable
electro-optic effect. Debye and Kleboth [81] examined a binary
liquid system and found that a field of 45,000 V/cm produced a
change of 0.015°C in the critical mixing temperature. The electro-
optic effect (modulation of the opalescence by an electric field)
apparently has not been studied.

VII.   EARLY ELECTRO-OPTIC STUDIES OF MACROMOLECULES

A desire to apply optical techniques to the measurement of
biopolymers led me to consider the extension of light-scattering
methods in Berkeley, California in the fall of 1948. While adapting and testing available equipment offered for this purpose by
Bruno Zimm, using tobacco mosaic virus (TMV), it occurred to me
that it would be intriguing to measure the anisotropy of the
scattering of light as a function of its polarization as the particles are oriented by an electric field. Flow birefringence of
TMV had already been studied, and Zimm reminded me that observing
birefringence between crossed polarizers is a very sensitive technique for detecting optical anisotropy. Having dealt with the
problem of achieving maximum sensitivity at the threshold of
detectable light-scattering signals in the presence of the shot
noise of a photomultiplier [82], it was obvious to me that the Kerr

technique, in which one measures changes in the intensity of the
transmitted light beam, offers an important advantage over electric
field light scattering, in which one measures the relatively weak
scattered light. Thus, before any scattering measurements, I decided to switch to electric birefringence studies of biopolymers.
Zimm and I set up crude cells using platinum electrodes, and we
applied sinusoidal fields of varying frequency to TMV solutions.
Shortly thereafter, our attention was called by Wendell Stanley to
earlier Kerr effect studies on solutions of TMV by Max Lauffer
[83], who employed 60 Hz sinusoidal fields and a polarizing microscope for average retardation measurements. Later I learned of
the elegant early study of colloids by the Kerr effect in sinusoidal
fields at varying frequencies by Errera, Overbeek, and Sack [84],
and still later, of a paper by Kuhn and Kuhn [84a] on "dragging"
birefringence of macromolecules in an electric field. A stimulating review which clearly emphasized the poor state of understanding
of the Kerr effect in disperse systems had been given by Heller
[85]. Lauffer's work on TMV showed that the field of frequency
dependences of the Kerr effect can be very complex and thus difficult to separate out, when observing only average light-intensity
measurements in sinusoidal fields. Thus it became clear, after
exploratory measurements, that measuring steady-state responses to
electric fields would be important for definitive studies. However, macromolecule solutions are often aqueous ones, and biological macromolecules are usually polyelectrolytes, so an applied
electric field causes energy dissipation and undesired heating as
well as electrophoresis. To observe buildup and decay, and to minimize the extraneous effects, I introduced short orienting pulses,
produced electronically. To further reduce electrophoresis, they
were given polarities which alternated [86]. Synchronized single
sweeps were not generally available on oscilloscopes at that time,
but in a few years the single-pulse method was introduced [87].

Unknown to me during my initial studies, Benoit was pursuing a suggestion of Ch. Sadron (Sadron, private communication, 1972; Benoit, private communication, 1972) to extend the interpretation of flow birefringence studies of macromolecules by measuring birefringence in electric fields. In 1949 Benoit published a note on pulsed technique using a motor-driven interrupter [88]. In another note [89], he presented an equation for the birefringence and its time dependence. That note, his 1950 Ph.D. thesis, and my 1950 paper with Zimm, all representing work completed by 1949, included derivation of the equation for the birefringence relaxation time, i.e., $\tau = 1/6\Theta$, which relates the birefringence field-free relaxation time, τ, to the rotational diffusion constant, Θ, for an axially symmetric particle. Benoit studied deoxyribonucleic acid (DNA) and vanadium pentoxide sols. I studied TMV solutions. He derived equations for the buildup of the birefringence. Zimm and I used square-wave fields and looked for frequency dispersion to get clues as to the origin of the orienting torque [86]; later, with Haltner, I introduced [87] the reversing pulse technique. Polarization of the counterion atmosphere was proposed as an important orienting mechanism for polyelectrolytes, and a theory of the Kerr constant for conducting systems came later (see Chaps. 3 and 19). DNA was shown to have a very strong electro-optic effect [90]. These were exciting developments as the power of the pulsed electro-optic technique was brought to bear in the study of macromolecular structures in solution.

Also unknown to us in 1948 was the then recent work of Kaye and Devaney [91] on the observation of the Kerr effect relaxation in viscous liquids, using an electronic pulse generator and an oscilloscope. Theirs was apparently the first direct observation of electro-optic relaxation this way. No doubt, had their interest been in macromolecules, they would have shared the experimental findings, the puzzlements, and the theoretical developments which are unfolded in the later chapters of this book.

But to Kerr himself, and to several others -- e.g., Raman and
Sirkar [92] and Kitchin and Mueller [93] -- we must give the credit
for having observed indirectly much earlier, in viscous liquids,
the effect of electro-optic relaxation, which gives rise to anomalous dispersion of the Kerr effect.

Kerr's work on electric birefringence of glass had shown a
relaxation time of the order of seconds [2]. In 1900 Abraham and
Lemoine [94] observed that the disappearance of the electric birefringence of liquids occurs in less than 10^{-8} sec -- a remarkable
experimental achievement in that day. Peterlin and Stuart [95],
in their important monograph on flow and electric birefringence,
had already treated the sine-wave dispersion behavior of macromolecules in terms of rigid models for macromolecules having dipole
moments and anisotropic electric and optical polarizabilities.

A group of investigators in the USSR also began developing
electro-optic methods for studying macromolecules in the late
1940s. Tolstoi and Feofilov [96] discussed the possibility of
improving the signals from a Kerr cell, then started pulse experiments [97] and studied some colloids [98]. Vollkenshtein and
Byutner [99] treated the theory of optically active molecules in an
external field. Thus, in the late 1940s there were workers in
three countries beginning to exploit the Kerr electro-optic relaxation method for macromolecule studies with photoelectronic
instrumentation. The basic theory and various electro-optic methods and their applications to macromolecules are summarized in
Chap. 3 and several succeeding chapters of this volume.

In this introductory historical sketch I have outlined some of
the early discoveries in several key subtopics of electro-optics.
More complete discussions are given in the following chapters.

REFERENCES

1.   J. C. Maxwell, A Treatise on Electricity and Magnetism, Oxford (1873).

2.   J. Kerr, Phil. Mag., (4) 50, 337 (1875).

3.   J. Kerr, Phil. Mag., (4) 50, 416 (1875).

4.   J. Kerr, Phil. Mag., (5) 8, 85 (1879).

5.   J. Kerr, Phil. Mag., (5) 8, 229 (1879).

6.   J. Kerr, Phil. Mag., (5) 9, 157 (1880).

7.   W. C. Röntgen, Phil. Mag., (5) 10, 77 (1880).

8.   J. Kerr, Phil. Mag., (5) 13, 153 (1882).

9.   J. Kerr, Phil. Mag., (5) 13, 248 (1882).

10.   J. Kerr, Phil. Mag., (5) 37, 380 (1894).

11.   J. Kerr, Phil. Mag., (5) 38, 144 (1894).

12.   P. Langevin, Radium, 7, 249 (1910).

13.   P. Langevin, Compt. Rend., 151, 475 (1910).

14.   J. E. H. Gordon, Phil. Mag., (5) 2, 203 (1876).

15.   O. J. Lodge, Phil. Mag., (5) 2, 353 (1876).

16.   J. J. Mackenzie, Ann. Physik M. Chem., (2) 238, 356 (1877).

17.   G. Quincke, Phil. Mag., 10, 537 (1880); Ann. Physik, 10, 536 (1880).

18.   H. Brongersma, Phil. Mag., (5) 14, 127 (1882).

19.   J. Kerr, Phil. Mag., (5) 3, 321 (1877).

19a.   J. Larmor, Phil. Trans., A190, 232 (1897); Aether and Matter, 1, 351 (1900).

20.   W. Voigt, Ann. Physik, 4, 197 (1901).

21.   R. Ladenburg and H. Kopfermann, Ann. Physik, (4) 78, 659 (1925).

22.   T. H. Havelock, Proc. Roy. Soc. (London), A80, 28 (1907); Phys. Rev., 28, 136 (1909); Proc. Roy. Soc. (London), A84, 492 (1911).

23.   A. Enderle, Dissertation, Freiburg (1912).

23a.   W. Voigt, Nadir. Kgl. Ges. Wiss. Göttingen, 577-93 (1912); Chem. Abstr., 7, 3914 (1913).

24.   F. Pockels, Radium, 10, 152 (1913).

25.   A. Cotton and H. Mouton, J. Chim. Phys., 10, 692 (1912).

26.   P. Debye, Physik Z., 13, 97 (1912).

27.   J. J. Thomson, Phil. Mag., 28, 757 (1914).

28.   P. Langevin, J. Phys., (4) 4, 678 (1905); Ann. Chim. Phys.,
       (8) 5, 70 (1905).

29.   M. Born, Ann. Physik, 55, 177 (1918).

30.   G. Szivessy, Handb. Physik, 21, S724 (1929).

31.   G. Briegleb and K. L. Wolf, Fortschr. Chem. Pnysik Physik.
       Chem., 21, 1-58 (1931).

32.   R. de Mallemann, Ann. Physik 2, 21 (1924); Compt. Rend., 193,
       523 (1931).

33.   P. Debye, Handb. Radiol., 6, 597, 760 (1925); Polar Molecules,
       Dover, New York (1929).

33a.   R. W. Wood, Physical Optics, 3rd ed., Macmillan, New York
       (1934), Chap. 21 and 22.

34.   J. W. Beams, Rev. Mod. Phys., 4, 133 (1932).

35.   C. V. Raman and K. S. Krishnan, Proc. Roy. Soc. (London),
       A117, 1 (1927).

36.   M. Born, Optik, Springer, Berlin (1933).

36a.   J. W. Beams, Rev. Sci. Instr., 1, 780 (1930).

37.   R. de L. Kronig, Z. Physik, 45, 458 (1927); 47, 702 (1928).

38.   M. Born and P. Jordan, Elementare Quantenmechanik, Springer,
       Berlin (1930).

39.   Th. Neugebauer, Z. Physik, 73, 386 (1932); 80, 660 (1933); 86
       392 (1933).

40.   R. Serber, Phys. Rev., (2) 43, 1003 (1933).

41.   J. H. van Vleck, Theory of Electric and Magnetic
       Susceptibilities, Oxford University Press, London (1932),
       p. 366.

42.   R. Gans, Ann. Physik, (4) 63, 97 (1921).

43.   P. Debye and H. Sack, 6, (ii) 69, 179 (1934).

44.   H. A. Stuart, Hand- und Jahrbuch der chemischen Physik, Vol.
       10, Part 3 (A. Eucken and K. S. Wolf, eds.) (1939).

45.   H. A. Stuart, Die Struktur des freien Molekuls, Springer,
       Berlin (1952), Chap. 7, p. 415.

46.   J. R. Partington, An Advanced Treatise on Physical Chemistry,
       Vols. 4 and 5, Longmans, Green, London (1953 and 1954).

47.   C. G. Le Fevre and R. J. W. Le Fevre, Rev. Pure Appl. Chem.,
       25, 261 (1955).

48.   A. Cotton and H. Mouton, Compt. Rend., 141, 317, 349 (1905).

49. J. C. Maxwell, Proc. Roy. Soc. (London), 22, 46 (1874); Ann. Phys., 151, 151 (1874).

50. G. Meslin, Compt. Rend., 136, 888, 930 (1903); J. Phys., 7, 856 (1908).

51. J. Chaudier, Compt. Rend., 137, 248 (1903); Ann. Chim. Phys., 15, 67 (1908); Compt. Rend., 149, 202 (1909).

52. P. Drapier, Compt. Rend., 157, 1063 (1913).

52a. Y. Bjornstahl, Ann. Physik, 56, 161 (1918).

53. C. T. O'Konski, K. Yoshioka, and W. H. Orttung, J. Phys. Chem., 63, 1558 (1959).

54. J. Kumamoto, J. C. Powers, Jr., and W. R. Heller, J. Chem. Phys., 36, 2893 (1962).

55. C. T. O'Konski and K. Bergmann, J. Chem. Phys., 37, 1573 (1962).

56. E. Lippert, Z. Elektrochem., 61, 962 (1957).

57. J. Czekalla, Z. Elektrochem., 64, 1221 (1960); Chimia (Aarau), 15, 26 (1961); J. Czekalla, W. Liptay,. and J. O. Meyers, Ber. Bunsenges. Physik, Chem., 67, 465 (1963); J. Czekalla and G. Wick, Z. Elektrochem., 65, 727 (1961).

58. W. Liptay, Z. Naturforsch, 20a, 272 (1965); W. Liptay and J. Czekalla, Z. Naturforsch, 15a, 1072 (1960); Z. Elektrochem., 65, 721 (1961).

59. H. Labhart, Adv. Chem. Phys., 13, 179 (1967); H. Labhart, Chimia (Aarau), 15, 20 (1961); Helv. Chim. Acta, 44, 457 (1961); Experientia, 22, 65 (1966); H. Labhart and G. Wagniere, Helv. Chim. Acta, 46, 1314 (1963).

60. W. Liptay, Mod. Quantum Chem., 3, 45 (1965).

60a. J. Czekalla and K. O. Meyer, Z. Physik. Chem. (Frankfurt), 27, 185 (1961).

61. G. Weber, J. Chem. Phys., 43, 521 (1965).

62. H. Siedentopf, Wiss. Mikrosk. M., 29, 1 (1912).

63. H. R. Kruyt, Kolloid-Z., 19, 161 (1916).

64. H. Freundlich, Z. Elektrochem., 22, 27 (1916).

65. P. Debye, J. Appl. Phys., 15, 338 (1944).

66. B. Zimm, J. Chem. Phys. 16, 1099 (1948).

67. K. A. Stacey, Light Scattering in Physical Chemistry, Academic, New York (1956).

67a. M. B. Huglin, Light Scattering from Polymer Solutions, Academic, New York (1972).

68. I. Tinoco, Jr., J. Phys. Chem., 60, 1619 (1956).

69.   N. Go, J. Chem. Phys., 43, 1275 (1965); J. Phys. Soc. Japan, 23, 88 (1967).

70.   S. J. Hoffman and R. Ullman, J. Polymer Sci. C 31, 205 (1970).

71.   A. D. Buckingham, Proc. Phys. Soc. (London), B69, 344 (1956).

72.   S. Kielich, IEEE J. Quantum Electronics, QE-4, 744 (1968).

72a. N. Bloembergen, Am. J. Phys., 35, 989-1023 (1967).

73.   R. Y. Chiao and J. Godine, Phys. Rev., 185, 430 (1969).

74.   W. Gordy, W. V. Smith, and R. R. Trambarulo, Microwave Spectroscopy, Wiley, New York (1953); C. H. Townes and A. L. Schawlow, Microwave Spectroscopy, McGraw-Hill, New York (1955).

75.   D. A. Dows and A. D. Buckingham, J. Mol. Spectry., 12, 189 (1964).

76.   A. D. Buckingham and D. A. Ramsey, J. Chem. Phys., 42, 3721 (1965).

77.   R. D. Conrad and D. A. Dows, J. Mol. Spectry., 32, 276 (1969).

78.   D. A. Haner and D. A. Dows, J. Mol. Spectry., 34, 296 (1970).

79.   J. M. Brown, A. D. Buckingham, and D. A. Ramsay, Can. J. Phys., 49, 914 (1970).

80.   R. Pecora, Ann. Rev. Biophys. Bioeng., 1, 257 (1972).

81.   P. Debye and K. Kleboth, J. Chem. Phys., 42, 3155 (1965).

82.   F. T. Gucker, Jr., C. T. O'Konski, H. B. Pickard, and J. N. Pitts, Jr., J. Am. Chem. Soc., 69, 2422 (1947).

83.   M. A. Lauffer, J. Am. Chem. Soc., 61, 2412 (1939).

84.   J. Errera, J. Th. G. Overbeek, and H. Sack, J. Chim. Phys., 32, 681 (1935).

84a. W. Kuhn and H. Kuhn, Helv. Chim. Acta, 27, 493 (1944).

85.   W. Heller, Rev. Mod. Phys., 14, 390 (1942).

86.   C. T. O'Konski and B. Zimm, Science, 111, 113 (1950).

87.   C. T. O'Konski and A. J. Haltner, J. Am. Chem. Soc., 78, 3604 (1956).

88.   H. Benoit, Compt. Rend., 228, 1716 (1949).

89.   H. Benoit, Compt. Rend., 229, 30 (1949).

90.   H. Benoit, Ann. Phys., 6, 561 (1951).

91.   W. Kaye and R. Devaney, J. Appl. Phys., 18, 912 (1947).

92.   C. V. Raman and S. C. Sirkar, Nature, 121, 794 (1928).

93.   D. W. Kitchin and H. Mueller, Phys. Rev., 32, 979 (1928).

94.   H. Abraham and J. Lemoine, J. Phys., (3) $\underline{9}$, 262 (1900).

95.   A. Peterlin and H. A. Stuart, Hand- und Jahrbuch der
      chemischen Physik, Vol. 8, Part 1B (A. Euken and K. L. Wolf,
      eds), Akademische Verlagsges., Leipzig, pp. 1-115 (1943),
      (Reprint, University of Michigan Press, Ann Arbor, 1948.)

96.   N. A. Tolstoi and P. P. Feofilov, Dokl. Akad. Nauk SSSR, $\underline{60}$,
      219 (1948); N. A. Tolstoi, Dokl. Akad. Nauk SSSR, $\underline{59}$, 1563
      (1948).

97.   N. A. Tolstoi and P. P. Feofilov, Zh. Eksperim, i Teor, Fiz.,
      $\underline{19}$, 421 (1949).

98.   N. A. Tolstoi and P. P. Feofilov, Dokl. Akad. Nauk SSSR, $\underline{66}$,
      617 (1949).

99.   M. V. Volkenstein and E. K. Byutner, Zh. Eksperim, i Teor.
      Fiz., $\underline{21}$, 1132 (1951).

Chapter 2

ELECTRIC BIREFRINGENCE IN GASES AND LIQUIDS

A. D. Buckingham

Department of Theoretical Chemistry
University of Cambridge
Cambridge, England

I.   INTRODUCTION

The application of an electric field to a fluid induces an anisotropy in its physical properties. The effect on the refractive index was first seen in Glasgow in 1875 by the Reverend John Kerr who observed birefringence in a piece of glass and in other solids subjected to a field [1]. Soon afterward, the effect was found by Kerr in liquids [2] and shown to be proportional to the square of the electric field strength [3].

A fluid experiencing a uniform electric field $E = E_z$ behaves optically like a uniaxial crystal, with its unique axis in the z direction parallel to the field. The Kerr constant B (which differs from the Kerr constant K defined in Chap. 1) is

$$B = (n_z - n_x)\lambda_0^{-1}E^{-2}$$

where n_z and n_x are the refractive indexes for light whose electric vectors are polarized in the z and x directions, respectively, λ_0 is the vacuum wavelength of the light, and E is the field strength in the fluid. The Kerr "constant" B varies with the pressure, temperature, and composition of the fluid, and with the wavelength λ_0; it is independent of the field strength E, at least for fields that are not strong enough to cause nonlinear dielectric polarization. The effect has yielded valuable information about molecular polarizabilities, both linear and nonlinear, and has been reviewed on several occasions [4-9].

By virtue of the Kramers-Kronig relationship between the real and imaginary parts of the electric susceptibility [10], there must be a difference in the absorption coefficients $k_z - k_x$ associated with the birefringence $n_z - n_x$. Thus, we have electric dichroism [11] to match the Kerr effect. Unlike the absorption coefficient, the refractive index is not negligible in transparent regions of the spectrum; both undergo rapid changes with frequency in the vicinity of absorption bands. Kerr spectra, analogous to magnetic rotation spectra [12], have been observed recently for the 3390-Å band of gaseous formaldehyde and the 3821-Å band of propynal by recording photographically the radiation transmitted by crossed polarizers at 45° to a strong electric field in the sample between the polarizers [13]. The anomalous dispersion of the Kerr constant in the vicinity of some sharp absorption bands enhances B sufficiently to cause significant transmission of radiation at that frequency; the differential absorption induced by the field also contributes to Kerr spectra [13]. The spectra are

simpler than absorption spectra (see Fig. 1) and the prominent
lines are those which show large Stark splittings. Kerr spectra
can also be recorded photoelectrically with a technique [13] which
yields the sign as well as the magnitude of B and which could be
applied to the broadbands of liquids and solutions.

If a beam of linearly polarized light propagates in a principal direction in an anisotropic medium, it becomes elliptically
polarized, and it is also rotated if there is absorption. Figure
2a illustrates the arrangement in a typical Kerr experiment. The
electric vector of a wave of angular frequency $\omega = 2\pi c/\lambda_0$,
propagating along the y axis and initially linearly polarized,
can be resolved into its z and x components:

$$E_z(\omega,t) = E_z^{(0)} \exp\left[\frac{-k_z\omega y}{c}\right]\cos\left[\omega t - \frac{n_z\omega y}{c}\right]$$

$$E_x(\omega,t) = E_x^{(0)} \exp\left[\frac{-k_x\omega y}{c}\right]\cos\left[\omega t - \frac{n_x\omega y}{c}\right]$$

(2)

where n_z, n_x and k_z, k_x are the refractive indexes and absorption coefficients for radiation whose electric vector is polarized
in the z and x directions, respectively. If the polarization
is at 45° to the z and x directions, as in Fig. 2b, the
amplitudes $E_z^{(0)} = E_x^{(0)} = E^{(0)}/\sqrt{2}$. In a birefringent medium,
$n_z \neq n_x$ and the wave is elliptically polarized with its electric
vector rotating about the direction of propagation, as in Fig. 2c.
In a dichroic medium, $k_z \neq k_x$ and the wave is anisotropically
absorbed, producing an optical rotation ϕ, as in Fig. 2d. The
effect of birefringence and dichroism together is to produce two
out-of-phase components of different amplitudes, as in Fig. 2e,
yielding elliptically polarized light whose major axis is rotated
from the original 45° position. Electric dichroism is considered
further in Chaps. 6 and 7.

It is convenient to introduce the phase difference δ between
the electric vectors E_z and E_x:

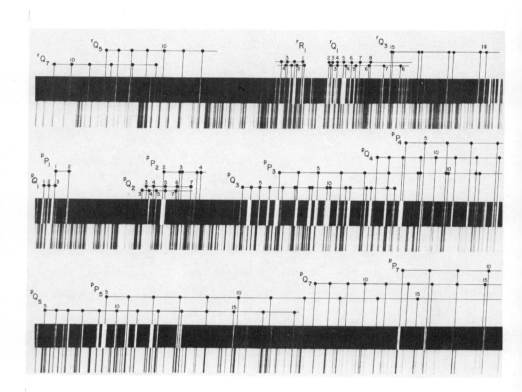

FIG. 1.   The Kerr spectrum (top) and the absorption spectrum (bottom) of the 339-nm band of formaldehyde vapor.   The notation is as follows: the superscripts p, q, r denote $\Delta|K| = -1$, 0, 1; P, Q, R denote $\Delta J = -1$, 0, 1; the subscripts 1, 2, 3, ... indicate the $|K|$ value of the lower state of the symmetric rotor which approximately describes H_2CO if J is not large [asymmetry splittings occur for large J and may be seen in $^PQ_3(10)$ and $^PQ_3(18)$]; the numbers on the lines indicate the positions of lines in the absorption spectrum;   some Kerr lines are shifted from the unperturbed frequencies by absorption. (Reproduced from Ref. 13 by courtesy of the National Research Council of Canada.)

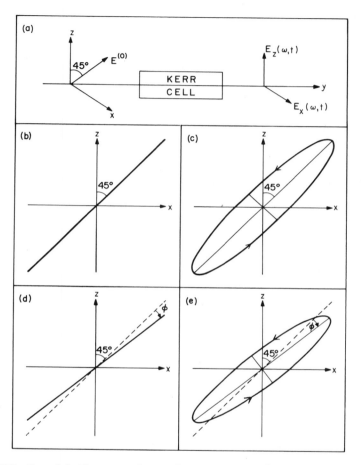

FIG. 2.   (a) The experimental arrangement in a Kerr experiment.
(b) Linearly polarized light propagating away from the viewer with
its electric vector at 45° or 135° to the x and z axes.
(c) Elliptically polarized light produced by a positive value of
$n_z - n_x$.   (d) Positive rotation of the plane of polarization produced by a positive value of $k_z - k_x$.   (e) Elliptically polarized
light, with its major axis rotated, produced by positive values of
$n_z - n_x$ and $k_z - k_x$.   (Reproduced from Ref. 13 by courtesy of the
National Research Council of Canada.)

$$\delta = \frac{2\pi(n_z - n_x)\ell}{\lambda_0} \tag{3}$$

where ℓ is the pathlength in the birefringent medium and δ is in radians. If complex refractive indexes are defined by

$$\hat{n}_z = n_z + ik_z$$
$$\hat{n}_x = n_x + ik_x \tag{4}$$

the complex Kerr constant is

$$\hat{B} = (\hat{n}_z - \hat{n}_x)\lambda_0^{-1}E^{-2} = \hat{\delta}(2\pi\ell)^{-1}E^{-2}$$
$$= B + iB' \tag{5}$$

where

$$B' = (k_z - k_x)\lambda_0^{-1}E^{-2} = \delta'(2\pi\ell)^{-1}E^{-2} \tag{6}$$

The dichroism δ' is related to the angle of rotation ϕ (see Fig. 2e) by [13]

$$\tan 2\phi = \frac{\sinh \delta'}{\cos \delta} \tag{7}$$

For small δ and δ',

$$\phi = \frac{1}{2}\delta' = \frac{\pi(k_z - k_x)\ell}{\lambda_0} \tag{8}$$

The sign convention for ϕ implicit in these equations is the opposite of that traditionally used in optical rotation studies [14]; however, it conforms to the natural choice that a positive rotation of the electric vector of the wave and the direction of propagation form a right-hand screw, and requires that a positive B' yield a positive rotation.

II.   THE MEASUREMENT OF ELECTRIC BIREFRINGENCE

The components involved in electro-optic birefringence measurements on gases and liquids are illustrated in Fig. 3.   The source may conveniently be a laser yielding continuous, parallel, monochromatic, polarized radiation, or it may be a thermal source coupled with a collimation system and a filter to select the appropriate wavelength.

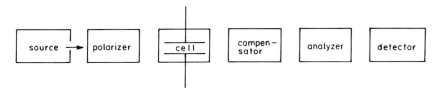

FIG. 3.   The basic components for measuring the birefringence induced in a fluid in the cell.

The polarizer should be of good quality and is normally required even when a "polarized" laser source is employed; Glan-Thompson polarizing prisms are particularly suitable for visible wavelengths, since they are very efficient and do not alter the path of the beam.

The polarizer should yield light linearly polarized at $45°$ to the uniform field   E   in the cell in a Kerr experiment.   In the nonuniform field $E'_{zz} = -E'_{xx}$ on the axis of a field-gradient or "four-wire" cell (see Sec. VII), the electric vector of the light wave initially makes equal angles with the   z   and   x   axes.

For gases the cell should be as long as possible, thereby increasing the induced phase difference and reducing the significance of "end effects" due to nonuniformities in the field at the extremities.   A length of about 1 m is practicable.

A multipass arrangement in which the light traverses the cell a number of times has obvious advantages, but the loss of polarization at each reflection must also be considered.   The cell should be rigid and made of noncorrosive materials.   Accurate temperature control of the cell is essential, and for gases a wide range of temperature should be available; low temperatures are particularly useful, since the induced birefringence of anisotropic molecules at constant density increases approximately as T^{-1} or T^{-2} (see Secs. V to VII).

The compensator is used to determine the birefringence induced in the light wave in its passage through the cell. Various types of compensator may be used and several have been discussed by Born and Wolf [15] and by Le Fèvre and Le Fèvre [8]. They may involve calibrated mechanical movement of one anisotropic solid over another, as in the Babinet-Soleil compensator, or the rotation of a wave plate as in the Szivessy-Dierkesmann and Brace compensators. The compensator may also comprise a reference fluid subjected to a field, and this is particularly appropriate when alternating voltages and "lock-in" detectors are employed. The reference compensator may be a Kerr cell containing nitrobenzene or carbon disulfide (the former cell tends to "age" through decomposition and conduction, and therefore requires frequent recalibration), or a quarter-wave plate and a Faraday cell comprising a liquid (such as water) or a glass rod on the axis of a solenoid. The quarter-wave plate converts the phase difference δ into a rotation $\frac{1}{2}\delta$, and this is compensated by an equal but opposite rotation induced in the Faraday cell by a magnetic field. It is usual to adjust the voltage on the standard Kerr cell or the current in the solenoid until a minimum amount of light is transmitted by the analyzer, which is crossed with respect to the polarizer. The detector in early work was normally the eye of the observer, but it is now usually a photomultiplier tube linked to the lock-in detector. An audiofrequency oscillator provides the potential for both the main cell and for the reference system, and the nulling voltage yields the birefringence of the fluid being studied.

Appropriate experimental arrangements are described in the review of Le Fèvre and Le Fèvre [8] and in the original literature [16]. If the fluid conducts to an appreciable extent, or if the relaxation properties are of interest, square-wave pulses of short duration are advantageous, and appropriate apparatus has been described by Benoit [17] and by O'Konski and Haltner [18] (see Chap. 3, IIIB).

Table 1 shows some representative values of the Kerr constants of simple gases at atmospheric pressure and at the frequency of the helium-neon laser red line at $\lambda_0 = 633$ nm; the numbers have been deduced from the results in papers cited in Table 4. Table 2 gives the Kerr constants of various simple liquids at room temperature and for yellow sodium light; the values were obtained from those given in the review article by Le Fèvre and Le Fèvre [8] and are of uncertain accuracy. The Kerr constants of polar liquids are found to depend critically on impurities, particularly those that increase the conductivity and, hence, reduce the effective field experienced by the liquid. It will be noticed that Kerr constants of gases and liquids cover an enormous range of values -- they are very small when the molecules are isotropically polarizable and large when they are polar and highly anisotropic. The values are negative when the polarizability in the direction of the dipole is less than the mean polarizability. An understanding of the actual values can be obtained by relating the Kerr constant to the molecular polarizability anisotropy and to the hyperpolarizabilities (see Sec. V).

III.   THE REFRACTIVE INDEX

For most purposes in optics, the effect of a medium on a light wave propagating in it can be described in terms of a complex refractive index $\hat{n} = n + ik$, where, as in Eq. (2), n determines the phase and k the intensity of the wave at any point. In the electric dipole approximation, $\hat{n}$ is related to the mean differential polarizability π, the extra moment induced in a molecule by the application of an additional weak unit electric field. We shall use a real field and moment to describe π. Suppose the

TABLE 1

Kerr Constants $B = (n_z - n_x)\lambda_0^{-1}E^{-2}$ of Gases at 1 atm
$(1.01325 \times 10^5 \text{ J m}^{-3})$ for Light of Wavelength $\lambda_0 = 633$ nm[a]

Gas	T(K)	$B(10^{-19} \text{ V}^{-2} \text{ m})$	$B(10^{-12} \text{ esu})$
He	296	0.024	0.022
Ne	296	0.047	0.042
Ar	296	0.53	0.48
Kr	296	1.3	1.1
Xe	296	3.5	3.2
CH_4	273.6	1.43	1.29
CF_4	288.8	0.70	0.63
SF_6	296	1.1	0.97
H_2	245.4	0.74	0.67
	332.4	0.48	0.43
N_2	248	3.63	3.26
	334	2.19	1.97
CO_2	252	30.8	27.8
	337	19.0	17.1
C_2H_6	255	5.13	4.61
	318	3.68	3.31
C_2H_4	262	16.7	15.0
	334	10.3	9.2
C_2H_2	254	26.1	23.5
	334	17.1	15.4
Cyclopropane	265	6.76	6.08
	318	5.30	4.77
CH_3F	250.8	107	96
	318.9	50.6	45.5
CH_2F_2	244.0	−17.5	−15.7
	303.2	−9.2	−8.3
CHF_3	245.5	−89.1	−80.2
	308.9	−40.8	−36.7

[a]The absolute accuracy of the B values is approximately
±5%.

TABLE 2

Kerr Constants $B = (n_z - n_x)\lambda_0^{-1}E^{-2}$
of Some Simple Liquids at 25°C for $\lambda_0 = 589$ nm[a]

Liquid	$B(10^{-16} V^{-2} m)$	$B(10^{-9} esu)$
Carbon tetrachloride	9.2 ± 0.1	8.3 ± 0.1
n-Hexane	5.2	4.7
Cyclohexane	6.0	5.4
n-Heptane	8.3	7.5
n-Octane	9.6	8.6
n-Decane	11.3	10.2
Carbon disulfide	339	314
Benzene	46.0 ± 1.0	41.0 ± 0.9
Toluene	86	77
Chlorobenzene	1,380	1,240
o-Dichlorobenzene	4,800	4,300
m-Dichlorobenzene	900	800
p-Dichlorobenzene	290	260
Nitrobenzene	44,000	40,000
Pyridine	2,270	2,040
Chloroform	-358	-322
Nitromethane	1,180	1,070
Paraldehyde	-2,600	-2,300

[a]Data from Ref. 8.

field, a vector, is

$$E_\alpha = E_\alpha^{(0)} \cos \omega(t - yc^{-1}) \qquad (9)$$

so that the time derivative of E_α is

$$\dot{E}_\alpha = -E_\alpha^{(0)} \omega \sin \omega(t - yc^{-1}) \qquad (10)$$

Then the dipole proportional to $E^{(0)}$ induced in a molecule in the field is

$$m_\alpha = \pi_{\alpha\beta}E_\beta + \pi'_{\alpha\beta}\,E_\beta\omega^{-1} \tag{11}$$

Thus, $\pi_{\alpha\beta}$ determines the α component of the dipole in phase with $\underline{E}$ induced in the molecule by the β component of $\underline{E}$; $\pi'_{\alpha\beta}$ determines the α component of the dipole out of phase with $\underline{E}$ induced by the β component of $\underline{\dot{E}}$. The energy absorbed by a sample which changes its dipole by $d\underline{m}$ in a field $\underline{E}$ is

$$dw = -d\underline{m} \cdot \underline{E} - \ldots \tag{12}$$

where the omitted terms involve the higher electric moments and the magnetic moments; the power absorbed is therefore

$$\dot{w} = -\underline{\dot{m}} \cdot \underline{E} - \ldots \tag{13}$$

From Eqs. (11) and (13) the mean power absorbed by a molecule, averaged over a full period of oscillation of the electromagnetic wave, in the electric dipole approximation, is

$$\overline{\dot{w}} = \frac{1}{2}\omega\pi'_{\alpha\beta}E_\alpha^{(0)}E_\beta^{(0)} \tag{14}$$

Thus, the symmetric part of $\underline{\pi}'$ is responsible for absorption, and $\underline{\pi}$ for refraction, of the wave.

The refractive indexes n_z and n_x for light with electric vector in the z and x directions is given in the electric dipole approximation by the Lorentz-Lorenz equation:

$$\frac{n_z^2 - 1}{n_z^2 + 2} = \frac{4\pi N}{3(4\pi\varepsilon_0)}\,\overline{\pi}_{zz}$$

$$\frac{n_x^2 - 1}{n_x^2 + 2} = \frac{4\pi N}{3(4\pi\varepsilon_0)}\,\overline{\pi}_{xx} \tag{15}$$

where ε_0 is the permittivity of free space ($4\pi\varepsilon_0 = 1.11265 \times 10^{-10}$ C V^{-1} m^{-1} = 1 esu), N is the number of molecules in unit volume, and $\overline{\pi}_{zz}$ and $\overline{\pi}_{xx}$ are mean polarizabilities averaged over the molecular translational and rotational motion and over internal vibrational and electronic states. In a liquid, it is usually sufficient to treat the translation and rotation classically, although, in a gas at low pressure, it is necessary to consider the discrete rotational states if the frequency is in or near an absorption band. In most cases, only molecules in the ground electronic state are present; some low-lying excited vibrational states may be populated, but their effect on the refractive index is generally small.

If the difference between n_z and n_x is small, as in electric birefringence experiments on fluids, Eqs. (15) yield

$$n_z - n_x = \frac{2\pi N}{9(4\pi\varepsilon_0)} \frac{(n^2 + 2)^2}{n} (\overline{\pi_{zz}} - \overline{\pi_{xx}}) \tag{16}$$

For a gas at low density, n_z and n_x are nearly equal to unity and

$$n_z - 1 = 2\pi N \overline{\pi_{zz}} (4\pi\varepsilon_0)^{-1}$$

$$n_x - 1 = 2\pi N \overline{\pi_{xx}} (4\pi\varepsilon_0)^{-1} \tag{17}$$

In the interpretation of optical activity, it is necessary to go beyond the electric dipole approximation and to consider the refractive index for left and right circularly polarized light [19, 20]. In the Faraday effect, optical activity is induced by a magnetic field B_y in the direction of propagation, and the rotation of the plane of polarization is proportional to $\overline{\pi'_{zx}} - \overline{\pi'_{xz}}$ which is linearly dependent on B_y [21].

IV. QUANTUM-MECHANICAL ASPECTS

Quantum-mechanical perturbation theory [22] leads to formulas for the polarizabilities $\underline{\pi}$ and $\underline{\pi}'$ of a molecule in the nth stationary state:

$$\pi_{\alpha\beta}^{(n)} = \hbar \sum_j \omega_{jn} f(\omega,\omega_{jn}) \, [<n|m_\alpha|j><j|m_\beta|n>$$

$$+ <n|m_\beta|j><j|m_\alpha|n>] = \pi_{\beta\alpha}^{(n)} \qquad (18)$$

$$\pi'^{(n)}_{\alpha\beta} = \hbar \sum_j \omega_{jn} g(\omega,\omega_{jn}) \, [<n|m_\alpha|j><j|m_\beta|n>$$

$$+ <n|m_\beta|j><j|m_\alpha|n>] = \pi'^{(n)}_{\beta\alpha} \qquad (19)$$

where the summation extends over all eigenstates j of the unperturbed system, including the continuum states, $\underline{m} = \sum_i e_i \underline{r}_i$ is the electric dipole moment operator, $\sum_i$ being a sum over all the charges e_i at positions $\underline{r}_i$ in the molecule; f and g are dispersion and absorption line-shape functions which are of the general form shown in Fig. 4. Forms for f and g, which are analytic and appropriate, are [21]

$$f(\omega,\omega_{jn}) = \frac{(\omega_{jn}^2 - \omega^2)}{[(\omega_{jn}^2 - \omega^2)^2 + \omega^2\Gamma_{jn}^2]} \qquad (20)$$

$$g(\omega,\omega_{jn}) = \frac{\omega\Gamma_{jn}}{[(\omega_{jn}^2 - \omega^2)^2 + \omega^2\Gamma_{jn}^2]} \qquad (21)$$

(a)   **(b)**

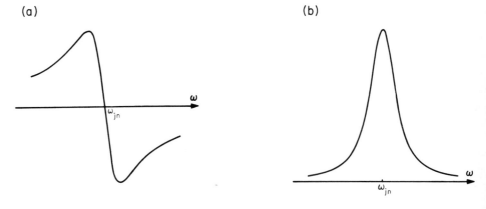

FIG. 4.   (a) A typical dispersion lineshape $f(\omega,\omega_{jn})$.
(b) A typical absorption lineshape $g(\omega,\omega_{jn})$. The dispersion function f is zero at $\omega = \omega_{jn}$ and g is approximately **symmetrical** about ω_{jn}.

where Γ_{jn} is approximately the full width at half-maximum height
of the $j \leftarrow n$ absorption line. In the region of the absorption
frequency ω_{jn}, the infinite summations in Eqs. (18) and (19) are
effectively replaced by single contributions -- a useful simplification. Far from an absorption band, g is zero and

$$f(\omega,\omega_{jn}) = (\omega_{jn}^2 - \omega^2)^{-1} \tag{22}$$

and for static fields $(\omega = 0)$, $g = 0$ and

$$f(0,\omega_{jn}) = \omega_{jn}^{-2} \tag{23}$$

The functions $|n>$ and $|j>$ may be eigenfunctions of the
full Hamiltonian of the rotating molecule, in which case, they are
of the form $|\tau_n J_n K_n M_n>$ and $|\tau_j J_j K_j M_j>$ where τ_n and τ_j
are the _internal_ electronic and vibrational states and the quantum
numbers J K M describe the overall rotation of the molecule, M
being the component of the total angular momentum in the direction
of the axis of quantization, which is conveniently chosen as the
direction z of the field in a Kerr effect experiment. In $\pi_{zz}^{(n)}$
the states n and j must have the same value of M, for
$\Delta M = 0$ in parallel polarization when the oscillating field is
parallel to z, and in $\pi_{xx}^{(n)}$ $\Delta M = M_j - M_n = \pm 1$ corresponding to
perpendicular polarization [23].

In a dense fluid, the rotational motion is normally treated
classically and the states $|n>$ and $|j>$ then refer only to the
internal degrees of freedom. The differential polarizability is
then referred to molecule-fixed axes. For example, in a molecule
with C_{2v} symmetry, there are three principal polarizabilities
$\pi_{11}^{(n)}, \pi_{22}^{(n)}$, and $\pi_{33}^{(n)}$, where 1, 2, and 3 are the orthogonal
principal axes of the molecule. In C_{2v} symmetry, the axes are
determined by symmetry, but, in general, this is not so and there
are six independent polarizabilities,

$$
\underline{\pi}^{(n)} = \begin{bmatrix}
\pi^{(n)}_{11} & \pi^{(n)}_{12} & \pi^{(n)}_{13} \\
\pi^{(n)}_{21} = \pi^{(n)}_{12} & \pi^{(n)}_{22} & \pi^{(n)}_{23} \\
\pi^{(n)}_{31} = \pi^{(n)}_{13} & \pi^{(n)}_{32} = \pi^{(n)}_{23} & \pi^{(n)}_{33}
\end{bmatrix}
$$

and **six** independent components of $\underline{\pi}'^{(n)}$. The symmetric tensor $\underline{\pi}^{(n)}$ is similar to the moment-of-inertia tensor and may be diagonalized by choosing principal axes 1, 2, 3; in that case, only three polarizabilities $\pi^{(n)}_{11}$, $\pi^{(n)}_{22}$, and $\pi^{(n)}_{33}$ need be specified, but six pieces of information are again needed in general, for the orientation of the principal axes in a molecule-fixed frame requires three angles for its specification. Thus, in a molecule like CHFCℓBr, whose point group is C_1, symmetry does not yield any information about the orientation of the principal axes in a frame determined by the C-X bonds. In a planar molecule such as LiOH, whose point group is C_s, the four polarizabilities π_{11}, π_{12}, π_{22}, π_{33} determine $\underline{\pi}$, where the 3-axis is at right angles to the plane, as do the three principal components together with the angle between one of the principal axes in the plane and the LiO bond.

The Kerr effect has its quantum-mechanical origin in the Stark shifts of the energy levels of molecules. For example, in the case of the sodium atom, there is a second-order Stark splitting of the excited $^2P_{3/2}$ states, the states with $M = \pm 3/2$ being higher in energy than those having $M = \pm 1/2$ by an amount proportional to $\Delta\alpha E_z^2$, where $\Delta\alpha$ is the anisotropy of the polarizability of the excited atom. Since the ground state is $^2S_{1/2}$ it is not split by the field E_z and the transitions contributing to $\overline{\pi_{zz}}$ ($\Delta M = 0$) occur at a lower frequency than three-quarters of those contributing to $\overline{\pi_{xx}}$ ($\Delta M = \pm 1$). Figure 5 illustrates the effect and the shape of the resultant Kerr dispersion and dichroism. Actually, the translational motion of the atoms, which causes the Doppler broadening, influences the line shape. There is a much weaker

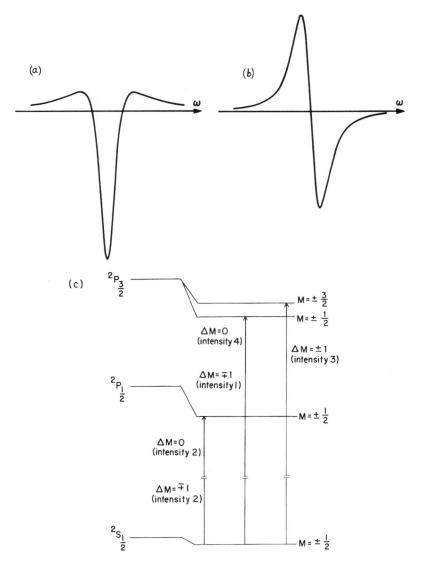

FIG. 5.   Kerr dispersion (a) and absorption (b) in the
vicinity of a transition exhibiting a second-order Stark splitting;
(c) energy-level diagram for the sodium atom; the shapes of (a) and
(b) near the $^2P_{3/2} \leftarrow {}^2S_{1/2}$ transition may be appreciated since
$\Delta M = 0$ transitions contribute to n_z and k_z, and $\Delta M = \pm 1$ to
n_x and k_x. The intensities shown apply in the absence of the
field; small changes are induced by the field.

Kerr effect near the transition frequency to the $^2P_{1/2}$ states;
the intensities of the parallel and perpendicular transitions are
affected differently by the field, producing a Kerr effect which
has been predicted to be approximately 1/2000 times the strength of
that near the $^2P_{3/2} \leftarrow {}^2S_{1/2}$ transition [24,25]. Early experimental work on the Kerr effect in sodium vapor was performed by
Kopfermann and Ladenburg [26], Bramley [27], and Gabler [28].

A first-order Stark effect, that is, a linear splitting of the
energy levels by E_z, can produce very large Kerr effects near
absorption frequencies. Figure 6 shows what could be seen near a

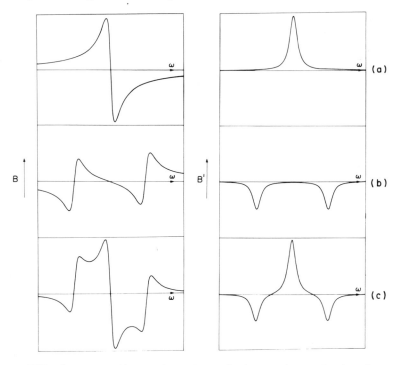

FIG. 6. Kerr dispersion B and absorption B' in the
vicinity of the $^pP_1(1)$ transition in a symmetric-top polar molecule. There is a first-order Stark splitting of the ground state
into three equally spaced lines having M = 1, 0, -1, and only a
second-order effect on the upper level. (a) Parallel polarization
($\Delta M = 0$ transitions from M = 0); (b) perpendicular polarization
with a negative sign ($\Delta M = \pm 1$ transitions from M = $\mp$1); (c) the
resultant (parallel minus perpendicular) birefringence and dichroism.

transition in which the ground state exhibits a large first-order
Stark splitting.

The energy levels may also be split by an electric field
gradient E', producing birefringence and dichroism proportional
to E'. Thus, in sodium vapor, the atoms in excited $^2P_{3/2}$ states
possess quadrupole moments which interact with E' to split the
level and produce a birefringence which is proportional to the
quadrupole moment of the atom in the excited $^2P_{3/2}$ state;
measurement of the birefringence or dichroism induced by E'
would therefore yield this interesting property [25].

In transparent regions the birefringence is much smaller than
near absorption frequencies, and there is no dichroism, but
measurements yield valuable information about the effect of the
electric field or field gradient on $\bar{\pi}$, and hence about the
polarizabilities and hyperpolarizabilities of the molecules in the
fluid. The remainder of this chapter is devoted to this aspect of
electric birefringence.

V. THE KERR EFFECT IN GASES AND LIQUIDS

The Kerr constant is proportional to $n_z - n_x$, and hence to
$\bar{\pi}_{zz} - \bar{\pi}_{xx}$, the difference in the molecular mean differential
polarizability parallel and perpendicular to the field E_z. In
fluids, this difference is an even function of E_z, since it must
be unaffected by a reversal of E_z. Normally, it is only the leading term in E_z^2 that is of interest.

The polarizability $\pi_{\alpha\beta}$ in a uniform field E may be written
as a power series in E:

$$\pi_{\alpha\beta} = \alpha_{\alpha\beta} + \beta_{\alpha\beta,\gamma} E_\gamma + \frac{1}{2}\gamma_{\alpha\beta,\gamma\delta}E_\gamma E_\delta + \cdots \qquad (25)$$

where $\alpha_{\alpha\beta}$ is the usual unperturbed polarizability tensor of a
molecule; the terms in the first and second hyperpolarizabilities
$\beta_{\alpha\beta,\gamma}$ and $\gamma_{\alpha\beta,\gamma\delta}$ describe the effect of the field on the

polarizability and give rise to <u>nonlinear optics</u> (see Chap. 12).
All tensors are symmetric in $\alpha\beta$ [see Eq. (18)] but since these
suffixes refer to time- or frequency-dependent quantities, and $\underline{E}$
is the external field, which is usually static or of very low frequency, the hyperpolarizabilities are not symmetric with respect to
interchange of γ or δ with α and β; $\gamma_{\alpha\beta,\gamma\delta}$ is symmetric in
$\gamma\delta$. From Eq. (25), with $\underline{E}$ in the z direction,

$$\pi_{zz} - \pi_{xx} = \alpha_{zz} - \alpha_{xx} + (\beta_{zz,z} - \beta_{xx,z})E_z$$

$$+ \frac{1}{2}(\gamma_{zz,zz} - \gamma_{xx,zz})E_z^2 + \ldots \quad (26)$$

At equilibrium at a temperature T

$$\overline{\pi_{zz}} - \overline{\pi_{xx}} = \frac{\displaystyle\sum_{\tau_n}\sum_{JKM} <\tau_n JKM | \pi_{zz}^{(n)} - \pi_{xx}^{(n)} | \tau_n JKM> \exp\left[-W_{\tau_n JKM} | kT\right]}{\displaystyle\sum_{\tau_n}\sum_{JKM} \exp\left[-W_{\tau_n JKM} | kT\right]} \quad (27)$$

where $W_{\tau_n JKM}$ is the energy of the quantum state $|\tau_n JKM>$ in the
field E_z; τ_n is a vibrational state of the nth electronic
state, and the states n and j in Eq. (18) for $\pi_{\alpha\beta}^{(n)}$ are
electronic states for fixed nuclear positions.

In many fluids at room temperature, only the ground electronic
state is populated and the rotational motion can be described
classically. In that case, both π and the potential energy u
are continuous functions of the orientation $\underline{\omega}$ of the molecule and

$$\overline{\pi_{zz}} - \overline{\pi_{xx}} = \frac{\int(\pi_{zz} - \pi_{xx}) \exp[-u/kT] d\underline{\omega}}{\int \exp[-u/kT] d\underline{\omega}} \quad (28)$$

where

$$u = u^{(0)} - m_\alpha^{(0)}E_\alpha - \frac{1}{2}\alpha_{\alpha\beta}^{(0)}E_\alpha E_\beta - \ldots \quad (29)$$

where $\underline{m}^{(0)}$ is the permanent dipole moment vector of the molecule,
and $\underline{\alpha}^{(0)}$ its static polarizability tensor. From Eq. (26), (28),

and (29), and the well-known results* for isotropic tensor averages
[29],

$$\overline{\pi_{zz}} - \overline{\pi_{xx}} = \frac{1}{30}E_z^2 \left[(3\gamma_{\alpha\beta,\alpha\beta} - \gamma_{\alpha\alpha,\beta\beta}) + \right.$$

$$+ \frac{2}{kT}(3\beta_{\alpha\beta,\alpha}m_\beta^{(0)} - \beta_{\alpha\alpha,\beta}m_\beta^{(0)})$$

$$+ \frac{1}{kT}(3\alpha_{\alpha\beta}\alpha_{\alpha\beta}^{(0)} - \alpha_{\alpha\alpha}\alpha_{\beta\beta}^{(0)}) +$$

$$\left. + \frac{1}{k^2T^2}(3\alpha_{\alpha\beta}m_\alpha^{(0)}m_\beta^{(0)} - \alpha_{\alpha\alpha}m_\beta^{(0)}m_\beta^{(0)}) \right]$$

$$(30)$$

If we introduce the anisotropy $\Delta\alpha_{\alpha\beta} = \alpha_{\alpha\beta} - \frac{1}{3}\alpha_{\gamma\gamma}\delta_{\alpha\beta}$, the
last two terms in Eq. (30) may be replaced by

$$\overline{\alpha_{zz}} - \overline{\alpha_{xx}} = \frac{E_z^2\Delta\alpha_{\alpha\beta}}{10kT} \left\{ \alpha_{\alpha\beta}^{(0)} + \frac{m_\alpha^{(0)}m_\beta^{(0)}}{kT} \right\}$$

The traceless symmetric tensor $\Delta\alpha_{\alpha\beta}$ is analogous to the field-
gradient tensor $-eq_{\alpha\beta}$ in nuclear quadrupole coupling.

Equation (30) is a very general result which is appropriate to
molecules of arbitrary symmetry; it simplifies in particular cases.

A. Molecules with No Permanent
Dipole Moment and with an Isotropic $\underline{\alpha}$

If $m^{(0)} = 0$ and $\alpha_{\alpha\beta} = \alpha\delta_{\alpha\beta}$, as for the inert gases and
tetrahedral and octahedral molecules (e.g., Ar, CH_4, SF_6)

---

*If $\underline{i}$, $\underline{j}$, $\underline{k}$ are unit vectors along the three orthogonal axes
of a Cartesian frame, then, for an isotropic average over all
orientations of the frame,

$$\overline{i_\alpha i_\beta i_\gamma i_\delta} = \overline{j_\alpha j_\beta j_\gamma j_\delta} = \overline{k_\alpha k_\beta k_\gamma k_\delta} = \frac{1}{15}\left\{ \delta_{\alpha\beta}\delta_{\gamma\delta} + \delta_{\alpha\gamma}\delta_{\beta\delta} + \delta_{\alpha\delta}\delta_{\beta\gamma} \right\}$$

$$\overline{i_\alpha i_\beta j_\gamma j_\delta} = \overline{i_\alpha i_\beta k_\gamma k_\delta} = \overline{i_\gamma i_\delta j_\alpha j_\beta} = \dots = \frac{1}{30}\left\{ 4\delta_{\alpha\beta}\delta_{\gamma\delta} - \delta_{\alpha\gamma}\delta_{\beta\delta} - \delta_{\alpha\delta}\delta_{\beta\gamma} \right\}$$

For example, $\overline{k_z^4} = \frac{1}{5}$, $\overline{k_z^2 k_x^2} = \overline{k_z^2 i_z^2} = \frac{1}{15}$, $\overline{k_z^2 i_x^2} = \frac{2}{15}$.

$$\overline{\pi}_{zz} - \overline{\pi}_{xx} = \frac{1}{30}E_z^2(3\gamma_{\alpha\beta,\alpha\beta} - \gamma_{\alpha\alpha,\beta\beta}) = \frac{1}{3}\gamma E_z^2 \qquad (31)$$

where the second equality defines γ. If dispersion is negligible, γ is a fully symmetric tensor and $\gamma = \frac{1}{5}\gamma_{\alpha\alpha\beta\beta}$, which, for a sphere such as an argon atom, reduces to $\gamma = \gamma_{1111} = 3\gamma_{1122}$ [22]. Thus measurements on $n_z - n_x$ lead through Eqs. (16) and (31) to accurate values of the second hyperpolarizabilities γ of isotropically polarizable molecules.

B. Axially Symmetric Molecules with a Center of Inversion

If a molecule has a center of inversion (e.g., H_2 and trans-glyoxal $\overset{O}{\underset{H}{\diagdown}}C - C\overset{H}{\underset{O}{\diagup}}$) it has no permanent dipole moment $\underline{m}^{(0)}$ or first hyperpolarizability $\underline{\beta}$ (for π in Eq. (25) and u in (29) must be unaffected by a reversal of $\underline{E}$, requiring all odd powers of $\underline{E}$ to be zero). If it is also axially symmetric (e.g., H_2 and CS_2)

$$\overline{\pi}_{zz} - \overline{\pi}_{xx} = E_z^2[\frac{1}{2}(\overline{\gamma_{zz,zz}} - \overline{\gamma_{xx,zz}} + \Delta\alpha\,\overline{P_2(\cos\theta)}] \qquad (32)$$

$$= E_z^2\left[\frac{1}{3}\gamma + \frac{\Delta\alpha\Delta\alpha^{(0)}}{15kT}\right] \qquad (33)$$

where θ is the angle between the molecular axis and E_z, and $P_2(\cos\theta) = \frac{1}{2}(3\cos^2\theta - 1)$. The importance of this equation is that it yields accurate values of the polarizability anisotropy $\Delta\alpha^{(0)} = \alpha_{\parallel}^{(0)} - \alpha_{\perp}^{(0)}$ (provided dispersion effects causing a difference between $\Delta\alpha$ and $\Delta\alpha^{(0)}$ are not large), and of γ when measurements of $n_z - n_x$ can be made on gases over a wide range of T [30-32]. Actually a 3-fold or higher axis of symmetry and a center of inversion, as in ethane and benzene, is sufficient for the applicability of Eq. (33).

C. Polar Molecules with an Axis of Symmetry

If a molecule is not centrosymmetric, but has an axis of symmetry (e.g., HF, HCN, OCS), there is one independent dipole, $m^{(0)} = m_3^{(0)}$, two polarizabilities, α_{33} and $\alpha_{11} = \alpha_{22}$ ($\Delta\alpha = \alpha_{33} - \alpha_{11}$), and three independent first hyperpolarizabilities, $\beta_{33,3}$, $\beta_{11,3}$ and $\beta_{13,1}$. Straightforward algebraic reduction leads to

$$\overline{\pi_{zz}} - \overline{\pi_{xx}} = \frac{1}{3}\gamma E_z^2 + \frac{2}{3}\beta\overline{P_1}E_z + \frac{3}{5}\Delta\beta\overline{P_3}E_z + \Delta\alpha\overline{P_2} \tag{34}$$

where $\beta = \frac{3}{5}(\beta_{33,3} - \beta_{11,3} + 3\beta_{13,1})$ and $\Delta\beta = (\beta_{33,3} - \beta_{11,3} - 2\beta_{13,1})$,

$$\overline{P_1} = \overline{\cos\theta} = \frac{m^{(0)}E_z}{3kT}$$

$$\overline{P_2} = \frac{1}{2}\left(3\ \overline{\cos^2\theta} - 1\right) = \frac{E_z^2}{15kT}\left(\Delta\alpha^{(0)} + \frac{(m^{(0)})^2}{kT}\right) \tag{35}$$

$$\overline{P_3} = \frac{1}{2}(5\ \overline{\cos^3\theta} - 3\ \overline{\cos\theta}) = \underline{0}(E_z^3)$$

Thus the term in $\Delta\beta$ does not contribute to the Kerr constant and may be dropped from Eq. (34). Hence

$$\overline{\pi_{zz}} - \overline{\pi_{xx}} = E_z^2\left[\frac{1}{3}\gamma + \frac{\Delta\alpha\Delta\alpha^{(0)}}{15kT} + \frac{2\beta m^{(0)}}{9kT} + \frac{\Delta\alpha(m^{(0)})^2}{15k^2T^2}\right] \tag{36}$$

In most polar molecules the final term is dominant so that a measurement of $n_z - n_x$ yields both the sign and magnitude of $\Delta\alpha$. In a typical case for small molecules

$$m^{(0)} \sim 10^{-18}\ \text{esu} = 1\ D = 3.336 \times 10^{-30}\ \text{C m}$$

$$\Delta\alpha \approx \Delta\alpha^{(0)} \sim \pm 10^{-24}\ \text{esu} = \pm 1.113 \times 10^{-40}\ \text{C}^2\ \text{m}^2\ \text{J}^{-1}$$

$$\beta \sim \pm 10^{-30}\ \text{esu} = \pm 0.3711 \times 10^{-50}\ \text{C}^3\ \text{m}^3\ \text{J}^{-2}$$

$$\gamma \sim 10^{-36}\ \text{esu} = 0.1238 \times 10^{-60}\ \text{C}^4\ \text{m}^4\ \text{J}^{-3}$$

and the ratios of the four contributions in Eq. (36) at room
temperature are 1:5:±15:±100. The plus and minus signs are introduced to emphasize that, while the contributions in γ and
$\Delta\alpha\Delta\alpha^{(0)}$ are positive, those in $\beta m^{(0)}$ and $\Delta\alpha(m^{(0)})^2$ may be
either positive or negative. Thus, in CH_3F, whose Kerr constant
is positive, the $\beta m^{(0)}$ term is negative; in CHF_3, whose Kerr constant is negative, $\Delta\alpha$ is negative and β is positive [33].

Accurate measurements of $n_z - n_x$ over a wide temperature
range in principle yield values for the terms in T^0, T^{-1}, and
T^{-2} in Eq. (30). In practice, the term in T^0 may be too small
to be determined in this way and it may be necessary to find γ
in other ways. One approximate approach is to use an additive
model in which $\gamma = \Sigma_{bonds}\gamma^{(bond)}$, just as mean polarizabilities α
may be approximated by adding bond polarizabilities [34]. This
model is appropriate to isotropic parts of tensors, including α
and γ; it may not be so reliable when applied to anisotropic
parts, like $\Delta\alpha$ and β.* Thus in studying the fluorinated
methanes, it was assumed that

$$\gamma_{CH_nF_{4-n}} = \frac{1}{4}\left[n\gamma_{CH_4} + (4 - n)\gamma_{CF_4}\right] \tag{37}$$

and this is helpful since γ_{CH_4} and γ_{CF_4} can easily be measured
since the molecules are tetrahedral [see Eq. (31)]. Alternatively,
if the molecules are rigidly held in a solid, γ may be measured
directly, for

$$\frac{d^2(\overline{\pi_{zz}} - \overline{\pi_{xx}})}{dE_z^2} = \frac{2}{3}\gamma \tag{38}$$

since the terms in T^{-1} and T^{-2} in Eq. (30) arise from
reorientation of the molecules in the field.

―――――――――

*Applequist et al [35] interpreted the polarizabilities of
molecules by ascribing an isotropic polarizability to each atom
and incorporating interaction of the induced atomic dipoles, as in
Silberstein's model [see Eq. (48)]. Agreement between experimental
and calculated mean polarizabilities is usually within 1 to 5%;
agreement for $\Delta\alpha$ is poorer, but is consistent with the view that
molecular anisotropy originates largely from atomic dipolar interactions.

VI. QUANTUM-MECHANICAL CORRECTIONS

The applicability of Eq. (28), and hence of (30), is dependent upon the assumption that the rotational motion is classical. While this may be reasonable in most fluids, it is not appropriate to light gaseous molecules at low densities. For them it is necessary to consider the effect of the field E_z on individual rotational states and then to average over all states with a Boltzmann weighting factor. For a rigid linear rotator in a Σ state (e.g., HCl) the averages in Eq. (35) must be replaced by [30]

$$\overline{\cos\theta} = \frac{m^{(0)}E_z}{3kT}\sigma^{-1}R_0^{-1} = \frac{m^{(0)}E_z}{3kT}\left(1 - \frac{1}{3}\sigma + \frac{2}{45}\sigma^2 + \ldots\right)$$

$$\frac{3}{2}\overline{\cos^2\theta} - \frac{1}{2} = \frac{E_z^2}{15kT}\left[\frac{1}{4}\Delta\alpha^{(0)}\left(1 + \frac{3S_0}{R_0}\right) + \frac{(m^{(0)})^2}{kT}\left(\frac{3}{2}\sigma^{-2} - \frac{1}{2}\sigma^{-1} - \frac{3}{2}\sigma^{-1}S_0\right)R_0^{-1}\right] \tag{39}$$

$$= \frac{E_z^2}{15kT}\left[\Delta\alpha^{(0)}\left(1 - \sigma_0 + \frac{8}{15}\sigma^2 + \ldots\right) + \frac{(m^{(0)})^2}{kT}\left(1 - \frac{5}{6}\sigma + \frac{31}{90}\sigma^2 + \ldots\right)\right]$$

where $\sigma = hcB_0/kT$ and B_0 is the rotational constant in wave numbers, R_0 is the rotational partition function,

$$R_0 = \sum_J (2J + 1)\,\exp\,[-(J^2 + J)\sigma] \tag{40}$$

$$S_0 = \sum_J \left[\frac{2J + 1}{(2J - 1)(2J + 3)} + \frac{4(2J + 1)}{(2J - 1)^2(2J + 3)^2\sigma}\right]\exp\,[-(J^2 + J)\sigma] \tag{41}$$

The rotational sums R_0 and S_0 can be evaluated in the form of asymptotic series [36], and the leading terms are given in equations (39). In the high-temperature limit $(\sigma \ll 1)$, Eqs. (39) reduce to the classical Eqs. (35), but significant corrections must be made for H_2 and other hydrides at room temperature (see Table 3). The correction for H_2 is particularly striking. The term in β survives only by virtue of the contribution of the ground rotational state $(J = 0)$; it is analogous to the quantum-mechanical expression for the dielectric polarization [37]. The

TABLE 3

Quantum Correction Factors at 300 K[a]

Molecule	H_2	D_2	N_2	Cl_2	HF	HCl
B_0 (cm^{-1})	59.34	29.91	2.00	0.243	20.55	10.44
$\frac{1}{4}(1 + 3S_0 R_0^{-1})$	0.755	0.867	0.990	0.999	0.906	0.951
$(\frac{3}{2}\sigma^{-2} - \frac{1}{2}\sigma^{-1} - \frac{3}{2}\sigma^{-1} S_0) R_0^{-1}$	--	--	--	--	0.921	0.959

[a]The polar factor is not shown for the homonuclear diatomics. B_0 is the rotational constant and $\sigma = hcB_0/kT$.

term proportional to $\Delta\alpha^{(0)}$ in $\frac{1}{2}(3\cos^2\theta - 1)$ in Eq. (39) is the same as that which appears in the formula for birefringence induced by an electric field gradient [38]; the effect of centrifugal distortion was also considered in that case [38], but the correction is minor, as it is in the Kerr effect.

VII. BIREFRINGENCE IN A FIELD GRADIENT

The symmetry of a fluid in a uniform field E_z requires the birefringence $n_z - n_x$ to be an _even_ function of E_z, and, therefore, to be proportional to E_z^2 under normal conditions. However, in a quadrupole cell, in which four wires parallel to the y axis are at $x = \pm a$, $z = 0$ and $x = 0$, $z = \pm a$, the induced birefringence on the axis must change sign on reversing the potential difference between adjacent wires, as it also does in a two-wire version (see Fig. 7). The birefringence is therefore _odd_ in E'. The field gradient must obey Laplace's equation $(E'_{\alpha\alpha} = E'_{xx} + E'_{yy} + E'_{zz} = 0)$ and is of the form

$$\underline{E}' = \begin{bmatrix} -E' & 0 & 0 \\ 0 & 0 & 0 \\ 0 & 0 & E' \end{bmatrix} \tag{42}$$

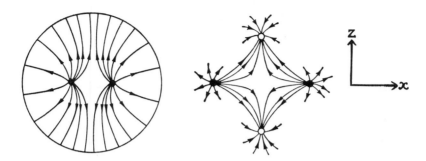

FIG. 7. The lines of force inside two-wire and four-wire cells for studying field-gradient-induced birefringence. The field gradient on the axis is of the form: $E'_{zz} = -E'_{xx}$, $E'_{yy} = 0$.

Equation (28) is still appropriate, but, in this case, the differential polarizability π must be written as a function of E' and (25) becomes

$$\pi_{\alpha\beta} = \alpha_{\alpha\beta} + \frac{1}{3}B_{\alpha\beta,\gamma\delta} E'_{\gamma\delta} + \cdots \tag{43}$$

where B is symmetric in $\alpha\beta$ and in $\gamma\beta$; in the static limit, B also describes the effect of a uniform field E on the molecular quadrupole moment Θ [39]:

$$\Theta_{\alpha\beta} = \Theta_{\alpha\beta}^{(0)} + A_{\alpha\beta,\gamma}E_\gamma + \frac{1}{2}B_{\gamma\delta,\alpha\beta}E_\gamma E_\delta + \cdots \tag{44}$$

The quadrupole moment is the expectation value of the traceless operator $\frac{1}{2} \Sigma_i e_i (3r_{i\alpha}r_{i\beta} - r_i^2\delta_{\alpha\beta})$ and is a measure of asymmetry in the second moment of the charge distribution. The potential energy analogous to Eq. (29) is

$$u = u^{(0)} - \frac{1}{3}\Theta_{\alpha\beta}^{(0)} E'_{\alpha\beta} - \cdots \tag{45}$$

and [39]

$$\overline{\pi_{zz}} - \overline{\pi_{xx}} = E'\left[b + \frac{2\alpha_{\alpha\beta}\Theta_{\alpha\beta}^{(0)}}{15kT}\right] = E'\left[b + \frac{2\Delta\alpha_{\alpha\beta}\Theta_{\alpha\beta}^{(0)}}{15kT}\right] \tag{46}$$

For molecules with a three fold or higher axis of symmetry,

$$\Theta^{(0)} = \Theta_{33}^{(0)} = -2\Theta_{11}^{(0)} = -2\Theta_{22}^{(0)} \quad \text{and}$$

$$\overline{\pi}_{zz} - \overline{\pi}_{xx} = E'\left(b + \frac{2\Delta\alpha\Theta^{(0)}}{15kT}\right) \tag{47}$$

The temperature-independent contribution b is equal to $\frac{2}{15}B_{\alpha\beta,\alpha\beta}$ together with terms related to the effect of a uniform field on the electric dipole induced by the field-gradient and by the magnetic field of the electromagnetic wave [40]; b is independent of origin, although $\underline{B}$ varies linearly with the position of the origin. What is the origin relative to which $\Theta^{(0)}$ is measured through Eqs. (16) and (47) in a dipolar species? In that case $\Theta^{(0)}$ varies linearly with the position of the origin $(\frac{d\Theta^{(0)}}{dR} = -2m^{(0)}$, where R is a distance along the axis of the molecule) so it is necessary to specify the appropriate origin, which is the point at which the anisotropy $\Delta\alpha$ in the polarizability may be considered to be located; this is the point at which the first moment of the polarizability anisotropy vanishes, i.e., at $\Sigma_i \Delta\alpha_i R_i = 0$ where the ith group in the molecule at R_i along the axis contributes an anisotropy $\Delta\alpha_i$ [34,35].

For a nondipolar molecule $m^{(0)} = 0$ and $\Theta^{(0)}$ is independent of the origin. If $\Delta\alpha$ is known, measurements of $n_z - n_x$ may readily yield reliable values for both the magnitude and sign of $\Theta^{(0)}$ [41]. For easily attainable fields and field gradients $(E_z \sim 10^6 \text{ V m}^{-1}, E_{zz}' \sim 10^9 \text{ V m}^{-2})\Theta E_{zz}'/\frac{1}{2}\Delta\alpha^{(0)}E_z^2 \sim 10^{-2}$, and the field-gradient-induced birefringence is generally smaller than that in the Kerr effect.

It is of interest that the temperature-independent distortion contribution b to the field-gradient-induced birefringence can be related to Faraday rotation constants; the relationship is rigorous for the inert gas atoms at very low optical frequencies [42].

VIII. ELECTRIC BIREFRINGENCE IN IMPERFECT GASES AND DENSE FLUIDS

Just as the Kerr effect in gases at very low densities yields the polarizabilities and hyperpolarizabilities of single molecules, so the effect in imperfect gases yields these properties of interacting pairs of molecules. Measurements are therefore a very sensitive probe of intermolecular forces and, like "dielectric virial coefficients" [43], are particularly sensitive to short-range angle-dependent potentials.

In the inert gases, $\Delta\alpha = 0$ for isolated atoms, but for pairs it varies with the internuclear separation R. In the Silberstein model [44]

$$\Delta\alpha = 6\alpha_0^2 R^{-3}(1 - 2\alpha_0 R^{-3})^{-1}(1 + \alpha_0 R^{-3})^{-1} \approx$$

$$\approx 6\alpha_0^2 R^{-3}(1 + \alpha_0 R^{-3} + 3\alpha_0^2 R^{-6} + \ldots) \qquad (48)$$

where α_0 is the polarizability of a free atom; only the longest range contribution to $\Delta\alpha$ is correct, since the interaction changes the intrinsic polarizability of the atoms by an amount which is proportional to R^{-6} at large separations [45]; overlap effects also influence $\Delta\alpha$ at short range [46]. Retention of the R^{-3} term alone in Eq. (48) gives reasonable agreement with observations on argon at 296 K and on CH_4 over a range of T, but leads to too large an effect in krypton and xenon and too small an effect in CF_4 and SF_6 [33,46]. The change in polarizability on molecular interaction also affects the polarization of scattered light [47, 48].

In dipolar gases there are several distinct contributions to the density dependence of the Kerr constant, and it is extremely difficult to find simple models of the polarizability and the potential that yield agreement with observation [33,49].

In dense fluids an expansion in powers of the density is inappropriate and little is known of the effects of molecular interactions on the Kerr effect. General statistical mechanical

TABLE 4

Polarizability Anisotropies, Hyperpolarizabilities, and
Quadrupole Moments (in esu)[a] Deduced from Electric Birefringence Measurements at 633 nm

Gas	$10^{24}\Delta\alpha$ (cm^3)	$10^{30}\beta$ (esu)	$10^{36}\gamma$ (esu)	$10^{26}\Theta(0)$ (esu)	$10^{38}b$ (esu)	Ref.
He	0	0	0.0217	0	-0.0140	60,42
Ne	0	0	0.051 ± 0.004	0	-0.0263	46,42
Ar	0	0	0.59 ± 0.04	0	-0.251	46,42
Kr	0	0	1.4 ± 0.1	0	-0.493	46,42
Xe	0	0	3.9 ± 0.3	0	-1.08	46,42
H_2	0.314 ± 0.005	0	0.28 ± 0.03	0.651	-0.21 ± 0.07	61,30,62,41
N_2	0.696 ± 0.006	0	0.7 ± 0.1	-1.4 ± 0.1	--	61,32,41
O_2	1.099 ± 0.007	0	--	-0.4 ± 0.1	--	61,41
CO_2	2.10 ± 0.01	0	4.5 ± 1.0	-4.3 ± 0.2	--	61,32,41
C_2H_2	1.86 ± 0.01	0	10.3 ± 2.2	--	--	61,32
C_2H_6	0.77 ± 0.006	0	1.9 ± 0.4	-0.8 ± 0.1	--	61,32,41
C_3H_6	-0.81 ± 0.01	--	4.0 ± 0.8	--	--	61,32

C_6H_6	-5.62 ± 0.05	0	6.5 ± 3	0	$--$	61,31
CS_2	9.6 ± 1	0	57 ± 5	0	$--$	31
CH_4	0	$--$	1.45 ± 0.07	0	-1.2 ± 0.3	33,41
CH_3F	0.20 ± 0.01	-0.5 ± 0.3	1.3 ± 0.1	$--$	$--$	33
CH_2F_2	-0.019 ± 0.002^b	-0.11 ± 0.03	1.1 ± 0.1	$--$	$--$	33
CHF_3	-0.243 ± 0.01	0.7 ± 0.2	0.9 ± 0.1	$--$	$--$	61,33
CF_4	0	$--$	0.75 ± 0.04	0	$--$	33
SF_6	0	0	1.2 ± 0.1	0	$--$	46

[a] The Values may be converted to SI units by using the following factors:
$\Delta\alpha$ 1 cm³ = 1.113×10^{-16} C² m² J⁻¹, β 1 esu = 0.3711×10^{-20} C³ m³ J⁻², γ 1 esu = 0.1238×10^{-24} C⁴ m⁴ J⁻³, $\Theta(0)$ 1 esu = 3.336×10^{-14} C m², [b] 1 esu = 0.3711×10^{-22} C³ m⁴ J⁻².

[b] $\Delta\alpha = 3/2 \, (\alpha_{33} - \alpha)$, where α_{33} is the polarizability in the direction of the dipole.

theories which examine the nature of the local field have been
developed [50-52], but these are neither entirely rigorous nor
simple; they show how the Kerr constant depends not only on the
pair distribution function, but also on the three-particle function
in dipolar fluids.

A simplified treatment of the effects of interaction in solutions and pure liquids on refractivity, dielectric polarization,
and the Kerr and Cotton-Mouton* effects has been described, in
which the effects of dipolar interactions are incorporated into
"correlation tensors" related to anisotropy in the distribution of
neighboring molecules. Solvent influences on the Kerr and Cotton-
Mouton effects may be large, even in nonpolar solvents [53].

Since the early days of dielectric studies, there has been a
search for a formula for the "effective" electric field (sometimes
called the internal field, or even the "infernal" field!) acting
on a molecule in a dense medium. The Lorentz [54] and Onsager [55]
fields have frequently been employed and may be very useful
approximations. However, it is now appreciated that no general
formula of this type exists and that the ratio of the effective
field to the actual field in the medium is different for quadratic
effects like the Kerr effect [which depend on $\overline{P_2(\cos \theta)}$] from its
value appropriate to dielectric polarization [which depends on
$\overline{P_1(\cos \theta)}$, which is linear in E_z -- see Eqs. (35)] [52,56,57].

Simple molecular models, such as an ellipsoid surrounded by a
continuum having the bulk properties of the medium [58,59], may be
useful in practice, particularly if the solute molecule is rigid
and large in relation to the solvent molecules; such a model is
employed in Chap. 3.

---

*
The Cotton-Mouton effect is the magnetic analog of the Kerr effect.

IX. SOME SELECTED RESULTS

The results of numerous Kerr effect measurements have been
recorded and discussed in review articles by Stuart [5] and
Le Fèvre and Le Fèvre [7,8]. In Table 4 we have assembled results
of some recent electric field and field-gradient birefringence
measurements on simple gases. The observations were made at room
temperature with a helium-neon laser as light source (λ_0 = 633.0
nm).

REFERENCES

1. J. Kerr, Phil. Mag., (4) 50, 337 (1875).

2. J. Kerr, Phil. Mag., (4) 50, 446 (1875).

3. J. Kerr, Phil. Mag., (5) 9, 157 (1880).

4. J. W. Beams, Rev. Mod. Phys., 4, 133 (1932).

5. H. A. Stuart, Die Physik der Hochpolymeren, Vol. 1, Springer,
    Berlin (1952), p. 415.

6. J. R. Partington, An Advanced Treatise on Physical Chemistry,
    Vol. 4, Longmans, Green, London (1953), p. 278.

7. C. G. Le Fèvre and R. J. W. Le Fèvre, Rev. Pure Appl. Chem.,
    5, 261 (1955).

8. C. G. Le Fèvre and R. J. W. Le Fèvre, Physical Methods of
    Chemistry, Part 3C, (A. Weissberger and B. W. Rossiter, eds.),
    Wiley-Interscience, New York (1972), p. 399.

9. A. D. Buckingham and B. J. Orr, Quart. Rev. (London), 21, 195
    (1967).

10. A. Abragam, The Principles of Nuclear Magnetism, Oxford
     University Press, London (1961), p. 93.

11. H. Labhart, Adv. Chem. Phys., 13, 179 (1967).

12. R. W. Wood, Physical Optics, 3rd ed., Macmillan, New York
     (1934), p. 729.

13. J. M. Brown, A. D. Buckingham, and D. A. Ramsay, Can. J.
     Phys., 49, 914 (1971).

14. M. Born and E. Wolf, Principles of Optics, 5th ed., Pergamon,
     New York (1975), p. 28.

15. M. Born and E. Wolf, Principles of Optics, 5th ed., Pergamon, New York (1975), Chap. 14.4.

16. A. D. Buckingham and R. L. Disch, Proc. Roy. Soc. (London), A273, 275 (1963).

17. H. Benoit, Ann. Phys. (Paris), 6, 561 (1951).

18. C. T. O'Konski and A. J. Haltner, J. Am. Chem. Soc., 78, 3604 (1956); S. Krause and C. T. O'Konski, J. Am. Chem. Soc., 81, 5082 (1959).

19. E. U. Condon, Rev. Mod. Phys., 9, 432 (1937).

20. I. Tinoco, Adv. Chem. Phys., 4, 113 (1962).

21. A. D. Buckingham and P. J. Stephens, Ann. Rev. Phys. Chem., 17, 399 (1966).

22. See, for example, A. D. Buckingham, Adv. Chem. Phys., 12, 107 (1967).

23. See, for example, C. H. Townes and A. L. Schawlow, Microwave Spectroscopy, McGraw-Hill, New York (1955), Table 4-4.

24. M. P. Bogaard, A. D. Buckingham, and B. J. Orr, Mol. Phys., 13, 533 (1967).

25. M. P. Bogaard and B. J. Orr, Mol. Phys., 14, 557 (1968).

26. H. Kopfermann and R. Ladenburg, Ann. Physik, (4) 78, 659 (1925); 79, 96 (1926).

27. A. Bramley, J. Franklin Inst., 205, 539 (1928).

28. F. Gabler, Phys. Z., 44, 108 (1943).

29. A. D. Buckingham and J. A. Pople, Proc. Phys. Soc. (London), 68, 905 (1955).

30. A. D. Buckingham and B. J. Orr, Proc. Roy. Soc. (London), A305, 259 (1968).

31. M. P. Bogaard, A. D. Buckingham, and G. L. D. Ritchie, Mol. Phys., 18, 575 (1970).

32. A. D. Buckingham, M. P. Bogaard, D. A. Dunmur, C. P. Hobbs, and B. J. Orr, Trans. Faraday Soc., 66, 1548 (1970).

33. A. D. Buckingham and B. J. Orr, Trans. Faraday Soc., 65, 673 (1969).

34. R. J. W. Le Fèvre, Adv. Phys. Org. Chem., 3, 1 (1965).

35. J. Applequist, J. R. Carl, and K.-K. Fung, J. Am. Chem. Soc., 94, 2952 (1972).

36. H. P. Mulholland, Proc. Cambridge Phil. Soc., 24, 280 (1928).

37. J. H. Van Vleck, The Theory of Electric and Magnetic Susceptibilities, Oxford University Press, London (1932), Chap. 7.

38. A. D. Buckingham and M. Pariseau, Trans. Faraday Soc., $\underline{62}$, 1 (1966).

39. A. D. Buckingham, J. Chem. Phys., $\underline{30}$, 1580 (1959).

40. A. D. Buckingham and H. C. Longuet-Higgins, Mol. Phys., $\underline{14}$, 63 (1968).

41. A. D. Buckingham, R. L. Disch, and D. A. Dunmur, J. Am. Chem. Soc., $\underline{90}$, 3104 (1968).

42. A. D. Buckingham and M. J. Jamieson, Mol. Phys., $\underline{22}$, 117 (1971).

43. H. Sutter, Specialist Periodical Reports, Dielectric and Related Phenomena, Vol. 1, Chemical Society, London (1972), p. 65.

44. L. Silberstein, Phil. Mag., (6) $\underline{33}$, 92, 521 (1917).

45. L. Jansen and P. Mazur, Physica, $\underline{21}$, 193 (1955).

46. A. D. Buckingham and D. A. Dunmur, Trans. Faraday Soc., $\underline{64}$, 1776 (1968).

47. M. Thibeau, B. Oksengorn, and B. Vodar, J. Phys. Radium, $\underline{29}$, 287 (1968).

48. W. M. Gelbart, Adv. Chem. Phys., $\underline{26}$, 1 (1974).

49. D. W. Schaefer, R. E. J. Sears, and J. S. Waugh, J. Chem. Phys., $\underline{53}$, 2127 (1970).

50. A. D. Buckingham and R. E. Raab, J. Chem. Soc., $\underline{1957}$, 2341.

51. J. M. Deutch and J. S. Waugh, J. Chem. Phys., $\underline{43}$, 2568 (1965); $\underline{44}$, 4366 (1966).

52. J. D. Ramshaw, D. W. Schaefer, J. S. Waugh, and J. M. Deutch, J. Chem. Phys., $\underline{54}$, 1239 (1971).

53. A. D. Buckingham, P. J. Stiles, and G. L. D. Ritchie, Trans. Faraday Soc., $\underline{67}$, 577 (1971).

54. H. A. Lorentz, The Theory of Electrons, Teubner, Leipzig (1909), p. 303.

55. L. Onsager, J. Am. Chem. Soc., $\underline{58}$, 1486 (1936).

56. C. W. Hilbers and C. MacLean, Mol. Phys., $\underline{16}$, 275 (1969).

57. A. D. Buckingham, MTP Int. Rev., Phys. Chem. (1), $\underline{2}$, 241 (1972).

58. Th. G. Scholte, Physica, $\underline{15}$, 437, 450 (1949).

59. C. J. F. Böttcher, Theory of Electric Polarization, 2nd ed., Elsevier, Amsterdam (1973).

60. A. D. Buckingham and P. G. Hibbard, Symp. Faraday Soc., $\underline{2}$, 41 (1968).

61. N. J. Bridge and A. D. Buckingham, Proc. Roy. Soc. (London) A295, 334 (1966).

62. D. E. Stogryn and A. P. Stogryn, Mol. Phys., 11, 371 (1966).

Chapter 3

ELECTRIC BIREFRINGENCE AND
RELAXATION IN SOLUTIONS OF RIGID MACROMOLECULES

Chester T. O'Konski

Department of Chemistry
University of California
Berkeley, California

Sonja Krause

Department of Chemistry
Rensselaer Polytechnic Institute
Troy, New York

I.   INTRODUCTION

Kerr effect data on gases and liquids provided important results in studies of molecular structure, as discussed in Chaps. 1 and 2. The theory is relatively complete and rigorous for dilute gases, where the complications of molecular interactions are absent. It is also in quite a good state for most liquids and dilute solutions of solutes in nonpolar solvents, but not so complete for polar molecules -- especially hydrogen bonding ones, which interact strongly in liquids.

Studies of macromolecules in solution by the Kerr effect began about two decades ago. In macromolecular solutions, the orientational relaxation times can readily be measured and they give valuable information about the size and shape of the macromolecules. Applications to biopolymer systems are providing new knowledge of the structures of polymers of importance in molecular biology and biochemistry. An aspect which has stirred deep interest was the discovery that the counterions of large polyelectrolyte macromolecules play a dominant role in the mechanism of orientation of such systems as nucleic acids, nucleoproteins, and other polyelectrolytes. In fact, the dielectric relaxation and the conductivity properties of solutions of polyelectrolytes have been shown to be strongly dependent upon polarization. This phenomenon was shown to be the dominant polarization mechanism of electro-optic effects in polyelectrolytes. This connection was established by a combination of dielectric and electro-optic relaxation studies which elucidated the polarization mechanisms. Electric birefringence studies since 1950 also have provided definitive information on dynamic properties through measurements of rotational diffusion constants. These circumstances provided the inspiration for this book.

In this chapter we shall deal with the Kerr electro-optic effect of solutions of rigid macromolecules. This furnishes a basis for treating other electro-optic effects in accord with the historical development. In Chap. 5, the more complex theory encountered with flexible macromolecules will be developed separately.

II.   THE STATIC ELECTRIC BIREFRINGENCE

A.   Theory of the Kerr Constant of Insulating Systems

Let us consider a rigid macromolecule of arbitrary structure in a solvent of uniform optical and electrical properties. The effects on the electric and optic properties of the solution which will be produced by this macromolecule may be described in terms of its electric dipole moment and a polarizability tensor, and its response to the applied electric field will be seen as an orientation effect. In this section we will deal with the low-frequency or static Kerr constant only. The dynamics of orientation will be discussed in Sec. III. The polarizability tensor is a well-known 3×3 matrix, and it can be diagonalized by proper choice of the Cartesian axis system, x, y, z. The dipole moment vector may always be resolved into three components, one along each of the respective axes, and we shall designate those μ_x, μ_y, μ_z. The hydrodynamic and the electric polarizability principal axis systems do not necessarily coincide.

In a dilute solution of macromolecules to which no field is applied, all possible orientations of any macromolecular axis system will be equally probable because of the random Brownian motion, and the solution will be optically isotropic. After an electric field, E, is applied and the system has come to a steady-state

distribution, we may calculate from the Boltzmann distribution
function the relative populations of molecules in the various
orientations, if we know how the energy varies with orientation.
Knowing how any one molecule in a given orientation will affect the
optical properties of the medium, we may calculate the optical properties of the solution as an integral over the entire population of
macromolecules in the ordered system, according to the Born-Langevin
orientation theory [1,2].

It is instructive to write out the expression for the birefringence of a dilute solution of rigid axially symmetric macromolecules in terms of the orientation distribution function, $f(\theta)$,
before considering more complicated expressions for macromolecules
of lower symmetry. Peterlin and Stuart's [3] equation for the
birefringence Δn may be written

$$\Delta n = \left[\frac{\pi C_v (g_a - g_b)}{n}\right] \int_0^\pi f(\theta)(3 \cos^2 \theta - 1) 2\pi \sin \theta \, d\theta \qquad (1)$$

Here $f(\theta)$ is the probability per unit solid angle of finding a
molecule with its symmetry axis at angle θ, with respect to the
direction of the applied electric field. The distribution function
is

$$f(\theta) = e^{-U/kT} \left[\int_0^\pi e^{-U/kT} 2\pi \sin \theta \, d\theta\right]^{-1} \qquad (2)$$

where U is the energy of the macromolecule as a function of
angle, C_v is the volume fraction of the particles, and the term
$(g_a - g_b)$ is the optical anisotropy factor which can be calculated
by application of an equation for ellipsoids due to Maxwell and
given by Peterlin and Stuart [3,4]. This equation is

$$g_j = \frac{1}{4\pi} \frac{n_j^2 - n^2}{1 + (n_j^2/n^2 - 1)A_j} \qquad (3)$$

where A_j is the depolarization factor, n is the refractive index of the solvent, and n_j $(j = a,b)$ are refractive indexes of
the particle for light with electric vector along the axes a and

b, respectively. The depolarization factor may be calculated
from the expression

$$A_j = \int_0^\infty \frac{abc}{2(j^2 + s)[(a^2 + s)(b^2 + s)(c^2 + s)]^{1/2}} \, ds \qquad (4)$$

Here, a, b, and c are the length of the semiaxes of the
particle, and s is a variable of integration. The integrals have
been tabulated for ellipsoids of revolution [5]. Graphs [6] and
special formulas also are available [7].

   To compute $f(\theta)$, we require an expression for the energy U
of the macromolecule as a function of angle. In insulating media,
this may be written

$$U = -\mu E B_a \cos \theta - \frac{1}{2}(\alpha_a - \alpha_b) E^2 \cos^2 \theta \qquad (5)$$

Here μ is the dipole moment, which is assumed to lie along the
symmetry axis of the macromolecule, designated by the subscript
$\underline{a}$, $\mu_b = \mu_c = 0$, and B_a is the internal field function which
expresses the ratio of the local field acting upon the macromole-
cule along the direction $\underline{a}$ to the component of the external field
along that axis. The polarizabilities of the macromolecule in excess of the solvent it replaces are α_a and α_b along the symmetry axis a, and the transverse axes, b = c, respectively. If
the particle can be represented as an ellipsoid in an insulating
homogeneous dielectric continuum, the internal field functions B_j
may be calculated from the relation

$$B_j = \left[1 + \left(\frac{\varepsilon_j}{\varepsilon} - 1\right) A_j\right]^{-1} \qquad (6)$$

in which ε and ε_j are the dielectric constants of the solvent
and of the ellipsoid along the axis j, respectively.

   From Eqs. (1), (2), and (5) one can readily obtain expressions
for the birefringence as a function of the applied electric field
and for the Kerr constant in weak fields, that is, U << kT. Except for the explicit inclusion of the internal field factor B_a

by O'Konski [8], the results were previously given by Peterlin and
Stuart [4] and Benoit [9]. The result for the specific Kerr constant, defined in Chap. 1, is

$$
K_{sp} = \frac{2\pi(g_a - g_b)}{15n^2} \left[\frac{\alpha_a - \alpha_b}{kT} + \frac{B_a^2 \mu^2}{k^2 T^2} \right] \tag{7}
$$

Here μ is not the vacuum dipole moment, but rather the dipole
moment of the solvated macromolecule in the solution, which is in
general different because of solvent orientation and polarization.

For the general ellipsoid with dipole moment components along
all three axes, the Kerr constant was obtained by Holcomb and
Tinoco [10] and it may be expressed as follows, after inclusion of
the internal field function B_a, B_b, and B_c, for the various
axes:

$$
\begin{aligned}
K_{sp} = & \frac{\pi}{15n^2 kT} [(g_a - g_b)(\alpha_a - \alpha_b) + (g_b - g_c)(\alpha_b - \alpha_c) \\
& + (g_a - g_c)(\alpha_a - \alpha_c)] + \frac{\pi}{15n^2 k^2 T^2} [(g_a - g_b)(B_a^2 \mu_a^2 - B_b^2 \mu_b^2) \\
& + (g_b - g_c)(B_b^2 \mu_b^2 - B_c^2 \mu_c^2) + (g_a - g_c)(B_a^2 \mu_a^2 - B_c^2 \mu_c^2)]
\end{aligned} \tag{8}
$$

The intermediate case of an ellipsoid of revolution (axes a and
b = c) with dipole moment components μ_a and μ_b was calculated
earlier [11].

The above equations apply for a homogeneous population of
macromolecules. If we are dealing with a heterogeneous population,
or a distribution of conformers, and are interested in properties
measured by techniques which are fast compared to the rate of
interchange among the conformations, we may, in the dilute solution
case, express the Kerr constant simply as a sum over various
species, that is,

$$
K_{sp} = \sum_i C_{v,i} K_{sp,i} \tag{9}
$$

where $C_{v,i}$ is the volume fraction of the species $\underline{i}$ and $K_{sp,i}$ is the specific Kerr constant of that species.

B.   Theory of the Kerr Constant of Conducting Systems

In this section we shall consider the Kerr effect of metallic suspensions, emulsions, and polyelectrolytes in solution. Many aspects of modern macromolecular science grew out of earlier studies of colloidal systems, and there is no clear line of demarcation between a colloidal suspension and a macromolecular solution. While it is important from the chemical point of view whether the units of a macromolecule are joined with covalent bonds or merely by intermolecular association forces, this is not necessarily an important consideration when dealing with the theory of electro-optics of rigid macromolecules, though it is relevant to flexible polymers (Chap. 5).

1. Metallic Suspensions

Several studies were made of metallic sols near the beginning of the century [12,13]. Recently, interest in the Kerr effect in metallic suspensions has been stimulated by the development of de- vices for controlling the phase and polarization of microwave radiation [14].

The theory for suspensions of metallic particles (e.g., gold sols), which are small compared to the wavelength of light, is essentially the same as that for micrometer-size particles in centimeter wavelength radiation fields, except for the computation of the optical anisotropy factors.

Metallic suspensions may be considered as highly conducting particles in an insulating medium. So far as the orientation effect is concerned, this simplifies the treatment which becomes similar to that for an insulating particle of infinite dielectric constant immersed in a medium of finite dielectric constant.

The excess electric polarizabilities of an anisotropic dielectric ellipsoid are given by the relation

$$\alpha_j = \frac{v}{4\pi} \frac{\varepsilon_j - \varepsilon}{(\varepsilon_j/\varepsilon - 1)A_j} \tag{10}$$

in which ε_j and ε are defined above, and v is the volume of the ellipsoid. Taking $\varepsilon_j \gg \varepsilon$, this leads to the following electric anisotropy term:

$$\alpha_a - \alpha_b = \frac{v\varepsilon}{4\pi}\left(\frac{1}{A_a} - \frac{1}{A_b}\right) \tag{11}$$

An exact treatment of the optical part of the problem is quite complex for metals, as in the Mie theory for the scattering of electromagnetic radiation from metallic particles at optical frequencies [15]. Here we shall assume that we are dealing with radiation of a frequency such that the particles may be regarded as perfect conductors. This approximation is expected to be useful for metallic colloidal particles at infrared and microwave frequencies.

The optical anisotropy factor is obtained from Eq. (3) by insertion of $n_j^2 \gg n^2$ with the result

$$g_a - g_b = \frac{n^2}{4\pi}\left(\frac{1}{A_a} - \frac{1}{A_b}\right) \tag{12}$$

Inserting Eqs. (11) and (12) into Eq. (8), and noting from Eq. (6) that $B_a = B_b = B_c = 0$, the expression for the Kerr constant of metallic suspensions becomes

$$K_{sp} = \frac{v\varepsilon}{240\pi kT}\left[\left(\frac{1}{A_a} - \frac{1}{A_b}\right)^2 + \left(\frac{1}{A_b} - \frac{1}{A_c}\right)^2 + \left(\frac{1}{A_a} - \frac{1}{A_c}\right)^2\right] \tag{13}$$

So far as we are aware, this straightforward extension of theory was not presented earlier [15a].

2.   Liquid Droplets

A Kerr effect will be produced in an aerosol or emulsion by the deformation of the liquid droplets by an electric field. Interfacial tension tends to keep the droplets spherical, but the electric free energy will, in general, be lowered by deformation from a spherical shape. The deformation of liquid droplets of arbitrary dielectric constant and conductivity in a suspending liquid of arbitrary dielectric constant and conductivity has been calculated [16] on the assumption that the Maxwell-Wagner boundary conditions [17] are applicable, and that the deformations are small, so the deformed particle may be approximated as an ellipsoid of revolution with symmetry axis along the field. The electric free energy of a conducting ellipsoid of revolution in a conducting medium has been calculated [16]. Electro-optic effects may be appreciable for aerosol or cloud droplets in atmospheric electric fields. Equations have been derived for the droplet deformation [18] and the birefringence produced by the deformed droplets can be calculated from the eccentricity and the refractive indexes of the particle and the surrounding medium. This is a case of form birefringence and the equations of Wiener [19] can be used.

To calculate the Kerr effect when the wavelength of the radiation is long compared to the diameter of the particle, one may use the expressions above for the optical anisotropy factor and simplify them for the case where the eccentricity is small. This would lead to the expressions for the Kerr constant, e.g., for cloud droplets at microwave frequencies. So far as we are aware, those calculations have not yet been carried to completion, although a microwave refractometer method for aerosols was considered by Thacher [Ref. 9 in Ref. 18].

Because of power dissipation effects and the necessity to apply rather large electric fields, the Kerr effect in emulsions would be of interest mainly for solvents of low electric

conductivity (e.g., water in oil suspensions) and droplets which
are micrometer size or larger.

3. Polyelectrolytes

By definition, polyelectrolytes are ionizable macromolecules,
and therefore, their solutions are electrically conducting.
Biological polyelectrolytes, such as the nucleic acids, proteins,
and polysaccharides, are ordinarily dissolved in solutions with
high ionic conductivity. Physiological solutions are around 0.15 M
salt. Both synthetic and natural polyelectrolytes may have local
ion densities considerably higher than that, and therefore the
conductivity effect is extremely important. In dilute aqueous
solutions of low salt content, a high-density polyelectrolyte such
as DNA may be regarded as a more or less uniformly charged
macromolecule with quite high surface-charge density arising from
the ionizable phosphate groups, surrounded by a swarm of counterions in a relatively insulating dielectric continuum. It was logi-
cal to expect, therefore, that the counterion atmosphere polarization effects would play a significant part in the orientation of
such macromolecules by electric fields [20]. This is because the
transport of ions profoundly influences the distribution of
electric fields near a macromolecule, and equations derived for
insulating systems are generally inapplicable. A comprehensive set
of experiments conclusively established that most of the orienting
torque in a model polyelectrolyte system, tobacco mosaic virus
(TMV), must be due to the counterion polarization effect [21].
Earlier birefringence studies on synthetic colloids had already led
to the suggestion that counterions were important in the orientation effect [22].

An important polyelectrolyte to consider would be a protein
molecule of nonuniform charge density (various groups) along a
flexible chain. It would require a very elaborate theory to treat
such a system, so we restrict ourselves now to rigid macromolecules
of uniform charge density.

A rigid polyelectrolyte, such as a short section of DNA or a virus, has a configuration of fixed charges, some of which bind counterions, and it will in general be surrounded in solution by a swarm of ions distributed diffusely according to the Poisson-Boltzmann distribution law [23]. Applying an electric field perturbs the counterion distribution. When the field is removed, the perturbation relaxes. The relaxation of the counterion distribution gives rise to frequency dependences of the Kerr effect, the dielectric constant, and the conductivity, all related to the counterion atmosphere relaxation time. In 1955 the relaxation time was calculated for a model system -- a sphere with a uniform surface conductivity produced by the mobile counterions [24]. Subsequently, a dielectric and conductivity theory based upon extensions of this model was developed by extending the Maxwell-Wagner theory to include the internal and external volume conductivities and the interfacial conductivity explicitly [5]. It was shown that a surface conductivity of value λ (ohms^{-1}) at the interface of a sphere is electrically equivalent to an internal conductivity, κ', of value $2\lambda/\underline{a}$ on a sphere of radius $\underline{a}$. In general, for anisometric particles, the surface conductivity will contribute different amounts to the effective conductivity along the different axes. Each axial conductivity may be written

$$\kappa_j = \kappa_j^0 + \kappa_j' \tag{14}$$

where the term κ_j^0 is the true volume conductivity along the axis j, κ_j' is the effective contribution along the axis j from the surface conductivity associated with the counterions and any mobile surface charges. The terms κ_j' were evaluated for various-shaped particles [5] and are summarized in Chap. 16 (Part 2). Prior to this theory the nearest treatment had been a calculation of the electrical conductivity and its frequency dependence for a suspension of spherical particles carrying a concentric shell of different electrical properties [25]. It can be shown through the

relationships between the complex dielectric constant and the
complex conductivity that this treatment gives equivalent results
in the limit of a negligibly thin shell of equal conductance [26].

Examples of physical situations which are expected to lead to
surface conductivity contributions have been discussed [5]. They
include ion atmosphere polarization [20] or the mobility of the
counterions of the ion atmosphere [5,24,27], the proton mobility on
the surface of proteins in solution [28], and electronic
semiconductivity. The external or "atmospheric" contributions from
mobile counterions appear to be of great importance for synthetic
polyelectrolytes and have been studied by various techniques [29-
33] in nucleoproteins [5,21] and nucleic acids [33-37].

To treat the static or low-frequency Kerr constant, we can
neglect not only the problem of the rate of the orientation of the
macromolecules because it is fast compared to one period at low
frequencies, but we may also ignore the dielectric dispersion and
treat the problem as one in the theory of steady-state conduction,
because the charge transport relaxation or ionic polarization
effects often are fast compared with the frequency. The range of
applicability of the Kerr constant theory derived with these
simplifying assumptions has been discussed in the original papers
[38].

Let us consider a general ellipsoid with axes a, b, and c,
having respective internal volume conductivities κ_a, κ_b, and κ_c
and dielectric constants ε_a, ε_b, ε_c. This assumes that the principal axis systems of the dielectric and conductivity tensors coin-
cide with the geometric axes. Considering that conditions of
steady-state electric conduction apply, and using the Maxwell-
Wagner boundary conditions, it was shown [38] that the equation for
the electric free energy of this ellipsoid in a solvent of dielectric constant ε and conductivity κ, subjected to an applied
field E is

$$U(\theta,\phi) = -E(\mu_a B_a \cos\theta + \mu_b B_b \sin\theta \cos\phi - \mu_c B_c \sin\theta \sin\phi)$$

$$- \left[\frac{\varepsilon vE^2}{8\pi}\right]\left[\cos^2\theta \left\{\left(\frac{\kappa_a}{\kappa} - \frac{\varepsilon_a}{\varepsilon}\right)B_a^2 + \frac{(\kappa_a - \kappa)B_a}{\kappa}\right\}\right.$$

$$+ \sin^2\theta \cos^2\phi \left\{\left(\frac{\kappa_b}{\kappa} - \frac{\varepsilon_b}{\varepsilon}\right)B_b^2 + \frac{(\kappa_b - \kappa)B_b}{\kappa}\right\}$$

$$\left.+ \sin^2\theta \sin^2\phi \left\{\left(\frac{\kappa_c}{\kappa} - \frac{\varepsilon_c}{\varepsilon}\right)B_c^2 + \frac{(\kappa_c - \kappa)B_c}{\kappa}\right\}\right] \qquad (15)$$

Here μ_a, μ_b, and μ_c are components of the permanent dipole
moment of the solvated macromolecule along the respective ellipsoid
axes, and v is the volume of the ellipsoid. The functions B_a,
B_b, and B_c are internal field functions expressed in terms of
electric conductivities as follows:

$$B_j = \left[1 + \left(\frac{\kappa_j}{\kappa} - 1\right)A_j\right]^{-1} \qquad (16)$$

and θ and ϕ are two of the Eulerian angles in the coordinate
system used to specify the orientation of the ellipsoid axes. The
dielectric constant and conductivity of the solvent are ε and κ.
From this expression one obtains a value for the specific Kerr constant which is

$$K_{sp} = \frac{2\pi}{15n^2}\left[(g_a - g_b)(P_a - P_b + Q_{ab})\right.$$

$$\left.+ (g_b - g_c)(P_b - P_c + Q_{bc}) + (g_c - g_a)(P_c - P_a + Q_{ca})\right]$$

where $P_i = \dfrac{B_i^2\mu_i^2}{k^2T^2}$, $\quad i = a, b,$ or c $\qquad (17)$

$$Q_{ij} = \varepsilon v\left\{(\kappa_i/\kappa - \varepsilon_i/\varepsilon)B_i^2 + (\kappa_i - \kappa)B_i/\kappa\right.$$

$$\left.- (\kappa_j/\kappa - \varepsilon_j/\varepsilon)B_j^2 - (\kappa_j - \kappa)B_j/\kappa\right\}/4\pi\kappa T$$

and other terms are defined above. In these equations the subscripts i and j label the quantities referring to the axes a,

b, or c. Coefficients of the form $(\kappa_i/\kappa - \varepsilon_i/\varepsilon)$ in Eq. (17) are proportional to the surface charges which accumulate, under the influence of an externally applied field, at the interface of two conductors having different conductivity and dielectric constant ratios; this is the essence of the Maxwell-Wagner polarization phenomenon. Details of the interpretation and the equations are presented in the original papers [5,38] and in Chap. 16 (Part 2).

The development of this theory was closely related to the interpretation of the high dielectric increments and the conductivity increments observed in solutions of polyelectrolytes [5]. Although theoretical extensions continue, and more complex calculations are needed for more general models, there appears to be widespread agreement now that the counterions are very important in electric polarization [32] and in electro-optic theories of polyelectrolytes.

C. Dispersion of the Kerr Constant in an Absorption Band

Buckingham [39] treated the dispersion of the Kerr constant of gases and liquids in the vicinity of an absorption band (see Chap. 2). In a dilute solution of macromolecules, the specific Kerr constant, K_{sp}, will vary with wavelength via the optical parameters, g_a, g_b, and g_c [Eqs. (3), (7), and (9)]. If the transition moment of the chromophore responsible for the absorption band can be assigned a direction with respect to the principal axes of the macromolecules, then the dispersion of the optical parameters g_a, g_b, and g_c can be calculated.

Conversely, a study of the dispersion of the Kerr constant can yield information on the structure of the macromolecule. For example, Orttung [40] studied the Kerr constant of horse oxyhemoglobin and methemoglobin at 546, 436, and 365 nm. He also calculated the expected Kerr constant at these wavelengths using the known spectra and structure of the macromolecule assuming

(a) orientation of the macromolecule with its long axis parallel
to the electric field, and (b) orientation with the short axis
parallel to the electric field. The data fitted case (b), and this
allowed a dipole moment parallel to the short axis of the macromolecule to be surmised.

Powers [41] studied the dispersion of the Kerr constant of
acridine orange-polyglutamic acid complex from 400 to 550 nm and
concluded that the long axis of the dye molecule is approximately
parallel to the long axis of the polymer helix in the complex.
The Kramers-Kronig relations between birefringence dispersion and
dichroism were developed by Kuball and coworkers [42] and were
applied to the DNA-proflavine complex by Houssier and Kuball [43].

D.   Experimental Results

The Kerr constants of some macromolecules of various types are
shown in Table 1. It is seen that there is a tremendous range of
K_{sp} values. The polyelectrolytes, especially the large rigid
ones, have the largest Kerr constants. The relatively rigid
polypeptide, poly-γ-benzyl-L-glutamate (PBLG) also has a very
large value, whereas the flexible coil polymer, polystyrene, has a
very low specific Kerr constant. Additional results for macromolecules are reported in Chaps. 15, 16, and 17 (Part 2). Clearly, the
constant is very sensitive to structure.

The synthetic polypeptide, PBLG is of special interest because
of extensive measurements of this system as a nonpolyelectrolyte
model for protein structure. TMV, a large, rigid, rodlike particle
3000 A long and 150 A in diameter, became a valuable model system
in studies of the mechanisms of polarization which produce the
electric field orientation effect. A very large Kerr constant was
found, which exceeded the value calculated from electrostatic
theory by a factor of 50 [21]. In addition, it was shown that the
dipole moment contribution to the Kerr effect is negligible. This

TABLE 1

Specific Kerr Constants of Some Macromolecules

Macromolecule	Mol. wt.	K_{sp} (esu)	Ref.
Tobacco mosaic virus (in distilled water, 0.049 g/liter, 25.5°C)	4.3×10^7	1.36×10^{-3}	44
Collagen (0.11 g/liter and 0.22 g/liter, in 2.9×10^{-3} N acetic acid, 25°C)	2.80×10^5	4.3×10^{-5}	45
Poly-γ-benzyl-L-glutamate (in ethylene dichloride, 0.1 to 4 g/liter)	3.5×10^5	1.3×10^{-5}	48
Sodium polyethylene sulfonate (in water, extrapolated to zero conc.)	2.4×10^4	1.0×10^{-5}	47
Poly-γ-benzyne-L-glutamate (in ethylene dichloride, 1 to 15 g/liter)	8.4×10^4	2.3×10^{-6}	49
Fibrinogen (pH 8.1, 2 g/liter, 25°C)	3.3×10^5	9.3×10^{-7}	48
γ-globulin (2.0 g/liter, in 10^{-3} M phosphate buffer at pH 7, 30°C	1.5×10^5	3.4×10^{-8}	46
Ribonuclease (9.8 g/liter, in 10^{-3} M phosphate buffer at pH 7, 30°C)	1.4×10^4	1.9×10^{-8}	46
Bovine serum albumin (in water, extrapolated to zero conc., 25°C)	6.7×10^4	1.70×10^{-8}	44
Polystyrene (in carbon tetrachloride, extrapolated to zero conc.)	2.5×10^5	8.7×10^{-14}	50

led to the recognition of the importance of counterion polarization effects and stimulated development of the counterion polarization theory [5]. The polyelectrolyte properties of TMV are known from the studies of the amino acid composition of the TMV protein, and from direct titration measurements. Thus, it was possible to obtain figures for the surface densities of counterions. Also, the specific optical properties of TMV are known from flow birefringence and electric birefringence saturation studies [44].

Comparison of the experimental specific Kerr constant with that calculated from a surface conductivity model [38] indicated satisfactory agreement in view of the diffuse nature of the counterion atmosphere and the necessary assumptions regarding counterion mobilities.

III.   ELECTRIC BIREFRINGENCE DYNAMICS

The rate at which macromolecules in a solution orient under the influence of an applied electric field determines the time dependence of the optical birefringence. It is well known from studies of nuclear magnetic resonance and dielectric dispersion in ordinary liquids that molecular reorientation processes of small molecules normally occur with lifetimes of 10^{-11} to 10^{-9} sec. Longer relaxation times are encountered with macromolecules in solution and the direct measurement of these relaxation times through electro-optic effects is now an important method for studying the sizes and shapes of macromolecules.

When an electric field is applied or changed across a solution containing macromolecules, there will be an accompanying change in the orientation distribution function, $f(\theta,t)$. Because of frictional and inertial effects, the orientation does not change instantaneously, so that transient effects in the birefringence will be observed. The orientation takes place about some center within the macromolecule, the center of hydrodynamic drag, and any motion about this center may be represented as the sum of rotations about three perpendicular axes. In general, some solvent molecules will be bound by the macromolecule, and for our purposes they may be regarded as part of the rigid hydrodynamic unit. The hydrodynamic properties may be described in terms of the frictional coefficients for rotation of the macromolecule about various axes through the center of drag, or three orthogonal axes for the principal frictional coefficients, or more conventionally, in terms

of the principal components of the rotational diffusion tensor.
These are related to the frictional coefficients by the Einstein
equation (see Chap. 4)

$$\Theta_{ii} = \frac{kT}{\zeta_{ii}} \tag{18}$$

where ζ_{ii} is a frictional coefficient, Θ_{ii} is the corresponding
rotational diffusion coefficient, or "constant," at temperature
T, k is Boltzmann's constant, and i labels one of the principal
axes.

We restrict the initial discussion to cylindrically symmetric
molecules. To calculate the time dependence of the birefringence
by the use of Eq. (1) we need to find the time dependence of the
orientation distribution function.

In general, one must consider the orienting torque due to the
electric field, the random Brownian motion which causes the particles to undergo a random walk in the orientation sense, the
friction between the particle and the surrounding medium, and
inertial effects. A straightforward calculation reveals that for
ordinary size macromolecules in common liquids, the inertial
effects are quite negligible; that is, a rotational motion is very
quickly dissipated as the molecule expends its inertial energy on
viscous drag in the surrounding medium. (See Appendix B.) Hence,
we will be concerned here with the formulation of the problem as
a Brownian diffusion process biased by the application of an electric field which produces a torque, M, acting on the macromolecule. The partial differential equation for the diffusional process may be written [9,20,51]

$$\frac{1}{\Theta} \frac{\partial f(\theta,t)}{\partial t} = \frac{1}{\sin \theta} \frac{\partial}{\partial \theta} \left[\sin \theta \left\{ \frac{\partial f(\theta,t)}{\partial \theta} - \frac{Mf(\theta,t)}{kT} \right\} \right] \tag{19}$$

where θ is the angle between the symmetry axis of the macromolecule under consideration here and the applied electric field, t
is the time and k, T, and $f(\theta,t)$ have the meanings stated
above. Θ is the diffusion constant for the reorientation of the

symmetry axes of the molecules by a rotation about their transverse
axes. This equation has been solved for situations in which
$M = 0$, corresponding to field-free relaxation, $M = M(\theta)$,
corresponding to a steady electric field, and also for $M = M(\theta,t)$
as in varying electric fields.

A. Field-Free Relaxation Theory and Results

It was shown [9,20] that in a static electric field, the
solution of Eq. (19) in the steady state may be expressed

$$f(\theta) = a_0 + a_1 P_1(\cos \theta) + a_2 P_2(\cos \theta)$$
$$+ a_3 P_3(\cos \theta) + \ldots \tag{20}$$

where the $P_n(\cos \theta)$ are the Lengendre polynomials of order n,
and a_0, a_1, and a_2 are constants. The term of this series
that contributes to the polarization tensor characterizing the
birefringence is the one involving $P_2(\cos \theta) = (3 \cos^2 \theta - 1)/2$.
Orthogonality of all other terms with respect to this term in the
integral of Eq. (1) leads to zero contributions from them. It also
was shown that the field-free decay of the birefringence followed
a first-order decay law which may be written

$$\Delta n = \Delta n^o e^{-t/\tau} \tag{21}$$

where the birefringence relaxation time, τ, is given by [9,20]

$$\tau = \frac{1}{6\theta} \tag{22}$$

B. The Electric Birefringence Relaxation Method and Apparatus

A convenient and versatile way of studying the electric,
optical, and hydrodynamic properties of macromolecules is the
transient electric birefringence or the electric birefringence
relaxation method. It was introduced at the University of

California at Berkeley in 1948, and about the same time in the
Soviet Union [52], and in France [9,53,54]. In the Berkeley work,
the inspiration for the method was a desire to study the sizes and
shapes of macromolecules in solutions by optical observations, and
the recognition of the need to understand better the puzzling
electro-optic behavior of colloidal systems observed in sinusoidal
fields of various intensities and frequencies [22,55-62]. Since
the systems to be investigated were aqueous solutions displaying
electric conductivity and electrophoresis, it was apparent that it
would be an advantage to apply short rectangular pulses, thus
minimizing these effects, and to observe the relaxation effects
directly. Experiments were carried out using an electronically
generated antisymmetric square wave of the kind illustrated in Fig.
1a [20].

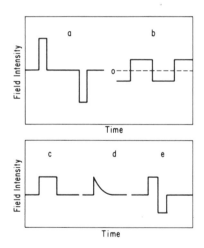

FIG. 1. Electric field waveforms for electro-optic experiments. (a) Continuously repeated pulses of opposite polarity (see
Ref. 5); (b) continuous square wave; (c) single pulse; (d) exponential pulse; (e) reversing field pulse.

Since pulses of equal and opposite polarity were applied, the
net electrophoretic motion of the particles was zero. The pulse
duration, amplitude, and the period between pulses was adjustable,

so that the duty cycle could be reduced to minimize the effects of
Joule heating of the solution. The early Strasbourg work was done
with a motor-driven rotating switch [9,53,54].

The first pulsed-field observations at the University of
California at Berkeley were made on TMV, which was a fortunate
choice because it was a rigid macromolecule which could be obtained
in homogeneous preparations, and it was large enough to permit
convenient measurements on an oscilloscope. It also turned out
that with TMV we could make measurements which separated the
relaxation effects associated with rotational diffusion from those
produced by the electrical mechanisms involved in orienting the
particles [21]. It was found [20] that the relaxation time for the
field-free decay of the electric birefringence of TMV was
approximately 0.6 msec, and that dilute solutions exhibit simple
exponential buildup and decay curves. Additional experiments were
conducted in which the field consisted of regular square waves,
shown in Fig. 1b (note that the field never is zero), and it was
found that the birefringence was substantially unchanged as the
frequency of the field was increased from values below to values
above the reciprocal of the observed birefringence time constant.
Also, there was no transient upon field reversal. These observations led to the proposal that the mechanism of orientation was not
primarily a permanent dipole moment coupling effect, as previously
supposed [56], but, rather, an induced polarization effect in the
dilute solutions [20]. More concentrated solutions of TMV showed
a complex buildup curve and a steady-state birefringence of negative sign. Values of steady-state optical retardation as a function of applied field strength at various concentrations are shown
in Fig. 2. These results led us to suggest that the rods may
orient across the direction of the electric field at higher
concentrations. This could be caused by a hydrodynamic orientation
effect, or by a small transverse dipole moment, which probably becomes relatively more important when the ion atmospheres overlap,
which should tend to reduce induced polarization orientation.

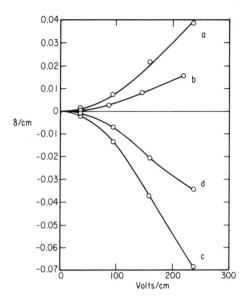

FIG 2. Peak values of optical retardation in radians per centimeter of solution versus applied field strength: (a) 0.073% TMV, 1.5×10^{-4} M buffer; (b) 0.073% TMV, 1.0×10^{-3} M buffer; (c) 1.4% TMV, 1.5×10^{-4} M buffer; (d) 2.75% TMV, 1.5×10^{-4} M buffer. (Reproduced from results given in Ref. 20.)

Subsequently, to minimize effects occurring as a result of Joule heating, the single-shot square pulse technique was introduced, and also an exponential pulse and reversing pulse, as illustrated in Fig. 1c, d, e [63-65]. The square pulse allows observations of the buildup of the birefringence under controlled conditions, as well as the decay, whereas the exponential pulse (Fig.1d) is particularly simple to obtain by the discharge of a condenser through the solution or through a resistor switched across it. With the exponential pulse, the field is decaying even as the birefringence is building up. A first-order kinetic analysis has been given [21]. Additional information can be obtained by applying a pulse and suddenly reversing its polarity, as shown in Fig. 1e. Besides the possibility of following the buildup and decay kinetics, one can observe the transient which may occur when

the electric field is rapidly reversed [21]. If the orientation is
due to permanent electric dipoles, then reversing the field will
cause the molecules which are preferentially oriented in a given
direction to diffuse until there is a preferential orientation in
the reverse direction. This will produce a transient in the electric birefringence. On the other hand, if the orientation is due
to an induced polarization which is fast compared to the rate of
reorientation of the macromolecules, then reversing the field will
at the same time reverse the polarization and leave the orienting
torque unchanged so that no transient will result [21]. Detailed
calculations of reversing field transients were carried out by
Tinoco and Yamaoka [66] and by Matsumoto et al. [67].

A schematic diagram of a typical pulsed electric birefringence
relaxation apparatus is shown in Fig. 3. White light from a
tungsten source, or monochromatic light, is brought to focus on a
slit by means of the condensing lenses, which have a diaphragm

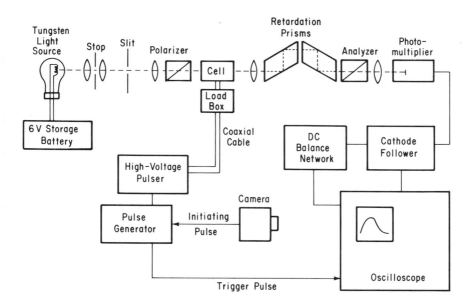

FIG. 3. Oscillographic electric birefringence apparatus (see
text for description). (Reproduced from Ref. 8.)

between them to control the aperture and convergence of the light
beam. The second lens, followed by a polarizer, brings the beam
to a slightly converging focus at the center of the cell which
generally contains plane-parallel platinum electrodes. A lens
beyond the cell picks up the slightly diverging beam and makes it
parallel. After this, it passes through a $\lambda/4$ retardation device
before it proceeds through the analyzer to a condensing lens, which
brings it to focus on a slit (not shown) in front of a photomultiplier tube. The polarizers used are conveniently of the Glan-
Thompson or of the Glan type, which do not divert the central ray
from the optic axis. In the instrument shown, the retardation
prisms are modified Fresnel rhombs, with a $\lambda/8$ retardation in
each prism. The two in sequence produce the $\lambda/4$ retardation, and
also restore the central ray to the original optic axis [68].

The photomultiplier tube converts the optical signal to a voltage drop produced across an anode load resistor, which is variable
in steps to permit adjustment of the sensitivity and the time constant of the instrument. Normally, it is the photomultiplier circuit time constant which determines the overall bandwidth of the
detecting system. Using either dc or ac coupling, the signal is
sent to a cathode follower located near the photomultiplier to reduce the effects of distributed capacitance, and then to a suitable
oscilloscope. A camera photographs the oscilloscope trace. The dc
balance network shown in Fig. 3 is used for photometric measurements of the transmitted light intensity and the stray light of the
optical system.

In the instrument shown, the sequence of measurement is
started when an operator triggers the shutter of a camera having
flash contacts, which produce an initiating pulse, energizing the
pulse generator. The pulse generator is actually an assembly of a
sweep generator and several pulse generators with adjustable de-
lays. This assembly produces trigger pulses which initiate the
oscilloscope sweeps, and it also triggers the high-voltage pulser
that applies a signal to the electrodes of the Kerr cell. In the

version of the instrument shown in the diagram, which was
developed for use on protein solutions, the high-voltage generator
is a thyratron which discharges a delay line consisting of coaxial
cable [68]. With a 550-ft cable and a high-voltage supply variable
to 16 kV, the circuit generated rectangular voltage pulses of 1.7
μsec duration and of amplitudes to 8 kV. By using a terminating
high-voltage resistor of 52 ohms, equal to the characteristic
impedence of the transmission line, it is possible to use a long
piece of coaxial line between the pulse generator and the Kerr cell
which is located close to the terminating resistor. With this
particular pulser the decay time of the rectangular pulses was
around 0.01 μsec to 50% and 0.1 μsec to 25%. The 10 to 90% rise
time was 0.05 μsec. Birefringence relaxation times down to 0.1
μsec were measurable. The electrodes and Kerr cell were described
in an earlier study [69]. With electrode separations of 0.2 cm,
fields up to 40 kV/cm have been achieved.

A hard tube pulse circuit, with variable pulse width from
around 10 μsec up to milliseconds and variable pulse amplitudes to
around 10 kV, also has been developed [70,71]. This pulser typi-
cally operates into Kerr cell impedances of 1000 ohms or higher,
and generates slower rising and decaying pulses, normally with time
constants of around 1 μsec. Jennings and Brown [72] devised a
pulser with similar characteristics and circuits were discussed by
Brown et al. [73], Yoshioka and Watanabe [74], and Stoylov [75],
among others.

The conventional Kerr cell arrangement has the polarizer
oriented so that the electromagnetic radiation is at 45° with respect to the polarizer. With this arrangement there is an increase
in light intensity, regardless of the sign of the birefringence
produced by the electric field, so that an independent measurement
is required to determine the sign. By using a quarter-wave
retardation device in the optical system as shown above, and rotating the analyzer from the crossed position, one introduces a
polarity so that the sign of the signal depends on the sign of the

birefringence. It was shown that this feature also increases the
sensitivity to small signals [20] by increasing the signal/noise
ratio.

The speed of response of the pulse detection system is determined by the frequency response characteristics of the phototube
circuit, the amplifiers, and the oscilloscope, if oscillographic
observations or photography is involved. In a well-designed instrument the frequency response of the oscilloscope is not generally the limiting factor; it is chosen fast enough to respond very
rapidly to the transients from the photodetector-amplifier system.
Ideally, only one of the instrument time constants limits the speed
of response; in our instruments this is the R-C combination at the
anode of the photomultiplier, including the distributed input
capacitance of the preamplifier (a cathode-follower stage), which
has a total value of around 30 pF. A switch selects high-frequency
resistors in steps, and this controls the limiting time constant
(30 nsec to 30 μsec) and thus the overall bandwidth, and the
sensitivity of the system. The use of a variable voltage supply to
a photomultiplier detector permits a wide variation of sensitivity
without changes of bandwidth. The signal/noise ratio, for a given
signal can, of course, be increased by using a larger time constant, because of the reduction of noise with decreased bandwidth
of the system. When the birefringence relaxation times are long
compared to the time constant, the latter can be neglected. Signals must be corrected for instrumental time constant, when R-C
becomes comparable to the relaxation times being observed.

To describe the response of the optical system, we may specify
the angular positions of the optical elements measured in the
clockwise direction looking down the light beam with respect to the
direction of the applied electric field. Thus, the field vector is
at zero degrees, the polarizer is at $\frac{\pi}{4}$ rad, and the retardation
prism will be oriented with its slow axis at $\frac{3\pi}{4}$. The angle of the
analyzer is given by χ. Then the change in light intensity, ΔI,
at the photomultiplier produced by an optical retardation δ is

given by

$$\frac{\Delta I}{I_o} = \frac{1}{2}[\sin (2\chi - \delta) - \sin 2\chi] \tag{23}$$

The relation between δ and the birefringence Δn is given in Eq. (2) of Chap. 1. I_o is the intensity when the analyzer is at the same orientation as the polarizer.

Optimum signal-to-noise is obtained at a value of χ determined by the stray light intensity of the optical system. In the present notation, it may be expressed in terms of I_s, the stray light transmitted with the analyzer in the crossed position [20]

$$\sin \left(\chi - \frac{3\pi}{4}\right) = \left(\frac{I_s}{I_o}\right)^{1/4} \tag{24}$$

The reason for the optimum angle is that the limiting noise in a properly designed optical system is the shot noise of the photomultiplier current due to background light. Rotating away from the crossed position increases the background light intensity; the shot noise rises as the square root of the intensity, but the signal increases more rapidly [20]. Values of the optical retardation δ, or of the birefringence Δn, are obtained from Eq. (23) using oscillographic measurements of the amplitude of the birefringence signal, ΔI. Stray light may arise from scattering and depolarization anywhere between polarizer and analyzer, and from any residual birefringence in that region. To reduce I_s/I_o to 10^{-4} or so, considerable attention must be given to clean optics. It helps to omit the lenses between polarizer and analyser entirely, and this is readily done when a laser source is used. Precautions are taken in the design of the equipment to preserve linearity between the light intensity and the photocurrent. By measuring the voltage drops across the anode resistor of known value, it is then a straightforward matter to obtain $\Delta I/I_o$ from a combination of oscillographic and potentiometric voltage measurements [21]. The value of I_o is normally obtained by rotation of the analyzer from the crossed position by a measured angle, r, which increases the light intensity by an amount $I_o \sin^2 r$. The relaxation data can be analyzed by methods discussed in Chap. 15.

C. Theory of the Buildup and Field Reversal Transients in Weak Electric Fields

Equations for the time dependence of the birefringence in a suddenly applied electric field were given by Benoit for the case of cylindrically symmetric macromolecules with dipole moment μ'_a along the symmetry axis, dipole orientation factors $P_b = 0$; $P = P_a = B_a^2 \mu_a^2/k^2 T^2$, and electric anisotropy term $Q = (\alpha_a - \alpha_b)/2kT$. The internal field function B_a has been introduced here in accord with the extended theory discussed above. For electric fields sufficiently weak so that the orientation energy is small compared to kT, Benoit obtained

$$\frac{\Delta n(t)}{\Delta n(\infty)} = 1 - \frac{3P}{2(P+Q)} e^{-2\Theta t} + \frac{P - 2Q}{2(P+Q)} e^{-6\Theta t} \tag{25}$$

where time is measured from the instant of application of the field. Equations for the less symmetric models were later derived by Holcomb and Tinoco [76].

For particles of the same hydrodynamic and polarizability symmetry, but with transverse as well as axial dipole moment components, Tinoco and Yamaoka [66] derived an expression for the birefringence transient produced by rapid reversal of the applied field. It may be written

$$\frac{\Delta n(t)}{\Delta n(\infty)} = 1 - \frac{3P_a}{P_a - P_b/2 + Q} e^{-2\Theta_b t}$$
$$+ \frac{6\Theta_b}{5\Theta_b - \Theta_a} \frac{P_b}{P_a - P_b/2 + Q} e^{-(\Theta_a + \Theta_b)t} \tag{26}$$
$$+ \frac{3P_a - 6\Theta_b P_b/(5\Theta_b - \Theta_a)}{P_a - P_b/2 + Q} e^{-6\Theta_a t}$$

Here Q is the same as Benoit's, and P_a and P_b can be obtained from Eq. (17) using Eq. (6) for the internal field function. Time is measured from the instant of field reversal.

They also derived equations for the case where the induced dipole polarization is not fast, for example, when the transverse and longitudinal counterion polarization relaxation times are of the same order of magnitude as $1/\theta$. The expressions are complex and are not reproduced here; they presented useful curves based upon numerical calculations over a wide range of parameters. Further examples have been presented elsewhere [67].

D.   Representative Data and Interpretation

Accurate measurements on homogeneous solutions of TMV monomer gave a 0.50 msec relaxation [63,64]. There was a discrepancy between this relaxation time, measured in dilute solutions, and the value calculated from the known dimensions of the rodlike macromolecule, using the Burgers equation for the hydrodynamic friction coefficient [64]. The constancy of the relaxation time with varying ionic strength showed that there is no great effect of the ion atmosphere, so far as rotational diffusion is concerned [21]. After examination of alternative possibilities, it was suggested that the hydrodynamic equation might not be sufficiently accurate. Later, Broersma made a study of the problem and decided that indeed a more extensive calculation than that of Burgers, based upon the Oseen hydrodynamic theory, was required for cylindrical macromolecules. He derived a more accurate expression for the frictional coefficients [77]. With it, a value of 305 nm was obtained for the length of the macromolecule from the birefringence data, compared with 298 nm for the dry length measured in an electron microscope [78], whereas the older Burgers equation had given about 342 nm. The excellent agreement between the improved theory and experiment was an early victory of the electric birefringence method [64,77,79]. Recent electric dichroism measurements [79a] gave $\tau = 0.49$ msec, verifying the earlier electric birefringence value to 2%, which corresponds to 0.7% in macromolecule length.

The Kerr constant of TMV turned out to be extraordinarily large compared with the value one would predict from electrostatic theory, and it was observed that the concentration of electrolyte had an appreciable effect on it [21]. These observations stimulated the development of the new theory (see Sec. II B) for electrically conducting systems such as polyelectrolytes. In rapidly reversing pulsed fields, no field reversal transient was observed in the common strain of TMV, confirming that the orientation was not due to a permanent dipole. This showed the significance of the ion atmosphere polarization. Later, a quantitative comparison between experiment and theory became possible [38] (see Chap. 19, Part 2). However, the Holmes Rib Grass strain of TMV was observed to give a large field reversal transient [80], indicating a permanent dipole moment for this case. Thus the homogeneous preparations of the crystallizable TMV played a key role in the development of electric birefringence relaxation technique and theory.

E. Dispersion of the Birefringence in Sine Wave Fields

The anomalous dispersion of the Kerr constant (a decrease of K with increasing frequency of the applied field) in sine wave fields was first observed by Raman and Sirkar [81]. It has been discussed for nonconducting dilute solutions of rigid molecules at low fields by a number of authors [4,21,82,83]. The molecules were assumed to have an axis of symmetry (rodlike or ellipsoidal) along which a dipole moment may or may not be present. Also, the principal axes of the dielectric tensor were assumed to lie along the molecular axes. Peterlin and Stuart [4], Benoit [82], and O'Konski and Haltner [21] considered monodisperse systems only, while Thurston and Bowling [83] also discussed polydisperse systems.

To calculate the birefringence in sine-wave fields, there is no change in the birefringence expression, Eq. (1), but the distribution function for the macromolecular orientation, Eq. (2), becomes time- and frequency-dependent. An additional factor appears in this expression, as discussed in connection with the light-scattering equations for rodlike particles in alternating electric fields by Plummer and Jennings [84], and in connection with the distribution function discussed by Stoylov [75].

Let the applied field be $E_o \sin \omega t$, where E_o is the maximum amplitude of the alternating electric field with angular frequency, ω. Then the calculated value of birefringence of the solution, Δn, at low field and at a time after turning on the field, t, which is long with respect to molecular rotation times, can be separated into a birefringence, $(\Delta n)_i$, due to the induced dipole moment alone, and a birefringence $(\Delta n)_p$, due to the permanent dipole moment only. The induced component is

$$(\Delta n)_i = (\Delta n)_{i,av} \left[1 \pm \frac{\cos (2 \omega t - \phi_i)}{(1 + 4 \omega^2 \tau^2)^{1/2}} \right] \qquad (27)$$

where $(\Delta n)_{i,av}$ is the average birefringence observed. The birefringence consists of two components, one of which is constant with time and the other of which alternates with twice the frequency of the applied field. Here again τ is the birefringence relaxation time of the molecule, and ϕ_i is the phase angle between the birefringence and the applied field, given by

$$\tan \phi_i = 2\omega\tau \qquad (28)$$

The value of $(\Delta n)_{i,av}$ is the same as that which would be found if a steady field equal to $E_o/\sqrt{2}$ or E_{rms}, were applied to the same system. When the frequency of the applied field becomes very high, only $(\Delta n)_{i,av}$ will be observed. At extremely high frequencies, such as the optical frequencies obtainable using high-power lasers, the induced dipole can probably no longer be assumed to vary instantaneously when the applied field changes, and the observed

birefringence, $(\Delta n)_i$, may drop below $(\Delta n)_{i,av}$. At lower
frequencies, $(\Delta n)_i$ fluctuates between maximum and minimum values
given by Eq. (27) with the numerator of the last term equal to ±1.

Experimental curves of such maximum and minimum values of
$(\Delta n)_i$ for a solution of TMV, a molecule with a negligible permanent dipole moment, are shown in Fig. 4. The above theory explains
the lower frequency region and does not treat the counterion
relaxation responsible for the "anomalous" dispersion around 0.1-1
MHz, which is explained in the last paragraph of this section.

The reason for the phase angle, ϕ, for the doubled frequency
component of the birefringence, and for the general variation of
the birefringence with frequency can be visualized as follows: At
normal frequencies, the induced dipole can be assumed to vary
instantaneously with the field intensity, fluctuating between zero
and a maximum magnitude which is the same for $\pm E_o$; it therefore
fluctuates between zero and its maximum absolute value twice during
each cycle of the applied field. At low frequencies, the molecular
orientation follows the applied field intensity, but lags behind it
because rotational diffusion takes a finite time. At high frequency, the molecular orientation time is so slow with respect to

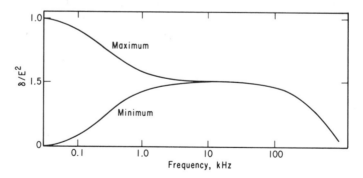

FIG. 4. Birefringence dispersion for a solution of TMV. At
low frequencies the macromolecules orient in phase with the field.
At 100-1000 Hz, they begin to lag, and above 10^4 Hz, they maintain
a steady-state orientation, which decreases because of ion-
atmosphere relaxation as frequency is increased. (Reproduced from
Ref. 21, courtesy of American Chemical Society Publications.)

changes in electric field strength, that the molecular orientation no longer changes with time and the observed birefringence equals that which would be observed in a steady field, as discussed above.

The birefringence due to a dipole moment along the molecular symmetry axis is

$$(\Delta n)_p = (\Delta n)_{p,0} \left[\frac{1}{1 + 9 \omega^2 \tau^2} + \frac{\cos (2\omega t - \phi_p)}{(1 + 9 \omega^2 \tau^2)^{1/2}(1 + 4 \omega^2 \tau^2)^{1/2}} \right] \quad (29)$$

Equation (29) takes no account of the birefringence due to the induced dipole moment; it is strictly a dipole term, and the dipole phase angle, ϕ_p, is given by

$$\tan \phi_p = \frac{5\omega\tau}{1 - 6\omega^2 \tau^2} \quad (30)$$

At any frequency, ω, the permanent dipole birefringence, Δn_p, itself consists of two components, one of which is constant with time, and the other of which alternates with twice the frequency of the applied field, just as in the case of an induced dipole. In the permanent dipole case, however, both the time-independent and the time-dependent components decrease to zero at high frequency. At lower frequencies, $(\Delta n)_p$ varies between maximum and minimum values given by Eq. (29) with the numerator of the last term equal to ±1. The value of $(\Delta n)_{p,o}$ is the birefringence that would be obtained in a steady field equal to E_{rms} for pure dipole orientation.

The behavior of the pure dipole $(\Delta n)_p$, as the frequency of the applied field changes, can be visualized as follows: At normal frequencies, the torque on the molecules is in one direction when the field is positive, and in the opposite direction when the field is negative. The molecules are induced to turn first in one direction, and then in the opposite direction during a single cycle of the applied field. The finite rotational diffusion time of the molecules again causes a time lag, and this produces a phase angle, ϕ_p, between the birefringence and the applied field. As the fre-

quency rises, the molecules can no longer follow the field, either
in magnitude or in direction, and the total birefringence, Δn_p,
falls to zero. At low frequencies, while the molecules are turning, there comes a time, twice in each cycle of the applied field,
when the molecules have oriented mainly perpendicularly to the
applied field, and the birefringence becomes negative. Also, twice
in each cycle of the applied field, namely, when $(2\omega t - \phi_p) = \pi/2$,
$3\pi/2, \ldots$, the birefringence will change its sign. This can also
be seen from Eq. (29).

Since a molecule with a dipole moment also has polarizability,
the total birefringence for such a molecule will be:

$$\Delta n = (\Delta n)_i + (\Delta n)_p \tag{31}$$

At low frequencies, $(\Delta n)_p$ predominates, while at very high
frequencies, only $(\Delta n)_i$ remains, in the form of $(\Delta n)_{i,av}$.

Thurston and Bowling [83] have analyzed the behavior of Δn
and the observed phase angle as a function of frequency for
different ratios of induced to permanent dipole moments. They also
showed how to determine τ from experimental data. When no dipole
moment is present, τ can be determined much more simply [21] from
the slope of a $\tan \phi_i$ vs. ω curve.

Thurston and Bowling [83], in effect, combined Eqs. (27) and
(29) to give

$$\Delta n = \Delta n_{st} + \Delta n_{alt} \cos (2\omega t - \phi_{alt}) \tag{32}$$

where Δn_{st} is the steady component and Δn_{alt} is the magnitude
of the alternating component of the birefringence having phase
angle ϕ_{alt}. Equation (32) is an alternate version of Eq. (31).
If the permanent dipole is along the axis of symmetry of the molecule, then

$$\Delta n_{st} = \frac{\Delta n_o}{r + 1}\left\{ r + \frac{1}{1 + 9\omega^2\tau^2} \right\} \tag{33}$$

where Δn_o is the low-frequency limiting value of Δn_{st}

$$\Delta n_o = (\Delta n)_{i,av} + (\Delta n)_{p,o} \tag{33a}$$

and

$$r = kT \frac{(\alpha_a - \alpha_b)}{\mu^2} \tag{33b}$$

where α_a, α_b, and μ are defined as for Eq. (7). The factor r (called P by Thurston and Bowling [83]) characterizes the relative magnitude of induced and permanent dipole moment effects in determining the orientation of the molecules. Strictly speaking, the factor μ^2 in Eq. (33b) should be $\mu^2 B_a^2$, as in Eq. (7).

Figures 5 and 6 show the frequency dependence of Δn_{st} for various values of r. Figure 5 shows the curves for values of "P" (of Thurston and Bowling) from 0 to ∞ and from -1 to $-\infty$, for which the sign of Δn_{st} does not change with frequency. Figure 6 shows the curves for values of r from 0 to -1, for which there is a reversal of sign with frequency for the steady

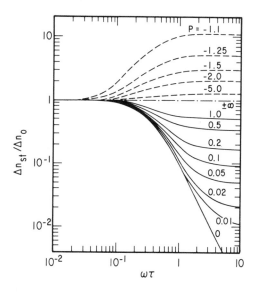

FIG. 5.   Frequency dependence of the steady component of the birefringence for $r \equiv P = 0$ to $+\infty$ and from -1 to $-\infty$. (Reproduced from results given in Ref. 83, by courtesy of Academic Press, Inc.)

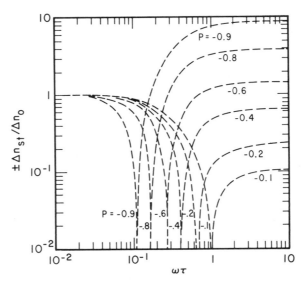

FIG. 6. Frequency dependence of the steady component of the birefringence for P = 0 to -1. The birefringence ratio is positive at the lower frequencies and negative at the higher frequencies. (Reproduced from results given in Ref. 83.)

component. Further discussion is given by Thurston and Bowling [83].

The above theory does not account for effects due to ionic relaxation, as no account of them was taken in the model.

The experimental setup for the study of Kerr effect in sine wave fields is very similar to that using square pulses. In studies of electrolyte solutions, it is often an advantage to use pulsed rather than continuous sine-wave fields [21,85] to minimize joule heating of the solutions. The birefringence and the phase angle, ϕ, have been measured using Lissajou's figures [82,86] or a lock-in amplifier [72,83].

Some of the polyelectrolytes studied, using alternating electric fields, have been TMV [21,82,83], bentonite [83,87], polyriboadenylic acid, and polyribouridylic acid [85]. Also the polypeptide poly-γ-benzyl-L-glutamate [88] was studied.

O'Konski and Haltner [21] and Thurston and Bowling [83] noticed that the birefringence of some materials, noticeably TMV, decreased at high frequency. O'Konski and Haltner [21] explained this anomalous dispersion in terms of relaxation of the ion atmosphere around the molecules (see Sec. II B; Fig. 4; and Chap. 19).

IV. HIGH FIELDS

When the applied electric field is intense enough, all systems will deviate from the Kerr law given in Chap. 1, because the degree of orientation of the macromolecules tends to approach an upper limit. Experimental measurements up into this saturation region are useful because they provide a basis for separately determining the electrical properties and the optical properties which are both involved in the Kerr constant. This method was reported in 1959. The theory of saturation of the Kerr effect was extended to include mixed permanent and induced dipole mechanisms [44]. Earlier theories were by Gans [89] who obtained a birefringence saturation expression for only permanent moment orientation, and by Peterlin and Stuart [3], who obtained an equation for only induced moment orientation. The method was tested and illustrated by measurements and analysis on TMV, the synthetic polypeptide, PBLG, and a synthetic polyelectrolyte, polyethylene sulfonate [44].

A. Theory of Saturation

For axially symmetric macromolecules it is convenient to define the degree of orientation Φ, given by

$$\Phi = \frac{1}{2}(3<\cos^2 \theta>_{av} - 1) \tag{34}$$

where θ is the angle defined in Eq. (1), which then may be expressed

$$\Delta n = \left(\frac{2\pi C_v}{n}\right)(g_a - g_b)\Phi \tag{35}$$

By use of Eqs. (2) and (5), Φ was calculated from Eq. (34) [44] with the result

$$\Phi = \frac{3\int_{-1}^{1} u^2 \exp(\beta u + \gamma u^2)\, du}{2\int_{-1}^{1} \exp(\beta u + \gamma u^2)\, du} - \frac{1}{2} \tag{36}$$

where we introduced new symbols

$$u = \cos\theta \tag{37}$$

$$\beta = \left(\frac{\mu B_a}{kT}\right)E \tag{38}$$

$$\gamma = \left[\frac{(\alpha_a - \alpha_b)}{2kT}\right]E^2 \tag{39}$$

At low fields the exponentials of Eq. (36) may be expanded in a power series and then the degree of orientation is

$$\Phi = \frac{(\beta^2 + 2\gamma)}{15} \tag{40}$$

in accord with the Kerr law, and with Eq. (7). Calculations based upon Eq. (36) were carried out [44] for the case where $\alpha_a - \alpha_b > 0$. The integrals were expressed in terms of a standard set of functions which is tabulated. Later, to explain the reversal of the sign of the birefringence of montmorillonite (polyelectrolyte clay mineral) particles in terms of this kind of theory, the case where $(\alpha_a - \alpha_b) < 0$ was calculated, and similar expressions were obtained [90]. In both contributions graphs and tables of Φ were given. Figure 7 shows the behavior of Φ as a function of $\beta^2 + 2\gamma$, which is proportional to E^2, for a wide variety of cases [74]. Matsumoto et al. [91] have published extensive tables of Φ and $15\Phi/(\beta^2 + 2\gamma)$ for various values of $\beta^2/2\gamma$. The latter function is very useful [44] for analyzing

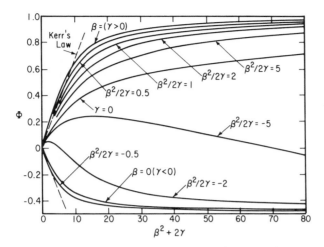

FIG. 7. Orientation factor Φ vs. $|\beta^2 + 2\gamma|$ for various values of $\beta^2/2\gamma$ (= $b^2/2c$). Deviations from Kerr's law occur at high field strengths. (Reproduced from results given in Ref. 74.)

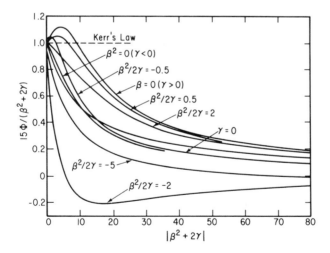

FIG. 8. $15\Phi/(\beta^2 + 2\gamma)$ vs. $|\beta^2 + 2\gamma|$ for various values of $\beta^2/2\gamma$ (= $b^2/2c$). These curves are useful for analyzing experimental data. (Reproduced from results given in Ref. 74.)

experimental results, and it is shown for several cases in Fig. 8.

The earlier saturation theories [89,3] treated permanent and induced moments as separate cases.

Calculations have been made [76] for a model in which both the electric and optical polarizabilities need not have any symmetry and the permanent dipole may have any orientation. The internal field functions defined above may be inserted into the dipole terms as before [8,38].

B. Experiments

The onset of electric birefringence saturation can be seen in the early experiments on TMV [9,20] and also in the results for DNA and vanadium pentoxide sols [9]. Benoit gave a first-order saturation term in the development of his equations, but his experimental saturation data were not quantitatively analyzed.

Experiments which extended into the region of strong electric orientation were performed on TMV, PBLG in dichloroethane solution, and sodium polyethylene sulfonate (NaPES) in water [44]. The saturation curve for TMV was consistent with the shape expected for an induced polarization orientation mechanism, in accord with earlier observations by other methods [21]. An improved fit of the data was obtained on the assumption of two macromolecular components. The optical anisotropy factor, $g_a - g_b$, was found to be 5.9×10^{-3}, and this was essentially unchanged upon extrapolating the data to an infinite field condition. The electric anisotropy term $\alpha_a - \alpha_b$, was found to be 3.3×10^{-14} cm^3 at pH 7 in 1.5×10^{-4} M phosphate buffer. This is in reasonable accord with other values discussed in the original paper, in view of the variation of the electric polarizability with conditions. The data on PBLG were consistent with a permanent dipole moment orientation mechanism, and led to a value for the dipole moment of 2700 D for the sample of molecular weight 195,000. For NaPES, the optical anisotropy factor was found to be 8×10^{-3}, and the orientation behavior was that of a permanent dipole system with

$\mu = 3.1 \times 10^3$ D for a sample of MW 24,000. The dipole moment
predominance was surprising in view of the polyelectrolyte nature
of the macromolecule and the expectation of a large counterion
polarization (see Chap. 3, 19), and it led to the suggestion that a
nonsymmetrical polyelectrolyte chain should have an intrinsic
configurational dipole moment [8,37,44].

The first measurements of a macromolecule which permitted
explicit determination of a dipole moment from the saturation
behavior of a mixed type, that is, with appreciable components
from both permanent and induced orientation effects, was collagen
[92]. Previous reports of permanent moments of polyelectrolytes
are generally unreliable. This macromolecule has a MW of 280,000,
a length of approximately 2800 Å, and is a triple helix. The optical anisotropy factor was found to be 1.7×10^{-3}, roughly in agreement with the value of 2.4×10^{-3} observed in flow birefringence
measurements [93]. The anisotropy of the electric polarizability
was found to be 2.7×10^{-15} cm^3, a large value as expected from an
ion atmosphere polarization in a rodlike polyion, and the permanent
dipole moment was estimated to be 1.5×10^4 D. For additional data
and interpretations in comparison with other literature the original reference should be consulted. This case was of special interest because earlier workers had no way of separating induced
and permanent contributions to the Kerr constant, which made reported dipole moments uncertain.

Shah and coworkers [87] observed that fractionated bentonite
suspensions (montmorillonite clay mineral) showed saturation behavior at modest electric fields (ca. 1 kV/cm) and exhibited a
reversal of the sign of electric birefringence at low electric
field intensities. It was proposed by one of us (C.T.O.) that the
saturation effects in a system involving both permanent and induced
polarization mechanisms might account for the birefringence reversal on the basis that the two orientation torques are perpendicular to each other, and the permanent dipole moment torque varies
linearly with the field, whereas the induced polarization torque

varies as the square of the electric field. Shah and coworkers [87] found evidence to support this idea by square-wave experiments, and subsequently by an interpretation based upon the extension of the saturation theory mentioned above [90]. The agreement between experiment and theory was quantitatively limited by the polydispersity of the preparations, but was qualitatively satisfactory.

Saturation effects have also been observed in the electric birefringence of a variety of systems by Yoshioka and coworkers. These include potassium polystyrene sulfonate in water, and dioxane-water mixtures [94], PBLG in mixed solvents [95], and potassium polystyrene sulfonate in various solvents [96,97]. Strong saturation effects also were observed in solutions of poly-(L-glutamic acid) (PLGA) in methanol-water mixtures [67,98] and also in a variety of other organic solvent systems. Electric birefringence saturation has also been studied in DNA and polynucleotides by Stellwagen, and is discussed in Chap. 18 (Part 2) [71]. In her experiments, and also in those of Yoshioka and coworkers, large deviations from the Kerr law were observed at <u>low</u> electric field strengths. These are discussed in Chap. 18.

The development of the electric birefringence saturation method has made the Kerr effect a much more efficient tool for studies of the electric and optical properties of macromolecules. As experimental methods are improved, many more biological systems currently of interest can be studied by this technique.

Saturation measurements are also important for the interpretation of dichroism experiments which are discussed in Chaps. 6 and 7.

C. Effects of High Fields on Buildup and Decay

In weak fields, the shape of a birefringence buildup curve is not the same as the decay curve when there is a permanent dipole moment, as illustrated in Fig. 9, but the buildup and decay curves for the induced moment case are symmetric as shown, resembling the curve of a condenser charge and discharge in a resistive circuit. This was shown theoretically in the equations of Benoit [9], and it was confirmed experimentally for TMV solutions [63]. However, when a field producing an orientation energy which is comparable to kT is applied, the buildup will be faster than the decay because of the presence of the torque during the buildup, as shown in Eq. (19).

It is difficult to solve the diffusion equation for arbitrary field strength, but at sufficiently high fields, the effect of rotational diffusion can be neglected as a first approximation, as was pointed out by Schwarz [99]. With this approximation, equations for the buildup of the birefringence were obtained [44] for three special cases:

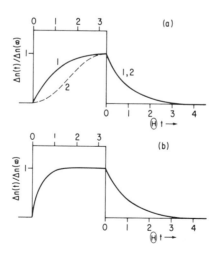

FIG. 9. Buildup and decay of birefringence. (a) 1, Induced moment, weak field; 2, permanent moment, weak field. (b) Induced moment, strong field.

(a) <u>Permanent dipole orientation, axial symmetry</u>. The torque is

$$M = -\mu E B_a \sin \theta \tag{41}$$

for a dipole moment μ along the symmetry axis a. The angular velocity becomes

$$\frac{d\theta}{dt} = - \left(\frac{\mu B_a}{kT}\right) \theta \, E \sin \theta \tag{42}$$

It was also shown [44] that the rise of the birefringence may be expressed

$$\frac{\Delta n(t)}{\Delta n(\infty)} = 1 + \frac{12e^x}{(e^x - 1)^2} - \frac{6xe^x(e^x + 1)}{(e^x - 1)^3} \tag{43}$$

where

$$x = \frac{2\mu B_a}{kT} \tag{44}$$

(b) <u>Induced dipole orientation, axial symmetry</u>. The torque is given by

$$M = -(\alpha_a - \alpha_b) E^2 \sin \theta \cos \theta \tag{45}$$

and the equation obtained [44] for the rise of the birefringence is

$$\frac{\Delta n(t)}{\Delta n(\infty)} = \frac{1}{1 - e^{-y}} \left(1 + \frac{1}{2} e^{-y} - \frac{3 \tan^{-1}\sqrt{e^y - 1}}{2\sqrt{e^y - 1}}\right) \tag{46}$$

where

$$y = 2(\alpha_a - \alpha_b) E^2 \theta t / kT \tag{47}$$

For the more general case of both permanent and induced dipole orientation, with axial symmetry, Nishinari and Yoshioka [100] have calculated the initial rise of the birefringence for arbitrary field strength. Their two equations for the initial buildup in the case of axial symmetry of the electrical properties (a $\neq$ b = c) may be written as follows:

$$\left[\frac{\Delta n(t)}{C_\alpha \Delta n(\infty)}\right]_{t\to 0} = \frac{\alpha_a - \alpha_b}{kT} E^2 \theta t$$

$$+ \left[\frac{\mu^2 B_a^2}{k^2 T^2} E^2 - \frac{3(\alpha_a - \alpha_b)}{kT} E^2 + \frac{(\alpha_a - \alpha_b)^2}{7k^2 T^2} E^4\right]\theta^2 t^2 \tag{48}$$

where we have introduced the coefficient C_α to cover the two
distinct cases

$$\alpha_a - \alpha_b > 0, \quad C_\alpha = \frac{2}{5} \tag{49}$$

$$\alpha_a - \alpha_b < 0, \quad C_\alpha = -\frac{4}{5} \tag{50}$$

Their equations for the orientation function Φ were carried to
higher order in both E and t.

O'Konski, Yoshioka, and Orttung [44] and Nishinari and
Yoshioka [100] showed that it is possible to determine the permanent dipole moment and anisotropy of polarizability separately from
suitable measurements in high fields. Nishinari and Yoshioka's method of separating permanent and induced dipole factors involves
experimental determination of initial slopes of the Δn vs. t
and the $\Delta n/t$ vs. t curves. The initial slope of the Δn vs.
t curve is zero [44] for pure permanent dipole orientation as
shown in Fig. 9. In the mixed case, the initial slope gives the
induced moment term.

(c) Decay of the birefringence from a steady state in
saturating fields. In a dilute system of rigid macromolecules
which is polydisperse, the decay is not exponential and the relative contributions of the species depend upon the field strength.
If we let ϕ_1 be the volume fraction of component i, and we
again consider axially symmetric macromolecules with permanent dipole moment along the symmetry axis of the polarizability ellipsoid, it has been shown that the birefringence decay may be
written [44]

$$\frac{\Delta n(t)}{\Delta n(0)} = \frac{\Sigma \ \phi_i (g_a - g_b)_i \Phi_i e^{-6\Theta_i t}}{\Sigma \ \phi_i (g_a - g_b)_i \Phi_i} \tag{51}$$

where Φ_i is the orientation function in the steady state for component i. In the limit of infinitely high field strength, all the Φ_i tend toward a limiting value of unity so that Eq. (48) reduces to

$$\frac{\Delta n(t)}{\Delta n(0)} = \frac{\Sigma \ \phi_i (g_a - g_b)_i e^{-6\Theta_i t}}{\Sigma \ \phi_i (g_a - g_b)_i} \tag{52}$$

Further, if the system is composed of macromolecules differing only in length and the axial ratio of each component is large, as for a synthetic polypeptide which is rigid, one obtains

$$\frac{\Delta n(t)}{\Delta n(0)} = \Sigma \ \phi_i e^{-6\Theta_i t} \tag{53}$$

Thus, the coefficient of the exponential function gives the value of the volume fraction for each component, and the analysis of birefringence decay from a saturated state will yield the weight distribution of polydisperse systems [44].

D.   Structural Transitions

It is well known that the electrostatic energy density in the immediate vicinity of a monovalent ion is great enough to produce strong ordering of the molecules of a polar solvent, e.g., water. Similarly, the polar groups of macromolecules such as proteins and nucleic acids, are strongly ordered into helical or other regular arrays by intermolecular and intramolecular forces such as the hydrogen bond, which are largely coulombic in nature. Thus, when high external electric fields are applied to macromolecules, there is a possibility that the perturbations would become great enough to alter the structure.

When there exist two possible configurations of a molecule, A and B, the relative populations at equilibrium will depend upon their free-energy difference. That difference must be of the order of RM per mole for both states to coexist at equilibrium. The concentrations of the two species are related by the equation

$$c_B = c_A e^{-(G_B - G_A)/RT} \tag{54}$$

Here c_A, c_B are the molar concentrations and G_A, G_B are the molar free energies. In an electric field the Gibbs free energies will be changed by increments which depend upon the electrical parameters. A structural transition in the macroscopic sense may occur if, for example, it happens that

$$G_A < G_B, \quad E = 0 \tag{55}$$

$$G'_A > G'_B, \quad E = E' \tag{56}$$

The increment of molar Gibbs free energy produced in each species by an applied field is calculable in principle, but requires treatment of the steady-state orientation distributions which will be different for both species in the general case. In high fields, the dielectric polarization of a dilute solution of macromolecules is not linear in the electric field intensity, so that a saturation theory of dielectric polarization would be required to evaluate the electrostatic free energy increments, $G'_B - G_B$ and $G'_A - G_A$. The following assertions should be useful in exploratory experimental studies.

If we have a macromolecular structure of zero dipole moment, it should become less stable than a polar alternative at sufficiently high fields, because the electrostatic free energy, g_E of a dipole μ in local field E_ℓ is given by

$$g_E = -\mu \cdot E_\ell = -\mu E_\ell \cos \theta \tag{57}$$

where θ is the angle between the direction of the dipole moment and the electric field. If all orientations of the dipole were equally probable, the mean value, $\bar{g}_E$, would be zero, but clearly they are not, and parallel orientations are favored [101,102] so that $\bar{g}_E$ is negative. Since this energy term would be zero for a nonpolar structure it is clear that polar structures tend to be favored in electric fields.

If μ is around 10^5 D, then it follows that, at room temperature, $\mu E = 5kT$ when $E = 600$ V/cm. Thus, if a nonpolar macromolecular structure were more stable than this highly polar one in zero field, by an increment of only 5 kcal/mole at 25°C (which is sufficient to keep the less stable form down to 1%) then applying a field of only 0.6 kV/cm could cause a conversion of 50% to the polar structure.

Polarizability contributions to the molecular free energies are expected to be less important than dipole moments in producing structural transitions, but they could be significant at very high fields when no polar structures are stable.

DNA, in the Watson-Crick structure, is an ordered apparently nonpolar macromolecule, except for nonaxial dipole components, which may not fully cancel depending upon the base sequence, in a long rigid helix. Electro-optic studies have been reported on DNA, RNA, and other synthetic polynucleotides, as discussed in Chap. 18 and elsewhere [37].

A dramatic effect in the pulsed birefringence signals of calf thymus DNA was discovered in fields of 10 to 20 kV/cm [71,103]. A typical result is shown in Fig. 10. It was found that the negative birefringence drops precipitously with the field still on at point A, but only at high fields. The birefringence in weak fields is normally monotonic in the time. The field required to produce the effect depended upon DNA concentration and buffer strength. See the figure caption for further information.

It was proposed that this high field effect is probably due to a structural transition in the DNA macromolecule, from the Wilkins "B" (or Watson-Crick) structure to the Wilkins "A" structure [104], or to some other unknown form, perhaps one in which the

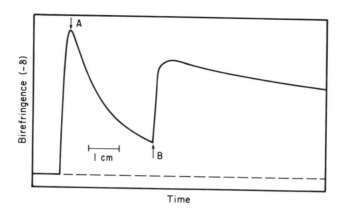

FIG. 10.   High field pulsed birefringence of calf thymus DNA
at 7.4 µg/ml in 10^{-4} M Tris buffer with a dc pulse of 21 kV/cm.
The sweep was 100 µsec/cm and the birefringence was negative, the
peak at A corresponding to $\delta = -0.047$ rad/cm.   The structural
transition commences at the point marked A, and the field was removed at B.   (Reproduced from results given in Ref. 103.)

bases are not perpendicular to the helix axis.   There were indica-
tions that the recovery is essentially complete after the field is
removed.   The high field birefringence experiment on DNA has been
repeated in this laboratory with similar results by P. Moser, and
independently, by M. Shirai, and by L. L. Mack.   Some observations
of changes in absorption of polarized UV light at high electric
fields have also been made at the University of California in
Berkeley and studies are continuing.

The theory of macromolecular structure transitions in electric
fields is highly relevant to problems of membrane biophysics.   It
was suggested that such transitions may be responsible for ion
transport control and the phenomenon of "excitability" associated
with the negative resistance characteristic of nerve membrane [103].

The possibility of electric field-induced effects in macromolecules has not been systematically studied from an experimental
point of view.   The kinetics of such effects are an untouched
subject except for one dielectric investigation on aggregation of

PBLG in solution, due to an electric field in which rates of deaggregation after the field was removed were studied [105]. Clearly, many interesting possibilities of this sort exist for biopolymer systems, and it is expected that as studied advance on biological mechanisms involving proteins and nucleic acids, electro-optic relaxation methods will lead to further observations of structural transformations which are biologically important, as well as physicochemically interesting.

V.   RELATIONSHIP OF ELECTRIC
BIREFRINGENCE THEORY TO THAT OF OTHER ELECTRO-OPTIC EFFECTS

The theories of electric birefringence summarized above involve the calculation of molecular orientation statistics in externally applied fields, and the evaluation of the static or time-dependent refractive index anisotropy of the medium in terms of the intrinsic molecular anisotropies. Theories of other electro-optic effects have, as one of their main elements, the orientation calculation, and then usually another kind of optical anisotropy needs to be evaluated. Usually, the optical effect can be expressed in terms of the degree of orientation function Φ defined above [Eq. (34)]. Once the relation between the observable and the degree of orientation function is worked out, it is often straightforward to write out the equations for any given electro-optic phenomenon originating in molecular orientation. Holcomb and Tinoco [76] and Stoylov [75] also recognized this. One should bear in mind that other than orientation effects may be involved, however. See, especially, Chap. 12 on nonlinear effects.

Thus, the advances in electric birefringence theory may be used as a basis for other theories of molecular electro-optics, as well as other effects, for example, anisotropy of conductivity, anisotropy of dielectric constant, and various magnetic phenomena. It is hoped that future researchers will take full advantage of the static and dynamic electric field orientation theory developed for

the Kerr effect, in predicting and interpreting electric field
phenomena of various types.

VI.   SUGGESTIONS FOR FUTURE STUDIES

Several topics for future work in the field of electric
birefringence of rigid molecules are suggested implicitly and
explicitly throughout this chapter. The purpose of this section
is to show a few of these more implicit possibilities more directly.

We have seen that the Kerr law is not always followed in
electric birefringence, and that this may be due to three causes,
(a) saturation of the orientation, (b) molecular structure changes
produced by high electric fields, and (c) a peculiarity of some
polyelectrolytes at low fields. Effect (c) is not yet fully understood and is a topic of continuing research. Effect (a) is quite
well understood, as discussed above in Sec. IV A. Structural
transitions (b) are highly significant in biology, and may be
studied by application of the powerful methods discussed in this
chapter.

Various aspects of theoretical interpretations would benefit
if equations for arbitrarily shaped particles with nonsuperimpos-
able principal axes of the electric and optical polarizability
tensors would be developed. Further work should also be done on
the theoretical treatment of the observed high-frequency dispersion
in the birefringence of polyelectrolyte systems (Sec. III C). The
interactions between translational and rotational processes need to
be reexamined; of interest here is Brenner's [106] work on coupling
between the translational and rotational Brownian motions of rigid
particles of arbitrary shape.

A number of the theoretical treatments reviewed above need to
be tested using well-characterized samples which conform to the

assumptions implicit in the theories. This can be quite a challenge, considering some of the assumptions. For example, one often assumes a dipole moment along the molecular symmetry axis which presupposes a symmetrical particle with a dipole moment in the proper direction.

Experiments on suspensions of conducting particles and liquid droplets were suggested earlier in this chapter. Hardly any experiments involving dispersion of the Kerr constant in an absorption band have been reported so far (see Sec. II C). This type of measurement could be useful for resolving certain macromolecular structure questions, and for assigning spectral absorptions to specific electronic or vibrational transitions.

Only a single investigation has been found on the Kerr effect in polymeric solids [107], and one study on polymer single crystals in suspension [108]. These suggest promising areas for future investigations in the solid state.

Current advances in electro-optic instrumentation for macromolecule studies include computerization of the apparatus for on-line data acquisition and processing; this is discussed in Chap. 15.

The decay process for electric birefringence in a chemically reacting system has been treated by Kobayasi [109], and this may be useful for the structural transition studies outlined in Sec. IV D.

The electric birefringence technique offers a very high sensitivity, which has not yet been fully exploited. DNA requires only around 5 $\mu g/cm^3$ of solution in a pathlength of 1 cm to achieve a signal-to-noise ratio of 20 or more, when applying a field pulse which saturates the orientation. With a commonly used electrode assembly in a spectrophotometer cuvette [110] the volume of solution required is about 2 cm^3. By using a laser source and a miniature cell, we estimate that the volume of the solution required could be reduced by a factor of 100 to 1000, which would make it possible to do experiments on a few nanograms of DNA.

Reducing the spacing between the electrodes also will facilitate the application of intense fields for saturation and structural transition experiments. Developments along these lines are currently underway in the University of California at Berkeley chemistry laboratories.

ACKNOWLEDGMENTS

This research has been supported in part from research grants as follows: To C. T. O.: Grant-in-Aid CA 12,540-02 from the National Cancer Institute, and to S. K.: Petroleum Research Fund Grant-in-Aid administered by the American Chemical Society. We thank Nancy Stellwagen, B. R. Jennings, and Koshiro Yoshioka for making suggestions leading to improvements in the manuscript.

REFERENCES

1. M. Born, Ann. Physik, $\underline{55}$, 177-240 (1918).

2. P. Langevin, ["Oeuvres scientifiques de Paul Langevin," p. 369-91, Service des publications du CNRS, Paris, 1950] Radium, $\underline{7}$, 249 (1910).

3. A. Peterlin and H. A. Stuart, Z. Physik, $\underline{112}$, 129-147 (1939).

4. A. Peterlin and H. A. Stuart, Hand- und Jahrbuch der chemischen Physik, Vol. 8, Part 1B (A. Eucken and K. L. Wolf, eds.), Akademische Verlagsges., Leipzig (1943), p. 1-115.

5. C. T. O'Konski, J. Phys. Chem., $\underline{64}$, 605-619 (1960).

6. J. A. Osborne, Phys. Rev., $\underline{67}$, 351 (1945).

7. H. C. van de Hulst, Light Scattering by Small Particles, Wiley, New York (1957).

8. C. T. O'Konski, Encyclopedia of Polymer Science and Technology, Vol. 9, Wiley-Interscience, New York (1968), p. 551.

9. H. Benoit, Ann. Phys., $\underline{6}$, 561-609 (1951).

10. D. N. Holcomb and I. Tinoco Jr., Biopolymers, $\underline{3}$, 121-133 (1965).

11. I. Tinoco, Jr., J. Am. Chem. Soc., $\underline{77}$, 4486-9 (1955).

12. C. Bergholm and Y. Björnståhl, Physik. Z., 21, 137-141 (1920).

13. Y. Björnståhl, Phil. Mag., 2, 701-732 (1926).

14. H. T. Buscher, R. M. McIntyre, and S. Mikuteit, IEEE Trans. Microwave Theory Techniques, 19, 950 (1971).

15. G. Mie, Ann. Physik, (4) 25, 377-445 (1908).

15a. C. T. O'Konski, note in preparation.

16. C. T. O'Konski and F. E. Harris, J. Phys. Chem., 61, 1172 (1957).

17. A. R. von Hippel, Dielectrics and Waves, Wiley, London (1954), pp. 228-235.

18. C. T. O'Konski and H. C. Thacher, Jr., J. Phys. Chem., 57, 955 (1953).

19. O. Wiener, Ber. Sächs. Ges. Wiss. (Math. Phys. Klg.), 62, 256 (1910).

20. C. T. O'Konski and B. H. Zimm, Science, 3, 113-116 (1950).

21. C. T. O'Konski and A. J. Haltner, J. Am. Chem. Soc., 79, 5634 (1957).

22. J. Errera, J. Th. G. Overbeek, and H. Sack, J. Chem. Phys., 32, 681-704 (1935).

23. S. A. Rice and M. Nagasawa, Polyelectrolyte Solutions: A Theoretical Introduction, Academic, New York (1961).

24. C. T. O'Konski, J. Chem. Phys., 23, 1959 (1955).

25. J. B. Miles and H. P. Robertson, Phys. Rev., 40, 583 (1932).

26. H. P. Schwan, Adv. Biol. Med. Phys., 5, 147 (1957).

27. J. P. McTague and J. H. Gibbs, J. Chem. Phys., 44, 4295-4301 (1966).

28. J. G. Kirkwood and J. B. Shumaker, Proc. Nat. Acad. Sci. U.S., 38, 855 (1952).

29. U. Schindewolf, Naturwissenschaften, 40, 435 (1953).

30. U. Schindewolf, Z. Elektrochem., 58, 697 (1954).

31. M. Eigen and G. Schwarz, J. Colloid Sci., 12, 181-194 (1957).

32. A. Katchalsky, S. B. Sachs, A. Raziel, and H. Eisenberg, Trans. Faraday Soc., 65, 77 (1969).

33. G. A. Johnson and S. M. Neale, J. Polymer Sci., 54, 241 (1961).

34. J. M. Neale and D. A. Weyl, Proc. Roy. Soc. (London), A291, 368 (1966).

35. S. Takashima, Adv. Chem. Ser., 63, 232-252 (1967).

36. S. Takashima, Biopolymers, 5, 899 (1967).

37. C. T. O'Konski, N. C. Stellwagen, and M. Shirai, Biophys. J., in press.

38. C. T. O'Konski and S. Krause, J. Phys. Chem., 74, 3243 (1970).

39. A. D. Buckingham, Proc. Roy. Soc. (London), A267, 27 (1962).

40. W. H. Orttung, J. Am. Chem. Soc., 87, 924 (1965); J. Phys. Chem., 73, 2908 (1968).

41. J. C. Powers Jr., J. Am. Chem. Soc., 88, 3679 (1966); 89, 1780 (1967).

42. H. G. Kuball, Z. Naturforsch., 22a, 1407 (1967); H. G. Kuball and R. Göb, Z. Physik. Chem. (Frankfurt), 62, 237; 63, 251 (1968); H. G. Kuball and D. Singer, Ber. Bunsenges. Physik. Chem., 73, 403 (1969); H. G. Kuball, W. Galler, R. Göb, and D. Singer, Z. Naturforsch., 24a, 1391 (1969).

43. C. Houssier and H-G. Kuball, Biopolymers, 10, 2421 (1971).

44. C. T. O'Konski, K. Yoshioka, and W. H. Orttung, J. Phys. Chem., 63, 1558-1565 (1959).

45. K. Yoshioka, and C. T. O'Konski, Biopolymers, 4, 499-507 (1966).

46. I. Tinoco, Jr., J. Am. Chem. Soc., 79, 4336 (1957).

47. A. Haschemeyer and I. Tinoco, Jr., Biochemistry, 1, 503 (1962).

48. S. Krause and C. T. O'Konski, Biopolymers, 1, 503 (1963).

49. P. Moser, P. G. Squire, and C. T. O'Konski, J. Phys. Chem., 70, 744 (1966).

50. C. G. LeFevre, R. J. W. LeFevre, and C. M. Parkins, J. Chem. Soc., 1958, 1468 (1958).

51. F. Perrin, J. Phys. Radium, 5, 497 (1934).

52. N. A. Tolstoi and P. P. Feofilov, Dokl. Akad. Nauk SSSR, 66, 617-620 (1949).

53. H. Benoit, Compt. Rend., 228, 1716-1718 (1949).

54. H. Benoit, Compt. Rend., 229, 30-32 (1949).

55. C. E. Marshall, Trans. Faraday Soc., 26, 173-189 (1930).

56. M. A. Lauffer, J. Am. Chem. Soc., 61, 2412-2416 (1939).

57. H. Mueller, Phys. Rev., 55, 508 (1939).

58. H. Mueller, Phys. Rev., 55, 792 (1939).

59. H. Mueller and B. W. Sakman, Phys. Rev., 56, 615-616 (1939).

60. H. Mueller and B. W. Sakman, J. Opt. Soc. Am., 32, 309-317 (1942).

61. F. J. Norton, Phys. Rev., 55, 668-669 (1939).

62. W. Heller, Rev. Mod. Phys., 14, 390-409 (1942).

63. A. J. Haltner, Ph.D. Thesis, University of California,
     Berkeley (1955), p. 108.

64. C. T. O'Konski and A. J. Haltner, J. Am. Chem. Soc., 78,
     3604-3610 (1956).

65. C. T. O'Konski and J. B. Applequist, Nature, 178, 1464-1465
     (1956).

66. I. Tinoco, Jr. and K. Yamaoka, J. Phys. Chem., 63, 423-427
     (1959).

67. M. Matsumoto, H. Watanabe, and K. Yoshioka, J. Phys. Chem.,
     74, 2182-2188 (1970).

68. S. Krause and C. T. O'Konski, J. Am. Chem. Soc., 81, 5082
     (1959).

69. R. M. Pytkowicz and C. T. O'Konski, Biochim. Biophys. Acta,
     36, 466-470 (1959).

70. C. M. Paulson, Ph.D. Thesis, University of California,
     Berkeley (1965).

71. N. C. Stellwagen, Configurations of Sodium Deoxyribonucleate
     and Sodium Phosphate in Solution, Thesis, University of
     California, Berkeley (1967).

72. B. R. Jennings and B. L. Brown, Eur. Polymer J., 7, 805 (1971).

73. B. L. Brown, B. R. Jennings, and H. Plummer, Appl. Opt., 8,
     2019 (1969).

74. K. Yoshioka and H. Watanabe, Physical Principles and Tech-
     niques of Protein Chemistry, Part A (S. J. Leach, ed.),
     Academic, New York (1969), pp. 335-67.

75. S. P. Stoylov, Adv. Colloid Interface Sci., 3, 45-110 (1971).

76. D. N. Holcomb and I. Tinoco, Jr., J. Phys. Chem., 67, 2691-
     2698 (1963).

77. S. Broersma, J. Chem. Phys., 32, 1626-1631 (1960).

78. R. C. Williams and R. L. Steere, J. Am. Chem. Soc., 73, 2075
     (1951).

79. A. J. Haltner and B. H. Zimm, Nature, 184, 265 (1959).

79a. F. S. Allen and K. E. Van Holde, Biopolymers, 10, 865 (1971).

80. C. T. O'Konski and R. M. Pytkowicz, J. Am. Chem. Soc., 79,
     4815 (1957).

81. C. V. Raman and S. C. Sirkar, Nature, 121, 794 (1928).

82. H. Benoit, J. Chim. Phys., 49, 517-521 (1952).

83. G. B. Thurston and D. I. Bowling, J. Colloid Interface Sci., 30, 34 (1969).

84. H. Plummer and B. R. Jennings, J. Chem. Phys., 50, 1033-1034 (1969).

85. S. J. Jakabhazy and S. W. Fleming, Biopolymers, 4, 793-813 (1966).

86. N. A. Tolstoi, A. A. Spartakov, and A. A. Trusov, Kolloid Zh., 28, 735-741 (1966).

87. M. J. Shah, D. C. Thompson, and C. M. Hart, J. Phys. Chem., 67, 1170 (1963).

88. V. N. Tsvetkov, Yu. V. Mitin, V. R. Glushenkora, A. Ye. Grischenko, N. N. Boitsova, and S. Ya. Lyubina, Vysokomolekul. Soedin., 5, 453 (1963).

89. R. Gans, Ann. Physik, 64, 481 (1921).

90. M. J. Shah, J. Phys. Chem., 67, 2215 (1963).

91. M. Matsumoto, H. Watanabe, and K. Yoshioka, Sci. Papers Coll. Gen. Educ. Univ. Tokyo, 17, 173-202 (1967).

92. K. Yoshioka and C. T. O'Konski, J. Polymer Sci., A-2, 6, 421 (1968).

93. H. Boedtker and P. Doty, Jr., J. Am. Chem. Soc., 78, 4267 (1956).

94. H. Nakayama and K. Yoshioka, Nippon Kagaku Zasshi (J. Chem. Soc. Japan), 85, 177-182 (1964).

95. H. Watanabe, K. Yoshioka, and A. Wada, Biopolymers, 2, 91-101 (1964).

96. H. Nakayama and K. Yoshioka, J. Polymer Sci., A-3, 813-825 (1965).

97. K. Kikuchi and K. Yoshioka, Rep. Prog. Polymer Phys. Japan, 10, 19-22 (1967).

98. M. Matsumoto, H. Watanabe, and K. Yoshioka, Biopolymers, 6, 905-915 (1968).

99. G. Schwarz, Z. Physik, 145, 563-584 (1956).

100. K. Nishinari and K. Yoshioka, Kolloid-Z. und Z. Polym., 235, 1189 (1969).

101. P. Debye, Polar Molecules, Dover, New York (1929).

102. H. Frölich, Theory of Dielectrics, Oxford University Press, London (1949).

103. C. T. O'Konski and N. C. Stellwagen, Biophys. J., 5, 607-613 (1965).

104. R. Langridge, H. R. Wilson, C. W. Hooper, M. H. F. Wilkins, and L. D. Hamilton, J. Mol. Biol., 2, 19-37 (1960).

105. M. Gregson, G. P. Jones, and M. Davies, Chem. Phys. Letters, 6, 215 (1970).

106. H. Brenner, J. Colloid Sci., 23, 407 (1967).

107. H. Pursey, Nature, 227, 834 (1970).

108. C. Picot, C. Hornick, G. Weill, and H. Benoit, J. Polymer Sci., C30, 349 (1970).

109. S. Kobayasi, Biopolymers, 10, 915, 2649 (1971).

110. R. M. Pytkowicz and C. T. O'Konski, Biochim. Biophys. Acta, 36, 466 (1959).

Chapter 4

THEORY OF ROTATIONAL DIFFUSION OF SUSPENSIONS

Don Ridgeway

Departments of Statistics and Physics
North Carolina State University
Raleigh, North Carolina

I. INTRODUCTION

The theory of rotational diffusion of a suspension of
particles deals with the random time evolution of rotational degrees of freedom of the suspension. As such, its results appear in
the theory of any experimental method, electro-optical or

hydrodynamic, in which contributions of individual particles to the
measured quantity depend either on the instantaneous orientations
of the particles or on time autocorrelations of orientation parameters. One of the most effective areas of application of such
methods has been the estimation of shapes and dimensions of rigid
macromolecules, primarily the proteins. It is the purpose of the
present chapter to discuss those parts of rotational diffusion
theory which usually underlie its role in the theories of these
methods.

We have chosen to present the subject within its context in
the theory of stochastic processes. For translational diffusion,
the comprehensive review of Chandrasekhar [1] covering all aspects
of this problem has long been available. The mathematical theory
of diffusion has been treated in detail in the recent monograph by
Nelson [2]. However, the fundamental distinction between differentiable spatial coordinates, such as the Cartesian coordinates of
the particle mass center, and kinematic representations of state,
such as the particle Euler angles, which separates the translational from the rotational problem, is serious enough to require separate treatment of rotational diffusion. In fact, it is precisely
in the point of greatest theoretical interest, the relationship
between the microscopic Ornstein-Uhlenbeck process and the
phenomenological diffusion equation, that this distinction expresses itself most directly. We shall not discuss specific aspects of solutions of the rotational diffusion equation and their
analytical properties, since the recent review of Favro [3] is
available. The form of application of these results in theories
of estimation of protein shape is displayed by Ridgeway [4,5],
specifically in the theory of Kerr (electric) birefringence of
protein solutions. Moreover, one cannot, within the scope of this
volume, present either the proper background (measure theory) or
the analytical formalism (theory of semigroups of operators on
function spaces) of the mathematical theory of diffusion. Nevertheless, we do feel that the physical meanings of the respective

parts of diffusion theory, their mutual interrelationships, and the assumptions underlying its results are particularly clear in even a qualitative discussion of diffusion theory in this context.

The thermal motion of suspensions is the subject of three relatively distinct areas of study. Since the diffusion tensor is defined in terms of coefficients of the diffusion equation, it is implicit in any description of the time behavior of a suspension which introduces a diffusion tensor that the aggregate or macroscopic behavior of the system is governed by a diffusion equation. Mathematical diffusion theory, as we shall use the term, is the study of the consequences of stochastic postulates about the microscopic motion of the system which are just sufficient to assure this property. It is the results of this area primarily that are employed in study of particle shape. The second area is rigorous theory of Brownian motion, which, as employed here, deals with a completely deterministic (to within assertion of a probability density function) dynamical theory of the microscopic motion of the suspension based on classical equations of motion for particles and solvent molecules. One important result in the great progress made in this area, particularly by Lebowitz and coworkers [6,7] is that the diffusion equation becomes an increasingly good description of time behavior as the solvent molecule-particle mass ratio decreases. This work has dealt almost exclusively with translational motions. The third area, study of Ornstein-Uhlenbeck (OU) processes, i.e., solutions of Langevin equations, is intermediate between the other two areas. It is closer to the theory of Brownian motion in being a dynamical theory, but closer to the mathematical theory of diffusion in that the particle-solvent coupling is defined by a stochastic postulate constructed to assure that the motion satisfy certain phenomenological laws. The OU processes have long been accepted as the correct stochastic basis for translational diffusion theory, and the Langevin equation, or its non-Markovian generalization, is the foundation of Kubo [8] linear response theory.

II. PHYSICAL MODEL

We consider first the physical aspects of the motion of a
polymer particle in thermal equilibrium with a suspending medium of
much smaller molecules. In the absence of external fields, the
forces on the particle primarily responsible for its motion are impulses received in the continuous rain of collisions from solvent
molecules, the mean frequency of which is by symmetry uniform over
the surface envelope of the particle if it is at rest, whatever its
shape. The integrated effects of these impulses are conveniently
described as representing two kinds of forces. The one kind is the
torque produced by vastly prolonged transient fluctuations in
collision frequency from the mean at different points on the particle surface. These fluctuations and their frequency are systematic
properties of the thermal motion of the solvent molecules. The
other kind of force is the dissipative torque experienced by any
particle in motion. It results from the fact that both the mean
collision frequency and the momentum imparted per colliding solvent
molecule of given normal velocity component are themselves larger
on the leading surfaces of the particle in a given motion than on
the trailing surfaces. Since the torque at a given surface element
for either kind of force is proportional to the distance of that
element from a rotation axis, both kinds of forces are dependent on
particle shape.

In describing the motion of such a system, one incorporates
these physical aspects as faithfully as he can into his formal
description of the system, recognizing that greater detail of description usually implies a more difficult analytical problem.
Clearly, the above picture requires an exact description of
particle-solvent molecule interactions upon collision. In the
theory of Brownian motion, for example, one most often assumes
that the particle is smooth and collisions elastic and specular.
In this case, rotational motion does not appear in the Brownian
problem because there is no way for the particle to exchange

rotational energy either with solvent or with translational
degrees of freedom. However, in order to remove analytical difficulties, work dealing with particle shape and its estimation usually simplifies this exact picture. The solvent molecules are
replaced by a viscous continuum, and their integrated thrust on the
particle by some randomly acting, continuous force. One then
assumes that the viscous resistance encountered by the particle in
its motion is that found from ordinary hydrodynamic laws of viscous
liquids, assigning to the medium the bulk viscosity of solvent and
to the particle its actual shape and size.

We shall use this physical model as the basis of most of the
following. Let us discuss, before so doing, two restrictions it
contains dealing with other forces, the interparticle interactions
and external fields, to which a particle in suspension might be
subjected. The assumption is made throughout that the suspensions
treated are sufficiently dilute such that we can consider the motions of the individual suspended particules to be mutually independent. It then suffices to study the random motion of a single
particule. The effects of interparticle interactions on Brownian
motion of dilute suspensions are taken into account explicitly by
Mazo [9]. Moreover, we direct our attention largely to the free
rotational motion of suspensions, i.e., motion in the absence of
externally applied fields. This restriction avoids discussion of
a number of interesting phenomena still requiring much theoretical
work, such as proton migration [10,11], which appear particularly
in interactions of proteins with external fields. It does not restrict applicability of the results to the field-free methods,
however. In fact, so long as the field interaction of a system is
linear in the external field strength, the appropriate theoretical
treatment of that system is through the formalism of linear response theory [8,12,13]; Berne and Harp [14] deal in particular with
application of this theory to experimental methods. The quantity required in this formalism to obtain the full dispersive
susceptibility in a given field problem is the time covariance

matrix of particular dynamical variables, depending on the nature
of the applied force and on the measured quantity, averaged over
the system in thermal equilibrium in the absence of external
fields. The covariance matrix which applies in linear response
theory in calculation of the rotational diffusion tensor arises
directly in study of free diffusion and will be displayed and so
indicated below. The relationship between linear response theory
and theories which include higher-order interactions is discussed
in particular by Bernard and Callen [15]. The concluding section
of this chapter is a brief discussion, without motivation of
assumptions, of the general theory of forced diffusion of suspensions.

Theory must provide two results if we are to estimate particle
shape from diffusion data. First, since a measured quantity is expressed as the expectation value of a state function, one needs the
instantaneous conditional orientation density, for given initial
distribution, with respect to which expectation values are defined.
Second, one must have a microscopic theory which relates the parameters which characterize diffusion data, viz., the elements of the
diffusion tensor, to those describing particle shape, the elements
of the particle viscous drag tensor in the present model. We treat
each of these in turn. The first is obtained from a stochastic
description of particle motion. The second is derived by relating
this description to the laws of motion and to physical equilibrium
distributions.

III.   STOCHASTIC PROCESSES

A.   Markov Processes

We define the instantaneous state of a particle to be its
angular orientation at that instant, independent of velocity, and
state space to be a suitable space in which distinct orientations

map as distinct points. State space may be, for example, a
subvolume of Euclidean 3-space, if orientation is described by the
Euler angles, or the surface of the unit ball in Euclidean 4-space,
if representation is in terms of the Cayley-Klein parameters. Denote by S the set, called the sample space, of all possible
trajectories a random particle might describe in time. We are concerned with characterization of the set of functions or random
process $\{x_t, \ t \geq 0\}$ mapping sample space, with a probability
structure we shall assign it, into state space such that for each
time t and fixed trajectory $\alpha \ \varepsilon \ S$, the point $x_t(\alpha)$ is the
state at time t of the particle on trajectory α. A set of values $\{x_t(\alpha), \ t \geq 0\}$ is called a sample path of the process. By
assuming throughout that the distributions in state space always
admit probability density functions, as we do, we avoid explicit
use of measure-theoretic notions and characterize the chance behavior of the particle orientation by density functions on state
space.

 We assume that the Brownian motion of the suspended particle
is a Markov process, i.e., that the conditional probability that
the trajectory of a particle pass through any fixed set of states
at particular times in the future, given the initial state, is
independent of motion before the initial state. This assumption is
based on the orders-of-magnitude difference in collision frequency
and in the frequency of local fluctuations significant to Brownian
motion in all cases of interest here, since it follows from this
difference that the effects of particle motion on the motion of
solvent molecules are dissipated almost instantaneously on the time
scale of Brownian motion. It is deviations from this assumption,
associated with inertial effects, that lead to dispersive viscous
behavior. A Markov process may be characterized uniquely by its
transition density, i.e., the function with values $p(x,t; \ y,s)$
which states the conditional density at orientation x, time t,
given the orientation y at an earlier time s, for all x,y
in state space and all times $t,s > 0$. The transition function of

a Markov process always satisfies as a composition rule the
Chapman-Kolmogorov equation, $p(x,t; y,s) =$
$\int p(x,t; z,u)p(z,u; y,s)\, dz$ for all $t \geq u \geq s \geq 0$, where
integration is over all of state space. The transition function
for a particle in thermal equilibrium with solvent in the absence
of external fields clearly depends on time only through the lapsed
time $t - s$ of a transition and not on t,x themselves, i.e., it
is temporally homogeneous. In this case, the Chapman-Kolmogorov
equation becomes $p(t + s,x,y) = \int p(t,x,z)p(s,z,y)\, dz$, where we
have written $p(x,t; y,s) = p(x,t - s; y,0) \equiv p(t - s,x,y)$, say.

In studying the time evolution of a system with an underlying
Markov process, one introduces two operators constructed from the
transition density of the process. First, let X be the set of
all bounded, integrable functions on state space. Define, for
each time t, the operator $T_t : X \to X$ by
$T_t f(x) \equiv \int f(y)p(t,y,x)\, dy$. It is observed that the integration
is over final states y of the transition function. Thus, T_t
transforms any function f belonging to X into the conditional
expectation $T_t f$ at time t of the function f given the initial
state x. Similarly, let Y be the set of all suitable density
functions on state space, and define, for each time t, the opera-
tor $U_t : Y \to Y$ by $U_t \rho(y) \equiv \int \rho(x,0)p(t,y,x)\, dx$. Now integration
is over initial states in the transition function. The operator
U_t transforms an initial probability density in state space,
$\rho = \rho(x,0)$, say, into the conditional density $\rho(y,t) \equiv U_t \rho(y)$
into which the system evolves in time t as the Markov process,
given the initial density. That transition function which generates a particular operator U_t is called its kernel or Green
function. Since one computes instantaneous expectation values of
state functions, i.e., $\langle f \rangle_t \equiv \int f(x)\rho(x,t)\, dx$, using the conditional density $\rho = \rho(x,t)$ for some given initial distribution, it
is the operator U_t or, equivalently, its Green function, which is
usually required in theories of physical methods. Most of modern
mathematical diffusion theory is based on studies of the sets of

operators $\{T_t\}$ and $\{U_t\}$ because of the applicability of the powerful methods of analytic semigroup theory to these sets. The greater attention has been given to T_t for reasons discussed by Feller [16, Sec. X.8].

B. The Diffusion Process

Let us now discuss properties of the first of two kinds of Markov processes applied to diffusion problems. A _diffusion process_ is a continuous Markov process with transition function which satisfy the following conditions: For all $\varepsilon > 0$, as $t \to 0$,

$$t^{-1} \int_{\|x - y\| > \varepsilon} p(t,x,y) \, dx \to 0$$

$$t^{-1} \int_{\|x - y\| < \varepsilon} p(t,x,y) (x_i - y_i) \, dx \to b_i(y)$$

$$t^{-1} \int_{\|x - y\| < \varepsilon} p(t,x,y) (x_i - y_i)(x_j - y_j) \, dx \to a_{ij}(y)$$

where $\|\cdot\|$ is any suitable norm on state space [cf. Ref. 16, Sec. X.4]. The first condition, called the _Lindeberg_ condition, has the physical meaning that large jumps become ever less likely as the transition time decreases. The remaining conditions characterize the first and second moments of the transition probabilities. Note in particular that the second moments, i.e., the _covariance_ elements, are proportional to the lapsed time t. We mention that Dynkin [17, Sec. 5.5] gives a more general definition of the diffusion process in terms of differential operators which characterize the random process and discusses the relationships between the two definitions.

The reason a random process satisfying the above conditions is called a diffusion process follows from the local differential behavior of the operators T_t and U_t associated with it. Consider a diffusion process with mean $b_i(x)$ and covariance matrix $a_{ij}(x)$ in an n-dimensional state space with points $x = (x_1, \ldots, x_n)$. The following results obtain, which were first announced in the famous paper of Kolmogorov [18]. Let f be any function in the domain X of T_t and let $g(x, t) = T_t f(x) \equiv \int f(y) p(t, y, x)\, dy$. Then g is a solution of the backward equation:

$$\frac{\partial g}{\partial t} = \frac{1}{2} a_{ij} \frac{\partial^2 g}{\partial x_i \partial x_j} + b_i \frac{\partial g}{\partial x_i}$$

where summation here and elsewhere is implied over all indexes repeated within a single term. Moreover, let $\rho = \rho(x)$ be any density function in the domain Y of U_t and for all $t > s$ let $\rho(x,t) \equiv U_t \rho(x) \equiv \int \rho(x) p(x,t; y,s)\, dy$. Then $\rho(x,t)$ is a solution of the forward equation:

$$\frac{\partial \rho}{\partial t} = \frac{1}{2} \frac{\partial^2}{\partial x_i \partial x_j} [a_{ij}(x)\rho] - \frac{\partial}{\partial x_i} [b_i(x)\rho]$$

(existence of the indicated derivatives is assumed). The forward equation is recognized to be the usual diffusion equation of physical phenomena, the coefficients a_{ij} and b_i being the diffusion tensor and the drift coefficients, respectively, of the phenomenon. It is noted that in translational diffusion, the function $\rho = \rho(x,t)$ is just the concentration at the point x at time t. Hence, under these conditions, all diffusion processes lead to solutions of the diffusion equation. Conversely, Wiener [19] showed that if the process is spatially as well as temporally homogeneous, i.e., if $p(t,x,0) = p(t,-x,0)$, so that the drift coefficients $b_i = 0$, then the only continuous solutions of the forward diffusion equation are the diffusion processes, called in this case Wiener processes. The free diffusion of a suspension is clearly spatially homogeneous. More general forms of the relationships between these differential equations and diffusion processes are given by Yosida [20, Chap. 13]. The principal tool employed by Favro [21] in his analysis of rotational diffusion of suspensions

is based on the result that the Green function p(t,x,y) of a
diffusion process itself satisfies a <u>forward</u> or <u>Fokker-Planck</u>
equation,

$$\frac{\partial p}{\partial t} = \frac{1}{2} \frac{\partial^2}{\partial x_i \partial x_j} [a_{ij}(x)p] - \frac{\partial}{\partial x_i} [b_i(x)p]$$

Favro [21] displays the derivation of the rotational diffusion
equation (his equation 2.6) from the above definition of a diffusion process for the spatially homogeneous case. The Markov
assumption is implicit in his equation 2.0, which defines the
operator U_t, the Lindeberg condition is his condition (c), and
his condition (b) characterizes the diffusion coefficients.
Favro's analysis is a generalization to rigid particles of arbitrary shape of the treatment by Furry [22] of the suspension of
spheres.

C. The Ornstein-Uhlenbeck Process

     Before discussing the second class of Markov process applied
to diffusion theory, we point out two aspects of the diffusion process which have led to study of this second class. The first aspect has to do with the velocity of a diffusion process, or, since
we are concerned with the spatially homogeneous case, of a Wiener
process. Wiener showed that with probability one the trajectory of
a Wiener process is not time differentiable and hence is of un-
bounded variation (infinite length). The first property means that
the ordinary definition of the velocity is inapplicable, the second that any reasonable definition introduced must lead to infinite
velocities. Qualitatively, one notes already in the definition of
the diffusion process that if the variance of the spatial distribution is proportional to the lapsed time t, then the root-mean-
square (rms) distance diffused is of order $t^{1/2}$, and the velocity
of order $t^{-1/2}$. The latter increases without bound as t tends

to zero. Evidently, neither of these is physically acceptable as a property of particle motion. We note that since the drift-free diffusion equation is equivalent to the Wiener process, rigorous application macroscopically of the diffusion equation does imply these inconsistencies in the microscopic motion. These conclusions all obtain for angular as well as linear velocities. The other aspect of the diffusion process which has required further study is that its definition does not provide enough connection with microscopic physical motion of the particle to relate its parameters to particle shape. In particular, one notes that the parameters of a diffusion process, as defined earlier, are sufficiently free of structure from the statistical and mechanical laws of nature as to contain no explicit dependence on either the temperature or particle mass.

Two approaches have been followed. In the first, employed by Einstein [23], who gave the first probabilistic derivation of the diffusion equation, assumes that the particle motion is a diffusion process. By studying the motion necessary to maintain an equilibrium distribution against diffusion streaming in the presence of a static applied field, Einstein related the viscous drag tensor and the diffusion tensor, coming from the drift and diffusion terms, respectively, to the temperature and particle mass appearing in the canonical density. Favro relates these quantities through an analogous treatment for the rotational case. This approach, which is indicated in the section on forced diffusion here, solves the second, but not the first, of the above objections. The other approach, followed in modern translational diffusion theory, introduces a different random process, the OU process, for particle motion at the outset. It removes the above difficulties with the particle velocities, yields the diffusion equation macroscopically except at very short diffusion times, and does not introduce interaction with external fields. Although the analogous treatment for the rotational case is at present incomplete, it is presumably the structure the theory eventually will take, as in the translational case. This fact is already accomplished in one

sense in application of the Kubo fluctuation-dissipation theorems
to rotational diffusion, for these are based on the OU process or
its non-Markovian generalization.

We introduce now the Ornstein-Uhlenbeck process [24,25,1,26].
Consider a space, called velocity space, in which distinct points
represent distinct velocities, i.e., rates of change of state parameters of a random particle. Let $\{B_t, t \geq 0\}$ be an n-dimensional Wiener process with values in this space having covariance
parameter matrix a_{ij}. The Markov process $\{v_t, t \geq 0\}$ which is
a solution of the system of stochastic differential equations

$$\frac{dv_i(t)}{dt} = -\gamma_{ij}v_j + \frac{dB_i(t)}{dt} \qquad (i = 1, \ldots, n)$$

is called an Ornstein-Uhlenbeck (OU) velocity process. It is
emphasized that sample paths of the processes $\{B_t\}$ and $\{v_t\}$ are
sequences of velocities, not states. This system of equations, an
n-dimensional Langevin equation [27], can have only formal significance because, as we have pointed out, the Wiener process $\{B_t\}$ is
not differentiable. It is well defined in terms of stochastic
integrals [cf. Refs. 28 and 29, Sec. 6.3]. If γ_{ij} is diagonal,
$\gamma_{ij} = \gamma_{(i)}\delta_{ij}$, say, then the equation has the immediate appearance
of the equation of motion of a particle of unit mass with viscous
drag tensor γ_{ij} in a viscous continuum subjected to a random
force dB_t/dt. The OU velocity process is a Gaussian process, with
the expressions for its mean and covariance given below. Its Green
function for the operator U_t on the space of velocity distributions is a solution of the forward (Fokker-Planck) equation [29,
Sec. 6.3]:

$$\frac{\partial p}{\partial t} = \frac{1}{2}\frac{\partial^2}{\partial v_i \partial v_j}(a_{ij}p) + \frac{\partial}{\partial v_i}(\gamma_{ih}v_h p)$$

The Green function tends asymptotically in the long-time limit to
the unconditional Maxwell-Boltzmann velocity distribution. If the
OU velocity process is integrable, i.e., if the process

$\{x_t, t \geq 0\}$ in state space is well defined by $x_t - x_o = \int_o^t v_s \, ds$, then the OU velocity process is closely related to a diffusion equation in state space. In fact, the process $\{x_t, t \geq 0\}$, called the associated OU process, is a Gaussian Markov process which converges in the long-time limit to a Wiener process [2, Theorems 9.3-5]. We illustrate one of the modes of convergence by the one-dimensional case. Let $\{B_t\}$ be a Wiener process with variance parameter $2\gamma^2 D$, and define the velocity process $\{v_t\}$ by $dv_t = -\gamma v \, dt + dB_t$. Suppose $\{v_t\}$ is integrable, and define $x_t = x_o + \int_o^t v_s \, ds$. Then the process x_t has mean $x_o + \gamma^{-1}(1 - e^{-\gamma t})v_o$ and variance $2Dt + \gamma^{-1}D(-3 + 4d^{-\gamma t} - e^{-2\gamma t})$. Clearly, these tend asymptotically to mean zero (if $v_o = x_o = 0$) and variance $2Dt$. But a Gaussian process with this mean and variance is a Wiener process with Green function the solution of the forward (Fokker-Planck) equation $\partial p/\partial t = D\partial^2 p/\partial x^2$. It follows, then, that if the OU process is well defined, it is natural to replace the Einstein description of particle motion as a diffusion process and to describe the motion as an OU process. This assures that the velocity exists and is continuous for all $t \geq 0$, since it is just the process $\{v_t\}$, and that the rms velocity is finite in the above sense in the limit of short time. Moreover, the system evolves in time in a way approximately described by a diffusion equation, the diffusion tensor appropriate to a given OU process being the covariance parameter matrix, to within a factor of 2, of the Wiener process to which it converges. We shall find below that this identification of an OU process and a diffusion process leads directly to an expression relating the viscous drag tensor and the diffusion tensor.

IV.   APPLICATION TO PARTICLE MOTION

A.   The Langevin Equation

Let us first show that the equation of motion for angular motion of the particle leads to a Langevin equation in the linear approximation.   Let I_{ij} be the particle inertia tensor referred to a given Cartesian particle-fixed frame and let ω_i be the angular velocity about the ith axis of that frame.   Euler's equation of motion is

$$I_{ij} \frac{d\omega_j}{dt} + \varepsilon_{ijk} I_{kl} \omega_j \omega_l = M_i$$

where ε_{ijk} is the completely antisymmetric 3-tensor, and M_i is the torque about the ith axis.   Set $M_i = -\gamma_{ij}\omega_j + (dB_i/dt)$, where γ_{ij} is the particle viscous drag tensor referred to the same axes, and $\{B_i(t)\}$ is a Wiener process independent of the motion.   Then

$$I_{ij} \frac{d\omega_j}{dt} = -(\gamma_{ij} + \varepsilon_{ijk} I_{kl} \omega_l) \omega_j + \frac{dB_i}{dt}$$

If the angular velocities are very small, then the quadratic term on the right is always small with respect to the linear term.   We neglect it, noting that the term dB_i/dt is not correlated with the neglected term.   Since the viscous drag tensor is always symmetric, we may, without loss of generality, assume that the particle reference axes have been selected to diagonalize it, say $\gamma_{ij} = \gamma_{(i)} \delta_{ij}$ (the sum convention not extending to indexes in parentheses).   Contraction with the inverse I_{ij}^{-1} of the inertia tensor yields

$$\frac{d\omega_i}{dt} = -I_{ik}^{-1} \gamma_{(k)} \omega_k + I_{ik}^{-1} \frac{dB_k}{dt}$$

With introduction of the definitions $\gamma'_{ik} = I_{ik}^{-1} \gamma_{(k)}$, $B'_i = I_{ik}^{-1} B_k$, noting that this conserves the Wiener property, we obtain finally

$$\frac{d\omega_i}{dt} = -\gamma'_{ik}\omega_k + \frac{dB'_i}{dt}$$

In this approximation, then, the angular velocity of the continuum model is an OU velocity process.

B. The Viscous Drag Tensor

We apply properties of the OU processes to this case. First, the velocity process $\{\omega_t\}$ exists and is finite for all t. It is a Gaussian process, with mean $<\omega_i(t)> = e^{-t\gamma'_{ij}}\omega_j(0)$ and covariance

$$r_{ij}(t,s) \equiv <\omega_i(t)\omega_j(s)> - <\omega_i(t)><\omega_j(s)> =$$

$$e^{-(t-s)\gamma'_{ih}} \int_o^s e^{-\tau\gamma'_{hk}} a'_{k\ell} e^{-\tau\gamma'_{j\ell}} d\tau$$

where $<dB'^2_i> = a'_{k\ell} dt$ [2, Theorem 8.2]. We have used the fact that the matrices γ_{ij} and I_{ij} are symmetric. This result leads to a relation expressing the viscous drag tensor γ_{ij} in terms of the covariance parameters $a_{k\ell}$ of the Wiener process producing the random force. Indeed, from the Gaussian property [e.g., Ref. 29, Sec. 2.3], the Green function for the velocity process is

$$p(t,\omega,\omega_o) = [(2\pi)^{-3} \det |r^{-1}_{ij}(t, t)|]^{1/2} \exp \{-\frac{1}{2}r^{-1}_{ij}(t, t) [\omega_i(t) -$$

$$- e^{-t\gamma'_{ih}}\omega_h(0)]$$

$$[\omega_j(t) - e^{-t\gamma'_{jk}}\omega_k(0)]\}$$

This tends asymptotically to

$$[(2\pi)^{-3} \det Q^{-1}]^{1/2} \exp (-\frac{1}{2} Q^{-1}_{ij}\omega_i\omega_j)$$

where $Q_{ij} \equiv \int_o^\infty e^{-t\gamma'_{ih}} a'_{hk} e^{-t\gamma'_{kj}}$ dt. But by definition, the OU
velocity process tends asymptotically to the unconditional Maxwell-
Boltzmann velocity distribution for thermal equilibrium, i.e.,
$[(2\pi kT)^{-3} \det I]^{1/2}$ exp $[- (1/2kT)I_{ij}\omega_i\omega_j]$. It follows that
$kTQ_{ij}^{-1} = I_{ij}$. Integration of this expression leads to the desired
result.

In the special case that the same choice of particle axes
diagonalizes the inertia tensor as well as the viscous drag tensor,
say $I_{ij} = I_{(i)}\delta_{ij}$, which is true for the circular disk and
(trivially) for the sphere, but may be true for more general shape
classes, these results simplify greatly. The Langevin equation becomes three uncoupled stochastic equations,

$$\frac{d\omega_i}{dt} = -\gamma'_{(i)}\omega_i + \frac{dB'_i}{dt} \quad (i = 1,2,3)$$

If $a'_{(i)}$ is the variance parameter of B_i, i.e.,

$<dB'_i(t)^2> = a'_{(i)}$ dt $= I_{(i)}^{-2} a_{(i)}$ dt, then $\omega(t)$ is Gaussian with

mean $e^{-t\gamma'_{(i)}}\omega_i(0)$ and variance

$$r_{ii}(t) = (a'_{(i)}/2\gamma'_{(i)})(1 - e^{-2\gamma'_{(i)}t}) \sim a_{(i)}/2\gamma_{(i)}I_{(i)}$$

From comparison with the Maxwell-Boltzmann distribution again, one
obtains $a_{(i)} = 2\gamma_{(i)}kT$, $k = 1,2,3$. We note that the elements of
the above covariance parameter matrix, which reduce to the variance
parameters in the special case, are the <u>autocorrelation functions</u>
which appear most often in Kubo formalism dealing with rotational
thermal motion.

C. The Diffusion Tensor

Before proceeding to the diffusion equation, we introduce the
diffusion tensor normally calculated in linear response theory.

The above forward equation for the Green function of the OU
velocity process is a diffusion equation with a drift term, with
diffusion tensor $D_{(\omega)ij} \equiv \frac{1}{2} a_{ij}$. In the case of the angular
velocity process constructed from the Wiener process B_t with
covariance parameter matrix a'_{ij}, the diffusion tensor is
$D_{(\omega)ij} \equiv \frac{1}{2} I^{-1}_{ih} I^{-1}_{jk} a_{hk}$. The relationship between it and the viscous drag tensor in the above special case is $D_{(\omega)(i)} = I^{-2}_{(i)} kT\gamma_{(i)}$.
These are the results displayed in Kubo [8]. It is critical to
note that this diffusion equation and diffusion tensor are defined
in terms of the velocities, not the spatial coordinates.

The usual diffusion equation, that in terms of which one defines the measured diffusion tensor, is referred to coordinates in
state space. We have seen that in the translational case, the OU
process converges to a diffusion process, thereby providing a
natural relationship between the parameters of the two processes.
Although attempts have been made [30,31] to apply the same results
to the rotational case, the theorems of diffusion theory in their
present form are not sufficiently general to include rotational
motion. Because of its importance here, we examine this fact in
some detail. Demonstrations of convergence of the OU process are
based on properties of the mean and covariance matrix of the process [cf. Ref. 2, proof of Theorem 9.3]. The above expressions for
the mean and covariance parameters of the OU process $\{x_t, t \geq 0\}$
are taken with respect to the probability structure assigned to the
Wiener process $\{B_t\}$ from which the associated OU velocity process
$\{v_t\}$ is constructed. This appears remarkable because $\{B_t\}$ and
$\{v_t\}$ are stochastic processes in velocity space, whereas $\{x_t\}$ is
a process in state space. Its justification depends on two facts.
First, since solutions of the Langevin equation are unique [25,
Theorem 8.1], to each Wiener trajectory there corresponds exactly
one velocity process which is a solution of the Langevin process.
It is therefore possible to consider the random process $\{v_t\}$ to
be defined on the sample space of $\{B_t\}$, with its probability
structure, and to compute expectation values with respect to this

structure. Second, if the definite integral $x_t - x_o = \int_o^t v_s\, ds$
by hypothesis well-defines the process $x_t - x_o$, then this process is a random process on the same sample space as the integrand
v_t, and its expectation values are to be taken with respect to
this same probability structure. In fact, the integral is a
continuous function of its integrand and hence a Baire function and
measurable [cf. Ref. 32, p. 108]. Now, the first fact applies to
the rotational as well as to the translational case, i.e., solutions of the rotational Langevin equation exist and are unique.
The second fact does not carry over, however. Indeed, whereas the
translational spatial coordinates are related to the velocities by
the above integral, there is no known coordinate system derivable
from the angular velocities ω^i in this way. In the absence of
such an identification, it is necessary, in defining the mean and
covariance parameters of the spatial process, to assign a
probability structure to the sample space of spatial trajectories
itself which cannot be inferred from the probability structure of
the original Wiener process or the OU velocity process. For
concreteness, let us consider this problem in terms of the Euler-
angle representation of orientation. Let $(\theta^1, \theta^2, \theta^3) = (\theta, \phi, \psi)$ be
the particle Euler angles. The transformations relating the θ^i
to the angular velocities ω^i are [33, Sec. 4.8]

$$\alpha^i_{\ j} = \left\| \begin{array}{ccc} \cos\theta^3 & \sin\theta^1\sin\theta^3 & 0 \\ -\sin\theta^3 & \sin\theta^1\cos\theta^3 & 0 \\ 0 & \cos\theta^1 & 1 \end{array} \right\|$$

$$\beta^i_{\ j} = \left\| \begin{array}{ccc} \sin\theta^3 & \cos\theta^3 & 0 \\ -\cos\theta^3/\sin\theta^1 & \sin\theta^3/\sin\theta^1 & 0 \\ \cot\theta^1/\cos\theta^3 & -\cot\theta^1/\sin\theta^3 & 1 \end{array} \right\|$$

where $\omega^i = \alpha^i_{\ j}\theta^j$ and $\theta^i = \beta^i_{\ j}\omega^j$. (In this discussion and in
that of Sec. IVF only, we employ subscripts and superscripts cor-
rectly in spite of their less familiar appearance; the superscripts
here are never powers.) The vectors $(\alpha^i_{\ 1}, \alpha^i_{\ 2}, \alpha^i_{\ 3})$, $(\beta^1_{\ i}, \beta^2_{\ i}, \beta^3_{\ i})$,
$i = 1, 2, 3$, are the co- and contravariant basis vectors of a proper

coordinate system, (π^i), say, iff the curl $\dfrac{\partial \alpha^i_j}{\partial \theta^k} - \dfrac{\partial \alpha^i_k}{\partial \theta^j} = 0$ for

all i,j,k, or equivalently iff the quantity $d\pi^i$ is an exact

differential, $d\pi^i = \alpha^i_j\, d\theta^j$. Since for example,

$\partial \alpha^1_1/\partial \theta^2 = 0 \neq \cos \theta^1 \sin \theta^3 = \partial \alpha^1_2/\partial \theta^1$, this condition is not

satisfied in the present case. This has two important consequences

for rotational diffusion theory. First, since $d\pi^i$ is inexact,

the integral $\displaystyle\int_{\pi^i(o)}^{\pi^i(t)} d\pi^i$ is path-dependent, i.e., the process

$\theta^i - \theta^i(0) = \displaystyle\int_o^t \omega^i(t')\, dt'$ depends not only on the limits of

integration $0,t$, but also on the particular trajectory in state

space along which the integral is taken [cf. Ref. 34, Sec. I.5].

Since velocities can be inferred from spatial trajectories at each

point by differentiation, the variable in the path integral

$\displaystyle\int_o^t \omega^i(s)\, ds$ is the path, not the integrand, and the sample space

with respect to which expectation values must be taken is the set

of paths, i.e., the set of spatial trajectories. The probability

structure of this set cannot be inferred from that of the Wiener

process in the absence of a postulate relating them and must be

assigned a priori on some other basis, as asserted. The second

consequence is that even if the OU process does converge to a dif-

fusion process in state space, one cannot expect the limiting dif-

fusion equation to have its standard appearance when referred to

the coordinates (π^i). The reason for this is that transformation

of an equation containing second or higher order partial deriva-

tives into a coordinate system with base vectors not satisfying the

above integrability conditions, called an anholonomic (or quasi-)

coordinate system, always introduces new terms with the quantity

$$\Omega^k_{ij} = \beta^m_i \beta^n_j \dfrac{\partial \alpha^k_m}{\partial \theta^n} - \dfrac{\partial \alpha^k_n}{\partial \theta^m}$$

called the object of anholonomity, which state the defining

condition on the $(d\pi^i)$ [35, Sec. II-10]. For example, transformation of Lagrange's equations for rotational motion into the anholonomic frame (π^i) leads to Euler's equations, with the new terms $\Omega^k_{ij}\omega^j \, dT/\partial\omega^k = \varepsilon_{ijk}I^k_h\omega^j\omega^h$ as used earlier here. This result is derived by Whittaker [36; Sec. II-30], with the notation $\gamma_{ijk} = \Omega^k_{ij}$. The standard form of Lagrange's equations, valid for any set of generalized coordinates, does not obtain precisely because the anholonomic coordinates are not by themselves an allowable parameterization of state space. The differential operator $D^{ij}\partial^2/\partial\theta^i\partial\theta^j$ in the diffusion equation is not covariant and would give rise to such terms, containing both D^{ij} and Ω^k_{ij}. Since the fundamental difference here is one of integrability, it is worthwhile to mention the application of invariant-integral techniques [37, Sec. 1,3] from group theory by Ivanov [38] in his derivation of the rotational diffusion equation as an asymptotic random-walk process in terms of elementary transformations of the rotation group. The Ivanov derivation is the rotational analog of the Kac [39] derivation of the translational diffusion equation.

In the absence of a proof, we take by assumption that an OU process does converge to a rotational diffusion process. We have mentioned earlier, and indicate in the section on forced diffusion, that forced diffusion theory leads to a relation between the diffusion tensor and the viscous drag tensor, in fact $D_{ij} = \gamma^{-1}_{ij}kT$, just under the requirement of stability of solutions for systems in equilibrium in static applied fields. We assume that the Langevin equation which converges to a given diffusion process is that in which the same viscous drag tensor appears in both processes. This is the assumption made before, e.g., by Hubbard [40].

D. The Green Function

Let us now discuss solution of the forward (Fokker-Planck) equation for the Green function of the rotational diffusion (Wiener)

process. We have seen that the instantaneous density
$\rho(x,t) = U_t\rho(x,0)$, where U_t is the operator constructed from
this Green function, and $\rho(x,0)$ is the initial orientation density, obtains exactly if it is assumed that the particle motion
underlying diffusion is a diffusion process and that it is a very
good approximation, except at short diffusion times, if the particle motion is an OU process. We have stated that theory must provide (a) the density function with respect to which instantaneous
expectation values of state functions describing experimental
measurements are to be taken, and (b) the relationship between
diffusion tensor and particle shape implied by that density function. These are satisfied, clearly, by the function $\rho = \rho(x,t)$
and the rotational Einstein relation $D_{ij} = \gamma_{ij}^{-1}kT$, respectively.
In application, the diffusion tensor is inferred from experimentally determined parameters. The viscous drag tensor has been calculated for some particle shapes, in particular the ellipsoid [41]
and the right circular cylinder [42]. These are generalizations
of the well-known relation $\gamma = 8\pi\eta r^3$ for the sphere of radius r
in a medium of bulk viscosity η. Ridgeway [5] gives computations
for preparation of nomograms for treatment of data with ellipsoids.

Since complete details of the solution itself can be found in
Favro's papers or, for the case of the ellipsoid, in the papers of
Perrin [43,44], we restrict ourselves to discussion of the form of
the solution and certain general points about the analysis. In
particular, we show that the solution provides exact expressions
for expectation values of the quantities which usually appear in
theories of methods which study rotational diffusion. Separation
of variables applied to the forward (Fokker-Planck) equation yields
two equations, one in the time, with solutions proportional to
$\exp(-\lambda^2 Dt)$, and one an eigenvalue problem in the spatial
coordinates. Values of the separation constant $-\lambda^2$ are acceptable which lead to single-valued solutions of the spatial equation,
i.e., eigenvalues. In the one-dimensional case, that of the infinite rotating cylinder, the spatial equation is formally

identical to the Schrödinger equation for the (trivial) 1-space
harmonic oscillator. The set $\{(2\pi)^{-1/2}e^{in\phi}\}$, $n = 0, \pm1, \ldots,$ is
a complete orthonormal (o.n.) basis of eigenfunctions, with
$\lambda^2 = n^2 = 0, 1, \ldots;$ D is the diffusion coefficient in the forward equation. Since the differential equation is linear, its
general solution is the superposition of solutions for all suitable
values of the separation constant, i.e., $p(t,\phi,\phi_o) =$
$\sum_{n=-\infty}^{\infty} A_n(\phi_o)e^{-in\phi}e^{-n^2Dt}$. We fix the arbitrary coefficients by the

regularity condition $p(0^+,\phi,\phi_o) = \delta(\phi - \phi_o)$. Noting that the

Dirac-δ density may be expanded $2\pi\delta(\phi - \phi_o) = \sum_{n=-\infty}^{\infty} e^{-in(\phi-\phi_o)}$,

we identify $A_n(\phi_o) = (2\pi)^{-1/2}e^{in\phi_o}$. The Green function becomes

$p(t,\phi,\phi_o) = \sum \psi_n(\phi_o)\psi_n^*(\phi)e^{-n^2Dt}$, where we have written

$\psi_n(\phi) = (2\pi)^{-1/2}e^{in\phi}$ for all n. In the three-dimensional case,
that of the rigid particle of arbitrary shape, separation of variables leads to the same time-dependent equation as the above and to
a spatial eigenvalue problem which is formally the same as the
Schrödinger equation for the asymmetric rigid rotator. A complete
set of eigenfunctions is known if and only if the particle possesses an axis of rotational symmetry. We denote the eigenfunctions
of this set $\{\psi_k\}$, where $\underline{k} = (j, m, n)$ specifies the three
indexes and ranges over the values $j = 0, 1, 2, \ldots,$ and
$m,n = -j, -j + 1, \ldots, j - 1, j,$ and use $\{\psi_k\}$ to denote the
general set. Let $L\Psi = \lambda\Psi$ be the spatial diffusion equation
arising from separation of variables, with differential operator
L. Favro expands the $\{\Psi\}$ in $\{\psi_k\}$-representation, selecting the
set $\{\psi_k\}$ such that L is diagonal with respect to indexes j,n
and satisfies the additional selection rule that the inner product
$(\psi_k, L\psi_{k'})$ vanishes unless m,m' have the same sign and are both
even or both odd. These rules partition the basis into finite sets
of noncombining eigenfunctions (cf. Ref. 45, Appendix 5] and reduce the secular determinant, formally of countably infinite

order, to a product of determinants of finite order. For small j,
the subdeterminants are of sufficiently small order to be solved
exactly by routine methods for the eigenvalues and thus for the
corresponding values of the separation constants. It is recalled
that, if a particular noncombining set has p elements, say,
then the subdeterminant from its elements is of rank p, with p
roots or eigenvalues. The corresponding eigenfunctions are p
o.n. linear combinations of its elements. The Fourier coeffi-
cients $(\psi_{\underline{k}}, \Psi)$ are obtained by inversion of the system of equa-
tions $(\psi_{\underline{k}'}, \Psi)[(\psi_{\underline{k}'}, L\psi_{\underline{k}}) - \lambda\delta_{\underline{k}\underline{k}'}] = 0$, where $\underline{k} = (j, m, n)$ and
$\underline{k}' = (j, m', n)$ range over the elements of the noncombining set.
The general solution of the forward equation is, as before,

$$p(t,x,x_o) = \sum_{\underline{k}} \psi_{\underline{k}}(x_o)\psi_{\underline{k}}^*(x)e^{-\lambda_{\underline{k}}t}.$$

E. Calculation of Time Averages

It is now very important in application that, although a complete o.n. basis of eigenfunctions of the general spatial diffusion
operator L is not known, nevertheless we can show that the above
procedure does yield exact expressions for expectation values of
any state function whose expansion in the basis has nonzero
coefficients only for terms of j small enough to permit solution
of the relevant subdeterminants. Indeed, let f be a state func-
tion with expansion $f = (f,\Psi_{\underline{n}})\Psi_{\underline{n}}$. In general, the λ and the
$\Psi_{\underline{n}}$ are unknown. The instantaneous expectation value $\langle f\rangle_t =$
$\int f(x)\rho(x,t) dx$. It suffices to treat the case with initial condition $\rho(x,0) = \delta(x - 0)$ [cf. Ref. 4]. Then $\rho(x,t) =$
$\int \rho(y, 0)p(t, x, y) dy = \int \delta(y - 0)p(t, x, y) dy = p(t, x, 0)$.

Thus, $\langle f\rangle_t = \int f(x)p(t, x, 0) dx = (f,\Psi_{\underline{k}}) \int \Psi_{\underline{k}}p(t, x, 0) dx =$

$$\sum_{\underline{n}} (f, \Psi_{\underline{k}})\Psi_{\underline{n}}(0)e^{-\lambda_{\underline{n}}t} \int \Psi_{\underline{k}}\Psi_{\underline{n}}^* dx$$

or $\langle f \rangle_t = \sum_n (f, \Psi_n) \Psi_n(0) e^{-\lambda_n t}$ because of the orthonormality of the

basis eigenfunctions. By hypothesis, for any given $\underline{k}$, $(f, \Psi_{\underline{k}}) \neq 0$
only if we can solve for all the eigenvalues of that noncombining
set and the corresponding eigenfunctions. But this is equivalent
to the requirement that all nonzero Fourier coefficients (f, Ψ_n)
and all the corresponding separation constants λ_n are known
exactly. The assertion follows from this.

It is important that this situation is the usual one in
practice. Let us discuss the particular case which occurs most
often. In the theory of an optical instrument depending on particle orientations in suspension, with or without externally applied
fields, the instrument, its fields and sensing devices, is related
to individual particles by transforming incoming signals to the
particle into instantaneous local particle coordinates and performing on outgoing signals the inverse transformation. The elements
of the matrix representation of the transformations are the direction cosines of the particle axes referred to the spatial frame.
For this reason, the quantities which appear in the theory of the
instrument are expectation values of products of the direction cosines. The transformations are displayed in terms of particle
Euler angles (θ, ϕ, ψ) and Cayley-Klein parameters $(\alpha, \beta, \gamma, \delta)$ in
Goldstein [33, Eqs. 4-46 and 4-63, respectively]. Favro parameter-
izes state space in terms of the following linear combinations of
Cayley-Klein parameters:

$$\Omega_1 = (2i)^{-1}(\beta + \gamma), \quad \Omega_2 = 2^{-1}(\beta - \gamma), \quad \Omega_3 = (2i)^{-1}(\alpha - \delta),$$

$$\Gamma = 2^{-1}(\alpha + \delta)$$

With these identifications, it is a straightforward problem to express products of direction cosines in terms of eigenfunctions if
the latter are known. Exact eigenfunctions for all m, n with
$j = 0, 1, 2$, and the corresponding eigenvalues are given in Favro's
paper. These suffice to expand all twofold products of the

elements of the transformation matrix. For example, the products
required for calculation of Kerr birefringence are the squares of
the direction cosines [4, Eq. 22]. Typical expansions are

$$c_{31}^2 = -\frac{\pi}{6}[2\psi(0,\ 0,\ 0) - \left(\frac{4}{5}\right)^{1/2}\psi(2,\ 0,\ 0)] -$$

$$- \left(\frac{\pi^2}{30}\right)^{1/2}[\psi(2,\ 0,\ 0) + \psi(2,\ 0,\ -2)]$$

$$c_{32}^2 = \frac{\pi}{6}[2\psi(0,\ 0,\ 0) - \left(\frac{4}{5}\right)^{1/2}\psi(2,\ 0,\ 0)] -$$

$$- \left(\frac{\pi^2}{30}\right)^{1/2}[\psi(2,\ 0,\ 2) + \psi(2,\ 0,\ -2)]$$

$$c_{33}^2 = \frac{\pi}{3}\psi(0,\ 0,\ 0) + \left(\frac{4\pi^2}{45}\right)^{1/2}\psi(2,\ 0,\ 0)$$

where we have written $\psi_{\underline{k}} = \psi(j,m,n)$, $k = (j,m,n)$.

F.   Relaxation of Kerr Electric Birefringence

As an example of application of the theory, we specialize the
results of the preceding section to the case of the relaxation of
Kerr birefringence. Consider as the physical system a suspension
of rigid particles of identical, but otherwise arbitrary, shape in
a viscous continuum. Initially, the suspension is in an oriented
state at equilibrium in a static uniform external electric field.
At a time denoted t = 0, the field is suddenly removed. We
wish to relate the loss of birefringence of the suspension in time
resulting from random rotational motion away from the initial
orientation to properties of particle shape. Conclusions in this
section are proven in Ridgeway [4,5], and details of the treatment
may be found there. The physical system is a generalization of
suspensions of axisymmetric particles, studied in Chap. 3.

Describe the instantaneous orientation of a particle at the
time t in terms of two Cartesian coordinate frames: one, the
reference frame, fixed spatially, the other, the particle frame,
comoving with the particle. Let a_{ij} be the cosine of the angle
formed by the ith reference axis and the jth particle axis at
time 0 and c_{ij} the cosine of the angle formed by the ith
particle axis at time 0 with the direction of the jth particle
axis at a later time t. Then the matrix (a_{ij}) formed from
these direction cosines transforms the particle from an orientation
with its axes coinciding with the reference axes into its orientation at time 0, and the matrix (c_{ij}) transforms its orientation
at time 0 to that at time t. Of course, the product matrix
$(a_{ik}c_{kj})$ transforms the reference axes directly into the particles
axes at time t. Use of the two matrices, rather than the product
matrix, separates the diffusion problem from the initial steady-
state problem, as we shall show.

The birefringence Γ of a suspension with respect to reference axes i and j, assumed fixed with respect to the measuring
instrument, is just the difference of the respective refractive
indices at the observed light wavelength for linearly polarized
light with propagation vector normal to the i,j plane and electric
vector parallel to i or j, i.e., $\Gamma = n_i - n_j$. Suppose that
particle axes are chosen to diagonalize the particle optical
polarizability tensor and denote by α_i the ith principal
polarizability at the observed wavelength. Then it is found that
the instantaneous Kerr birefringence of the suspension with respect
to spatial axes 3 and 1 is [4, Eq. 16]

$$\Gamma = (N/2n_o) < \sum_i \alpha_i [\sum_j (a_{3j}^2 - a_{1j}^2)c_{ji}^2 + 2 \sum_{j>k} \sum (a_{3j}a_{3k} - a_{1j}a_{1k})c_{ji}c_{ki}] >$$

where N is the density of mass centers, assumed uniform, and n_o
is the refractive index of the suspending medium. The indicated
expectation value is taken with respect to the instantaneous

rotational orientation density function. Because of the linearity
of the integral, this expression is a linear combination of terms
of the form $<a_{hj}a_{hk}c_{ji}c_{ki}>$. Since free diffusion is a spatially
homogeneous process, the elements of the matrix (a_{ij}) are
stochastically independent of those of (c_{ij}), so that the prod-
uct simplifies as $<a_{hj}a_{hk}c_{ji}c_{ki}> = <a_{hj}a_{hk}><c_{ji}c_{ki}>$. The expectation values $<a_{hi}a_{hk}>$ are taken with respect to the canonical
density formed from the Hamiltonian of the field-particle interaction and will not concern us further. We obtain the expectation
values $<c_{ji}c_{ki}>$ from diffusion theory.

Since all c_{ij} satisfy the initial condition $c_{ij} = \delta_{ij}$,
calculation of the expectation values $<c_{ji}c_{ki}>$ requires only
knowledge of conditional transition probabilities from origin,
i.e., the densities $p(t,y,0)$. The analysis of the preceding section therefore applies directly. We have, first, that

$$<c_{ps}c_{qs}> = \sum_{\underline{n}} (c_{ps}c_{qs}, \psi_{\underline{n}}) \psi_{\underline{n}}(0) e^{-\lambda_{\underline{n}}t}.$$ Second, all expansions of

pairwise products of direction cosines in terms of the Favro
eigenfunctions, like those of the c_{3s}^2 displayed above, involve
only terms with $j \leq 2$. Since all such eigenfunctions and the
corresponding eigenvalues $\lambda_{\underline{n}}$ are known, the expectation values of
these products can be evaluated exactly. In fact, define the
quantities $D = 3^{-1} \sum_{i} D_{(i)}$ and $P^2 = 3^{-1} \sum_{i>j} \sum D_{(i)}D_{(j)}$. If the
choice of particle axes diagonalizes both the diffusion tensor and
the viscous drag tensor, so that the Einstein-Favro relation
$D_{(i)} = kT\gamma_{(i)}^{-1}$ obtains, then substitution of the final expressions
for the expectation values into that for the birefringence yields

$$\Gamma = A_+ e^{-6\theta_+ t} + A_- e^{-6\theta_- t}$$

where the relaxation constants $\theta_{\pm}$ may be written either as
$\theta_{\pm} = D \pm (D^2 - P^2)^{1/2}$ in terms of the rotational diffusion constants, or as

$$\Theta_{\pm} = (kT/3) [\sum_i \gamma_{(i)}^{-1} \pm (\sum_i (\gamma_{(i)}^{-1})^2 - \sum_{i>j} \sum \gamma_{(i)}^{-1} \gamma_{(j)}^{-1})^{1/2}]$$

in terms of the particle viscous drag constants. The coefficients A_t are time-independent functions of the particle polarizabilities and the matrix elements a_{ij} carrying the dependence on initial orientation.

One concludes, then, that the expression for the time-dependent decay of birefringence is of the simple form of a sum of two relaxation terms. The relaxation constants, i.e., the coefficients of t in the exponents, depend only on the frictional properties of the particle and the temperature. They are, in particular, independent of initial orientation and (therefore) of electric peoperties of the particle or orientation mechanism. Since no restriction on particle shape has been introduced into the analysis, these are general results. As a corollary, then, it follows that Kerr relaxation curves cannot provide an effective basis for distinguishing classes of particle shape, e.g., whether a particle is cup-shaped or ellipsoidal, since one predicts at most two relaxation times for all shapes, however irregular.

Because of the observation experimentally of exponential relaxation curves, it is of great interest to obtain conditions on the suspended particle under which the expression for the birefringence reduces to a single term [5]. A sufficient condition for one of the coefficients $A_{\pm}$ to vanish identically is that the suspended particle possess cylindrical optical and geometric symmetry about one of its axes. This is also found to be necessary condition unless the polarizabilities and shape parameters satisfy a fixed physically unlikely algebraic relation. An exponential relaxation curve could arise formally, apart from a zero coefficient, if the relaxation constants were equal. However, this equality, which occurs only for the spherical particle, is accompanied by the identity $A_+ = A_- = 0$, the condition for no Kerr birefringence. Thus, with the one algebraic exception, the

relaxation curve is exponential if and only if the particle
possesses a symmetry axis such that its geometric cross sections
perpendicular to that axis are circles. Note in particular that
the condition allows the radii of the circular cross sections to
vary arbitrarily along the length of the axis, and that no restriction is required on the particle electric properties, such as
permanent charge distribution (or dipole moment orientation) within
the particle.

These results apply in many different ways in evaluation of
experimental data (cf. Chap. 3) for discussion and references).
As a first example, isoionic bovine serum albumin displays an
exponential relaxation curve. Two particle shapes have been
proposed for this protein, a prolate spheroid [46] and a pearl-
necklace trimer with collinear centers [47]. From the above condition, one cannot distinguish qualitatively between these two
structures, since both are axisymmetric, although one can rule out
either a nondegenerate ellipsoid, on the one hand, or a trimer with
noncollinear centers, on the other. As a second example, one notes
that whereas a relaxation curve with more than two exponential
terms always implies, for rigid particles, that the suspension is a
mixture of particle types, a sum of two terms can arise either from
a mixture of two distinct axisymmetric particle types or from a
monodisperse suspension of an asymmetric particle. The hemocyanin
of the Roman snail, Helix pomatia, has a doubly exponential relaxation curve. It has been treated both as a mixture of species [48]
and, for illustrative purposes, as an asymmetric ellipsoid [5].
Finally, the Kerr relaxation method should find wide use in study
of the relationship between protein structures in the crystalline
state, as obtained from crystallographic methods, and the structures in aqueous solution. Whereas the relaxation curve is not,
by itself, a good method for distinguishing between shape classes,
it is applicable to any particle shape, however irregular, and
may within a given shape class be very sensitive to small changes
in shape parameters for fixed particle volume. If the viscous drag

tensor has been calculated for the postulated structure, then the
relaxation times can be written directly. This is the case, for
example, for the right circular cylinder and its application to the
tobacco mosaic virus [49]. If expressions for the elements of the
viscous drag tensor are not known, they can always be measured
with scaled models of the postulated structures and applied to
calculate the relaxation constants, since the relevant hydrodynamic
equations are independent of scale. Such measurements have been
made recently, for example, by Douthart and Bloomfield [50] for the
bacteriophage T2.

G. Diffusion in External Fields

We conclude with a brief summary of treatment of the diffusion
equation for forced diffusion. The additional assumptions, beyond
that of a diffusion process for random behavior, are stated as they
occur. We consider again state space to be the unit ball in
Euclidean 4-space, parameterized now by the particle Euler angles
following the discussion by Perrin [43]. State space is, therefore, a Riemannian space with fundamental tensor g_{ij} having
diagonal elements $1/4$ and off-diagonal elements zero except for
$g_{23} = g_{32} = (1/4) \cos \theta^1$. The random force and external torque
lead to probability current densities $D_{ij} \, \partial\rho/\partial\theta^j$ and $\rho \, d\theta^i/dt$,
respectively, where $\rho = \rho(x,t)$ is the probability density at the
point $x = (\theta^1, \theta^2, \theta^3)$ on the hypersphere at time t. If we assume
that these two currents superimpose, then the continuity equation
for probability density is

$$\frac{\partial\rho}{\partial t} = g^{-1/2} \frac{\partial}{\partial\theta^j} \left[g^{1/2} \left(D^{jk} \frac{\partial\rho}{\partial\theta^k} + \rho \frac{d\theta^j}{dt} \right) \right]$$

where $g^{1/2} \equiv (\det|g_{ij}|)^{1/2} = (1/8) \sin \theta^1$ [cf. Ref. 51, p. 32].
If we neglect inertial effects, then the torque produced by the
external field is approximately at equilibrium with the viscous

drag, hence $-\gamma_{ij}\omega^j = M_i$ and $d\theta^i/dt = -\beta_j^i(\gamma^{-1})^{jk}M_k$, where M_i is the torque due to the external fields and (β_j^i) is the transformation matrix given earlier. We have finally

$$[\frac{\partial}{\partial t} - \frac{1}{g^{1/2}}\frac{\partial}{\partial\theta^j}\ (g^{1/2}D^{jk}\frac{\partial}{\partial\theta^k})]\rho = -\ [\frac{1}{g^{1/2}kT}\frac{\partial}{\partial\theta^j}\ (g^{1/2}\beta_k^j D^{k\ell}M_1)]\rho$$

where we have identified $D^{ij} = (\gamma^{-1})^{ij}kT$, which is necessary if ρ is to be the canonical distribution in the static field case [21]. This equation has been treated by Perrin [43] and Debye [52]. In the absence of external fields, $M_i = 0$, the right-hand side vanishes, and the left-hand side is just the free diffusion equation treated heretofore. The equation is of the form of a standard perturbation problem in quantum mechanics in which exact eigenfunctions are known (in principle) for the unperturbed case. The equation can be used to construct a hierarchy of equations to treat field interactions of any order by standard methods (cf. Ref. [53, Sec. XVI 2]. This approach has formally been followed by Kirkwood [54] and Favro [3].

REFERENCES

1. S. Chandrasekhar, Rev. Mod. Phys., 15, 1 (1943).

2. E. Nelson, Dynamical Theories of Brownian Motion, Princeton University Press, Princeton, N.J.

3. L. D. Favro, in Fluctuation Phenomena in Solids (R. E. Burgess, ed.), Academic, New York (1965), pp. 79-102.

4. D. Ridgeway, J. Am. Chem. Soc., 88, 1104 (1966).

5. D. Ridgeway, J. Am. Chem. Soc., 90, 18 (1968).

6. J. Lebowitz and E. Rubin, Phys. Rev., 131, 2381 (1963).

7. J. Lebowitz and P. Résibois, Phys. Rev., 139A, 1101 (1965).

8. R. Kubo, Prog. Phys., 19(I), 255 (1966).

9. R. M. Mazo, J. Stat. Phys., 1, 89 (1969).

10. J. G. Kirkwood and J. B. Shomaker, Proc. Nat. Acad. Sci. U.S., 38, 855 (1952).

11. W. Scheider, Biophys. J., 5, 617 (1965).

12. R. Kubo, J. Phys. Soc. Japan, 12, 570 (1957).

13. R. Zwanzig, Ann. Rev. Phys. Chem., 16, 67 (1965).

14. B. J. Berne and G. D. Harp, Adv. Chem. Phys., 17, 63 (1970).

15. W. Bernard and H. B. Callen, Rev. Mod. Phys., 31, 1017 (1959).

16. W. Feller, An Introduction to Probability Theory and Its Applications, Vol. 2, Wiley, New York (1966).

17. E. B. Dynkin, Morkov Processes, Vol. 1, Academic Press, New York (1965).

18. A. N. Kolmogorov, Math. Ann., 104, 415 (1931)

19. N. Wiener, J. Math. Phys. MIT, 2, 131 (1923).

20. K. Yosida, Functional Analysis, 2nd ed., Springer, New York (1968).

21. L. D. Favro, Phys. Rev., 119, 53 (1960).

22. W. H. Furry, Phys. Rev., 107, 7 (1957).

23. A. Einstein, Ann. Phys., 17, 549 (1905).

24. G. E. Uhlenbeck and L. S. Ornstein, Phys. Rev., 36, 823 (1930).

25. J. L. Doob, Ann. Math., 43, 351 (1942).

26. M. C. Wang and G. E. Uhlenbeck, Rev. Mod. Phys., 17, 323 (1945).

27. P. Langevin, Compt. Rend., 146, 530 (1908).

28. K. Ito, Mem. Am. Math. Soc., No. 4 (1951).

29. J. L. Doob, Stochastic Processes, Wiley, New York (1953).

30. W. A. Steele, J. Chem. Phys., 38, 2404; 2411 (1963).

31. K. Mishima, J. Phys. Soc. Japan, 31, 1796 (1971).

32. M. Loève, Probability Theory, 3rd ed., Van Nostrand, Princeton, N.J. (1963).

33. H. Goldstein, Classical Mechanics, Addison-Wesley, Reading, Mass. (1950).

34. F. Klein and A. Sommerfeld, Ueber die Theorie des Kreisels, Teubner, Stuttgart (1965).

35. J. A. Schouten, Ricci-Calculus, 2nd ed., Springer, Berlin (1954).

36. E. T. Whittaker, A Treatise on the Analytical Dynamics of Particles and Rigid Bodies, 4th ed., Cambridge University Press, London (1937).

37. I. M. Gel'fand, R. A. Minlos, and Z. Y. Shapiro,
     Representations of the Rotation and Lorentz Groups and Their
     Applications, Pergamon, Macmillan, New York (1963).

38. E. N. Ivanov, Sov. Phys. JETP, 18, 1041 (1964).

39. M. Kac, Am. Math. Monthly, 54, 369 (1947).

40. P. S. Hubbard, Phys. Rev., 131, 1155 (1963).

41. D. Edwardes, Quart. J. Math., 26, 70 (1892).

42. S. Broersma, J. Chem. Phys., 32, 1626 (1960).

43. F. Perrin, J. Phys. Radium, 5, 497 (1934).

44. F. Perrin, J. Phys. Radium, 7, 1 (1936).

45. J. C. Slater, Quantum Theory of Molecules and Solids, Vol. 1,
     McGraw-Hill, New York (1963).

46. S. Krause and C. T. O'Konski, J. Am. Chem. Soc., 81, 5082
     (1959).

47. V. A. Bloomfield, Biochemistry, 5, 684 (1966).

48. R. M. Pytkowicz and C. T. O'Konski, Biochim. Biophys. Acta,
     36, 466 (1959).

49. C. T. O'Konski and A. J. Haltner, J. Am. Chem. Soc., 78, 3604
     (1956).

50. R. J. Douthart and V. A. Bloomfield, Biochemistry, 7, 3912
     (1968).

51. L. P. Eisenhart, Riemannian Geometry, Princeton University
     Press, Princeton, N.J. (1925).

52. P. Debye, Polar Molecules, Dover, New York (1929).

53. A. Messiah, Quantum Mechanics, Vol. 2, Wiley, New York (1965).

54. J. G. Kirkwood, J. Polymer Sci., 12, 1 (1954).

APPENDIX I: THE VISCOUS DRAG TENSOR

The phenomenon of viscosity is the result of the tendency of a
liquid to resist relative motion of its parts and steady conversion
by the liquid of the kinetic energy of such motion into heat as the
motion progresses. If a particle is rotated in a liquid initially
at rest, and if, as is usually assumed in hydrodynamical theories,
there is no slippage of liquid at the particle surface, then the
motion of the particle produces relative motions throughout the

liquid. Since the internal energy of the particle and liquid
together must be conserved, the appearance of heat throughout the
liquid must be compensated by a decrease in the kinetic energy,
and the angular velocity, of the particle unless work is done on
the particle by external forces to maintain the motion. This decrease appears formally as the result of a retarding force, the
viscous drag, on the particle.

Hydrodynamical theory postulates that it is possible to express these effects in terms of stresses at the particle surface.
In fact, one defines the <u>stress tensor</u>, $\Phi_{ij} = -p'\delta_{ij} + 2\mu e_{ij}$,
such that the resultant force on a surface element dS is
$\Phi_{ij}dS_j$, where (dS_i) is a vector of length dS in the direction
of the outward normal to dS, δ_{ij} is the Kronecker-δ, p' the
constant hydrostatic pressure and μ the constant coefficient
of viscosity. The tensor $e_{ij} = 2^{-1}(\frac{\partial q_j}{\partial x_i} + \frac{\partial q_i}{\partial x_j})$ is the symmetric
rate-of-strain tensor, with q_i the liquid velocity at each point
in the direction of the Cartesian reference coordinate x_i. The
stress tensor is the sum of two terms, then, a hydrostatic term,
which can lead only to a stress normal to the surface at each
point, and a viscosity-dependent term, which in general produces
tangential as well as normal stresses.

The velocity field (q_i), from which the components of the
rate-of-strain tensor are formed, is determined by the Navier-
Stokes equation $\rho\frac{dq_i}{dt} = \mu\nabla^2 q_i - \frac{\partial p}{\partial x_i}$, where $p = 3^{-1}\Phi_{ii}$ and ρ is
the liquid density. In principle, if one could solve this equation
for appropriate boundary conditions, he could then calculate the
rate-of-strain tensor (e_{ij}) directly, obtain the stress tensor
(Φ_{ij}) by substitution into its definition, and determine the
viscous force of rotation by evaluation of the integral $\int \Phi_{ij}dS_j$
taken over the closed particle surface. The 3×3 diagonal
matrix with components equal to the viscous drags for steady
rotation at unit angular velocity about the three respective principal axes of the particle is called the <u>viscous drag tensor</u>.

   Unfortunately, the Navier-Stokes equation has remained
refractory to solution except in a few special cases because of the
nonlinearity of the derivative $\dfrac{dq_i}{dt} \equiv \dfrac{\partial q_i}{\partial t} + q_j \dfrac{\partial q_i}{\partial x_j}$. Two methods of
approximation have been employed. In the first, one simply neg-
lects the nonlinear term $q_j \dfrac{\partial q_i}{\partial x_j}$ and solves the resulting linear
equation. The second method of approximation, introduced by Oseen
[1] and elaborated particularly by Burgers [2], linearizes the
Navier-Stokes equation by superimposing onto the velocity field of
the unperturbed liquid motion the velocity field due to forces on
the liquid required to maintain the motion of the particle. The
maintaining forces are introduced in the rotational problem as
expressions for torques about the axes of rotation containing
arbitrary constants which can be specialized to satisfy the boundary condition of no slippage at the particle surface in a liquid
of infinite extent. The superimposed field is described in terms
of an object now designated as the Oseen interaction tensor.

   The Oseen method is more readily adaptable to treatment of
particles of different shapes particularly because of a second
approximation it permits, the replacement of the rigorous condition
of no surface slippage with a weaker condition on slippage, e.g.,
no average slippage over the surface. The first method has been
applied to obtain the viscous drag on a rotating sphere (cf. the
textbook discussion of Sommerfeld [3; p. 315f], the spheroid [4],
and the completely asymmetric ellipsoid [5]. The Oseen method has
been employed to treat the rotation of the right circular cylinder
of finite length by Burgers [2] and, to a better degree of approximation both along the length of the cylinder and at the ends, by
Broesma [6]. It has been applied to collections of small spheres
of different shape envelopes by Kirkwood [7], Riseman and Kirkwood
[8], and by Felson and Bloomfield [9].

REFERENCES

1. C. W. Oseen, Neuere Methoden und Ergebnisse der Hydrodynamik, Akademische, Leipzig (1927).

2. J. M. Burgers, Second Report on Viscosity and Plasticity, Ch. III, Amsterdam Academy of Sciences, Nordemann (1938).

3. A. Sommerfeld, Mechanik der Deformierbaren Medien, 4. Aufl., Akademische, Leipzig (1957).

4. G. B. Jeffery, Proc. London Math. Soc., 14, 327 (1915).

5. D. Edwardes, Quart. J. Math., 26, 70 (1892).

6. S. Broersma, J. Chem. Phys., 32, 1626 (1960).

7. J. G. Kirkwood, J. Polymer Sci., 12, 1 (1954).

8. J. Riseman and J. G. Kirkwood, in Rheology, Theory and Applications (F. R. Eirich, ed.), Vol. I, Ch. 13, Academic, New York (1956).

9. D. P. Felson and V. A. Bloomfield, Biochemistry, 6, 1650 (1967).

Chapter 5

FLEXIBLE POLYMERS

Robert L. Jernigan*

Physical Sciences Laboratory
Division of Computer Research and Technology
National Institutes of Health
Bethesda, Maryland

Douglas S. Thompson**

Engineering Research and Development Division
Engineering Physics Laboratory
E. I. du Pont de Nemours and Co., Inc.
Wilmington, Delaware

*Current affiliation: Laboratory of Theoretical Biology, National
Cancer Institute, National Institutes of Health, Bethesda, Maryland.
**Current affiliation: Department of Chemistry, Hampden-Sydney
College, Hampden-Sydney, Virginia.

159

I. ELECTRO-OPTICS OF FLEXIBLE POLYMERS

Flexible polymers in solution are oriented by an externally
applied electric field. In addition to orientation, as occurs with
rigid rods, the relative abundances of configurations are usually
perturbed. These small conformational changes can affect the
molecule's average electric birefringence, dichroism, and, of
course other properties. Overall orientation and changes in internal freedom arise from the action of the electric field on both
the permanent and the induced dipole moments. In addition, birefringence and dichroism depend on detailed chemical structural
features, such as bond angles and the energies of hindered internal
rotations.

Previously, the theory of optical properties of flexible
macromolecules has lagged behind experiment. Now, however, mean-
ingful experiments are needed in order to test current theories,
especially for electric birefringence and electric dichroism. For
other optical properties of chain molecules not treated here, such
as depolarization of light scattering, strain optical effects, and
flow optical effects, the experimental situation is more satisfactory. In those cases, direct connections between experiments and
detailed molecular theories have been established.

The first half of this chapter presents equilibrium theories
for electric birefringence (Kerr effect) and electric dichroism;
the second half discusses the dynamic aspects of these properties.
All theories presented are appropriate only for isolated polymer
molecules in solution. Concentration and solvent effects can
usually be corrected by experimental means. We assume throughout
that the total molecular quantity (dipole moment, polarizability
tensor, or absorptivity tensor) is given as a simple sum of contributions from each repeating chain unit. The validity of additivity
of chain group dipole moments in flexible polymers is known. In

the case of polarizability tensors, only recently has their
additivity (the so-called valence-optical scheme) been verified
[1]. The plethora of bond polarizabilities that have been
suggested for common bond types has been responsible for much of
the confusion about this principle. Proposed polarizabilities have
not been derived for the conditions of ideal dilute polymer solutions. Unfortunately, calculations of the Kerr effect with a detailed molecular theory, such as presented below, require
specification of optical, in addition to static, polarizability
tensors for each type of chain bond.

The other optical properties that are most similar to electric
birefringence and electric dichroism are flow-induced birefringence
and dichroism. Only the types of externally applied force fields
differ. The similar measurement of the strain optical properties
is somewhat less widely applicable because of the necessity of
forming a covalently bonded network from the polymer. For both
flow and strain optical properties, it is possible to assume that
the major molecular polarizability axis lies in the direction of
the applied strain or flow. The situation for electro-optic properties is not as simple in general. The distribution of group
dipole moments and polarizable bonds within the molecular structure
serves to determine the molecule's average orientation with respect
to the electric field. Only if the molecular dipole moment and
major polarizability axis coincide with the end-to-end vector can
this vector then be designated as the principal molecular axis.
For most types of molecules these axes are not coincident.

Both electric birefringence and electric dichroism depend on
the molecules' response to the electric field through their static
polarizability tensors and dipole moments. In addition, electric
birefringence depends on the optical polarizability tensor, and
electric dichroism is related to the tensor of absorption coefficients.

II. CONNECTION BETWEEN OPTICAL
PROPERTIES AND INTERNAL CONFIGURATIONS

The first connection between optical properties and internal configurations of flexible chain molecules was established by Kuhn and Grün [2]. They related the strain birefringence to the mean square end-to-end distance, $<h^2>_s$. The brackets $< >_s$ indicate a configurational average in the presence of this external field. Their expression for the average molecular optical anisotropy of a flexible polymer is

$$<\Delta\alpha^0>_s = \left(\frac{3}{5}\right)(\alpha_1^0 - \alpha_2^0)\left(\frac{<h^2>_s}{Nb^2}\right)$$ (1)

This chain is composed of N identical links (bonds or groups of bonds), each with a length b and an optical polarizability anisotropy equal to $(\alpha_1^0 - \alpha_2^0)$. The polarizability tensor of the link is assumed to have cylindrical symmetry. This result for the mean molecular optical polarizability anisotropy $(\Delta\alpha^0 = \alpha_\parallel^0 - \alpha_\perp^0)$ was obtained at the limit of complete flexibility for a chain of freely jointed links [2]. More realistic models yield similar expressions [3]. Flow dichroism has also been treated for the freely jointed chain model by Callis and Davidson [4]. Their similar result is

$$<\Delta A>_s = \left(\frac{3}{5}\right)(\epsilon_1 - \epsilon_2)\left(\frac{<h^2>_s}{Nb^2}\right)$$ (2)

where the link has an absorption coefficient ϵ_1 along the link and ϵ_2 perpendicular to it. This expression for the mean absorption coefficient anisotropy requires that the molecular absorption coefficient tensor be cylindrically symmetric.

In contrast, an electric field does not necessarily force polymers to align their end-to-end vectors parallel to the field. The alignment will depend on the locations within the chain of

polar and polarizable groups. Below, the effect of the electric
field on chain configurations is developed in general.

III.   ORIENTATION BY AN ELECTRIC FIELD

The molecular energy of a polar molecule in the presence of
an external electric field E along the x direction is given by

$$U_E = U_0 - \mu_x E - \left(\frac{\alpha_{xx}}{2}\right)E^2 \tag{3}$$

Hyperpolarizability terms with higher powers of the field strength
have been ignored (see Chap. 2). U_0 is the field free configurational energy, μ_x is the permanent dipole moment component along
the direction of the applied field, and α_{xx} is the corresponding
component of the static polarizability tensor. In order to perform
averages with these energies, a partition function is defined.

$$Z_E = (8\pi^2)^{-1} \int \cdots \int \exp\left(\frac{-U_0}{kT}\right)$$

$$\left[1 + \frac{\mu_x E}{kT} + \frac{\mu_x^2 E^2}{(2k^2 T^2)} + \frac{\alpha_{xx} E^2}{(2kT)} + \cdots\right] d\Omega \tag{4}$$

This expansion of the field dependent Boltzmann factor is usually
possible because $E/kT \ll 1$. Integration is over internal molecular coordinates as well as the external angles which specify overall spatial orientation. Each term in Eq. (4) is merely a
field-free average. For small electric fields we truncate the
series to obtain

$$Z_E = Z_0 \left[1 + \frac{\langle\mu_x^2\rangle E^2}{(2k^2 T^2)} + \frac{\langle\alpha_{xx}\rangle E^2}{(2kT)}\right] \tag{5}$$

The field-free partition function is just
$Z_0 = (8\pi^2)^{-1} \int \cdots \int \exp(-U_0/kT) \, d\Omega$. The angle brackets denote

the average over all internal molecular coordinates. Spatial
symmetry permits this to be reexpressed as [3,5]

$$Z_E = Z_0\left\{1 + \left[\frac{<\mu^2>}{kT} + Tr<\alpha>\right]\left(\frac{E^2}{6kT}\right)\right\} \tag{6}$$

where Tr indicates trace of the matrix. This partition function
is appropriate for averaging any property in the presence of small
electric fields. It includes two field-dependent terms. The first
is important only for polar molecules, in which case it is usually
dominant. The other depends on the magnitude of the molecule's
static polarizability. Because this partition function is utilized
in forming all subsequent equilibrium average configurational properties, these two types of terms may be observed in many of the
following expressions. As a simple example, the squared dipole
moment in an electric field can be averaged with this partition
function. The result is [6]

$$<\mu^2>_E = \frac{Z_0}{Z_E}\left\{<\mu^2> + \left[\frac{<\mu^4>}{kT} + <\mu^2 Tr\alpha>\right]\frac{E^2}{(6kT)^2}\right\} \tag{7}$$

The field-free averages required on the right side of this equation may be obtained by use of a model for the internal degrees of
freedom of the macromolecule. Several of these models will be
discussed in detail in the next section.

IV.   KERR CONSTANT AND EQUILIBRIUM FIELD-FREE AVERAGES [3,5,7]

By means of the Lorentz-Lorenz equation the difference in
refractive indexes parallel and perpendicular to the electric
field can be expressed in terms of the difference between the
polarizabilities parallel and perpendicular to the field [8],

$$\Delta n = \left(\frac{2\pi}{9}\right)\left[\frac{(n^2 + 2)^2}{n}\right]N_A\left(\frac{c}{M}\right)\Delta\alpha^0 \tag{8}$$

where N_A is Avogadro's number, M is the polymer molecular

weight, n is the average refractive index, and c is the
concentration in grams per cubic centimeter. Derivation of this
equation requires that $\Delta n \ll n$.

The molar Kerr constant may be expressed as

$$K_m = \left(\frac{4\pi}{3}\right)\frac{N_A \Delta\alpha^0}{E^2}$$ (9)

Combining Eqs. (8) and (9) yields an expression for the electric
birefringence,

$$\Delta n = \frac{(n^2 + 2)^2 cE^2 K_m}{6nM}$$ (10)

Treatment of the Kerr constant of flexible polymer solutions
has usually [5,7] proceeded from these equations; however, a better
expression for K_m was derived by Buckingham [9] and later applied
to polymer solutions by Dows [9]. The difference between the results obtained by the two methods is a simple numerical factor.
Results for K_m corresponding to Buckingham's method can be obtained from those presented below [based on Eqs. (9) and (10)] by
multiplying by 1/9.

Before considering detailed configuration models, we must
average $\Delta\alpha_E^0$ over the external orientation angles. If x is
the direction of the electric field, then by the use of the partition function in Eq. (4), we obtain

$$\Delta\alpha^0 = (1/2)(Z_0/Z_E)[<(\alpha_{xx}^0 - \alpha_{yy}^0)\mu_x^2>_{ext}(E/kT)^2$$
$$+ <(\alpha_{xx}^0 - \alpha_{yy}^0)\alpha_{xx}>_{ext}E^2/kT]$$ (11)

First, the two complicated expressions in angle brackets must be
averaged over external Euler angles for a fixed internal configuration in order to eliminate the coordinate dependence indicated by
the x and y subscripts. The first term averaged over external
coordinates is

$$
\langle(\alpha_{xx}^0 - \alpha_{yy}^0)\mu_x^2\rangle_{ext} = \sum_{p=1}^{3} \alpha_p^0 \mu^2 [\langle\cos^2 T_x^p \cos^2 T_x^\mu\rangle_{ext}
$$

$$
- \langle\cos^2 T_y^p \cos^2 T_x^\mu\rangle_{ext}]
$$

(12)

where the principal components of α^0 are α_1^0, α_2^0, and α_3^0. The angle T_x^p is the angle between coordinates x and α_p. The external coordinate average of this expression is

$$
\langle(\alpha_{xx}^0 - \alpha_{yy}^0)\mu_x^2\rangle_{ext} = \left(\frac{1}{15}\right)[3\underline{\mu}^T\underline{\alpha}^0\underline{\mu} - \mu^2 Tr\underline{\alpha}^0]
$$

(13)

Vectors and matrices are indicated by underbars; transposition is denoted by a superscript T. For the other required term we find

$$
\langle(\alpha_{xx}^0 - \alpha_{yy}^0)\alpha_{xx}\rangle_{ext} = \left(\frac{1}{15}\right)[3Tr(\underline{\alpha}^0\underline{\alpha}) - Tr(\underline{\alpha})Tr(\underline{\alpha}^0)]
$$

(14)

Substitution of these terms back into Eq. (11) and averaging over internal degrees of freedom yield

$$
\langle\Delta\alpha^0\rangle = \left(\frac{1}{30}\right)[3\langle\underline{\mu}^T\underline{\alpha}^0\underline{\mu}\rangle - \langle\mu^2 Tr\underline{\alpha}^0\rangle
$$

$$
+ 3kT\langle Tr(\underline{\alpha}^0\underline{\alpha})\rangle - kT\langle Tr\underline{\alpha}^0 Tr\underline{\alpha}\rangle]\left(\frac{E}{kT}\right)^2
$$

(15)

This is a result of general validity restricted only to small electric field strengths. The averages on the right may be calculated in the absence of the electric field. Equation (15) may be simplified somewhat by introducing traceless polarizability tensors,

$$
\hat{\underline{\alpha}} = \underline{\alpha} - \overline{\alpha}I_3
$$

(16)

The mean polarizability is $\overline{\alpha}$ and the unit matrix of order 3 is I_3. Incorporation of Eq. (16) into Eq. (15) yields

$$
\langle\Delta\alpha^0\rangle = \frac{1}{10}\left[\langle\underline{\mu}^T\hat{\underline{\alpha}}^0\underline{\mu}\rangle\left(\frac{E}{kT}\right)^2\right.
$$

$$
\left. + \langle Tr(\hat{\underline{\alpha}}^0\hat{\underline{\alpha}})\rangle\left(\frac{E^2}{kT}\right)\right]
$$

(17)

This result holds in the absence of intermolecular interactions
and is independent of a model for internal configurations. The
treatment of polar molecules requires only the first term; whereas
all molecules necessitate inclusion of the other term which represents an induced effect. Experimentally it should be possible to
distinguish between the two terms in Eq. (17) by taking advantage
of their different temperature dependences. However, for strongly
polar molecules, this would not be possible because the first term
would dominate. Models appropriate for performing the averages
indicated in Eq. (17) over internal molecular coordinates must now
be delineated.

V. CONFIGURATIONAL MODELS AND AVERAGE KERR CONSTANTS

A variety of models [3,8,10] have been developed to perform
averages over internal degrees of freedom in polymer chains.
Usually these methods are applied to treat such simpler properties
as the square of the end-to-end distance. Several treatments of
the more complicated averages required in Eq. (17) have appeared
[3,7,9,11-13]. Here we will consider only two of these models and
will emphasize the most realistic one, the rotational isomeric
state model. If desired, results for other less general models can
be obtained easily as limits [3] of this model.

We assume that the molecular dipole moment and the molecular
polarizability tensor may be expressed as sums of the contributions
from each of the N units.

$$\underline{\mu} = \sum_{i=1}^{N} \underline{\mu}_i$$

and

$$\underline{\hat{\alpha}} = \sum_{j=1}^{N} \underline{\hat{\alpha}}_j \tag{18}$$

All terms in the sums must, of course, be expressed in the same coordinate system. These principles have been verified for several types of molecules [1,3]. Substitution of these expressions into Eq. (17) yields

$$<\Delta\alpha^0> = \frac{1}{10}\left[\sum_{i,j,k} <\underline{\mu}_i^T\hat{\underline{\alpha}}_j^0\underline{\mu}_k>\left(\frac{E}{kT}\right)^2\right.$$

$$\left. + \sum_{i,j} <Tr(\hat{\underline{\alpha}}_i^0\hat{\underline{\alpha}}_j)>\left(\frac{E^2}{kT}\right)\right]$$

(19)

The simplest model is the freely jointed chain. This is a random flight chain composed of identical links joined by universal joints. All internal configurations have equal internal energies. The model does not include the more realistic features of rigid bond angles and internal hindered rotations. Nor does it permit incorporation of the variations in molecular structure manifested in chain molecules with various chemical structures. However, the results for a freely jointed chain are instructive. The molar Kerr constant from Eqs. (19) and (9) averaged with this model [9,11] is

$$K_m = \frac{4\pi}{45kT} N_A N\Delta a^0\left[\left(\frac{m^2}{kT}\right) + \Delta a\right]$$

(20)

where m is the dipole moment of each link and Δa^0 is the optical anisotropy of a link. The dependence on N can be observed from Eq. (19); the only nonzero terms in the sums are those in which all indexes are the same, i.e., i = j = k and i = j, respectively. This leads, for both quantities, to N identical terms. A linear dependence of the Kerr constant on chain length will thus be observed for all long completely flexible chain molecules. Application of this result to real molecules requires specification of an artificial equivalent chain. It is somewhat arbitrary how one chooses a number of real bonds to correspond to one equivalent freely jointed link. Also, dipole moment and optical anisotropy of the equivalent link must be specified. We can avoid such problems by utilizing a more detailed molecular model.

For the rotational isomeric state model, the bond properties
required are chosen unambiguously from the properties of small
molecules.

It is also appropriate to apply Eq. (19) to treat rigid rods.
In this case all chain units are correlated. The result is [9]

$$K_m = \frac{4\pi}{45kT} \, N_A \Delta a^0 N^2 \left[\left(\frac{m^2}{kT} \right) N + \Delta a \right] \tag{21}$$

Comparison of Eqs. (20) and (21) indicates that the Kerr constant
for rigid rods exhibits a higher order molecular weight dependence
than for flexible chains.

Volkenstein [12] has presented a theory by Gotlib for nonpolar
molecules typified by polyisobutylene. This theory, in addition to
being approximate and cumbersome, includes hindered rotations only
in the approximation that each bond rotates independently of all
others. The importance of including the neighbor dependences of
bond rotations has been amply demonstrated for a variety of polymer
molecules [3].

Flexibility in polymer molecules arises principally through
rotations about backbone bonds. Usually these rotations are restricted. The probabilities of any rotational angle depend upon
the rotational positions assumed by its neighbors. The degree of
rotational restriction and neighbor interdependence directly
determine the chain's flexibility. Distortions of bond angles have
little influence on average equilibrium properties. In the present
rotational isomeric state model the bond angles are taken to be
fixed. Thus configurations are completely specified by the set of
backbone bond rotation angles, ω_2, ω_3, ..., ω_{N-1} (see Fig. 1).
In this model only a small number, ν, of rotational angles about
each bond are permitted. These are chosen to coincide with the
minima in energy. Usually the angular dependence of the energy is
steep; consequently, replacement of a continuous angular probability with ν individual probabilities located at the energy minima
is valid for the purpose of calculating average equilibrium properties. This model has been successfully applied to treat many
different properties [3].

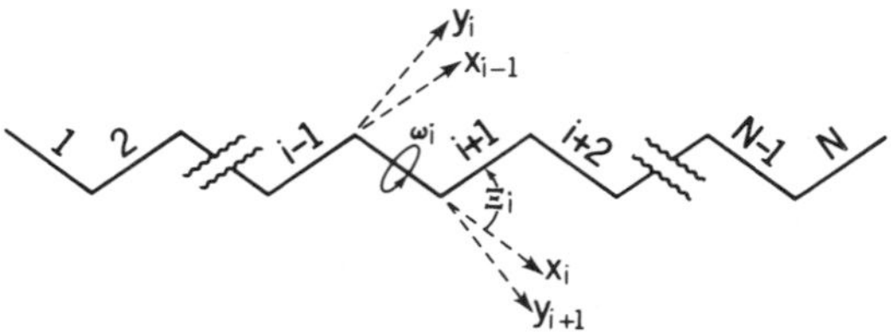

FIG. 1. Portions of molecular backbone. Bond numbering
system is indicated. Several coordinate axes as defined in the
text are shown.

A partition function is defined in terms of statistical
weights for each of the ν rotation states. Accounting for the
neighbor dependence of the rotational state energies has normally
proven to be essential [3]. The statistics of the chains require
knowledge of the array of energies,

$$
\underline{U}_i = \begin{bmatrix}
u_{11} & u_{12} & \cdots & u_{1\nu} \\
u_{21} & u_{22} & \cdots & u_{2\nu} \\
\cdot & & & \cdot \\
\cdot & & & \cdot \\
\cdot & & & \cdot \\
u_{\nu 1} & u_{\nu 2} & \cdots & u_{\nu\nu}
\end{bmatrix}
\tag{22}
$$

The energy u_{pq} is the pair energy, in the absence of an electric
field, for the configuration with backbone bond $i - 1$ in rotational state p and bond i in rotational state q minus the
singlet energy for the configuration with bond $i - 1$ in state p.
A statistical weight matrix is formed from Eq. (22) by transforming
the energies into Boltzmann factors,

$$
\underline{V}_i = \begin{bmatrix}
\exp\left(\dfrac{-u_{11}}{kT}\right) & \exp\left(\dfrac{-u_{12}}{kT}\right) & \cdots & \exp\left(\dfrac{-u_{1\nu}}{kT}\right) \\
\cdot & \cdot & & \cdot \\
\cdot & \cdot & & \cdot \\
\cdot & \cdot & & \cdot \\
\exp\left(\dfrac{-u_{\nu 1}}{kT}\right) & \exp\left(\dfrac{-u_{\nu 2}}{kT}\right) & \cdots & \exp\left(\dfrac{-u_{\nu\nu}}{kT}\right)
\end{bmatrix}
\tag{23}
$$

A partition function which includes the sums of the statistical weights for all permitted configurations is given by

$$Z_o = \underline{J}^* \underline{V}_2 \cdots \underline{V}_{N-1} \underline{J} \tag{24}$$

where $\underline{J}^*$ is the $1 \times \nu$ row $(100\ldots0)$ and $\underline{J}$ is the $\nu \times 1$ column $(11\ldots1)$. This partition function can be used directly to perform equilibrium averages of most polymer properties. Results for this model rely on the selection of energies u_{pq} in Eq. (22). Judicious combination of experimental results for rotational energy functions in small molecules, calculations of semiempirical energy functions, and use of space-filling models permit the energies to be chosen with confidence. Molecules whose rotational isomeric state chain statistics have been described include polymethylene, polyoxymethylene, higher polyoxyalkanes, polydimethylsiloxane, polyphosphate, polymeric sulfur and selenium, polytetrafluoroethylene, polyamides, polyesters, polypeptides, polyisobutylene, polybutadiene, polyisoprenes [3], polysaccharides [14], and polynucleotides [15]. Any sequence of monomer types may be treated by forming the product in Eq. (24) with the appropriate sequence of different types of matrices $\underline{V}$. This model is for isolated molecules with no long-range interactions. These conditions are met in dilute polymer solutions at the θ condition [3,15].

Before averaging the quantities in Eq. (17) with this partition function, the chain geometry must be specified. Figure 1 shows the supplement of bond angle Ξ_i and the rotational angle ω_i. A coordinate system is specified for each backbone as follows: the axis x_i is taken to lie along bond i, y_i is in the plane of bonds $i-1$ and i, its direction so chosen as to yield a positive projection on x_{i-1}, and the position of z_i is dictated by the right-handedness of the coordinate system. The matrix which transforms a property expressed in coordinate system $i + 1$ to its representation in the ith system is

$$
\underline{T}_i = \begin{bmatrix}
\cos \Xi_i & \sin \Xi_i & 0 \\
\sin \Xi_i \cos \omega_i & -\cos \Xi_i \cos \omega_i & \sin \omega_i \\
\sin \Xi_i \sin \omega_i & -\cos \Xi_i \sin \omega_i & -\cos \omega_i
\end{bmatrix} \tag{25}
$$

Averaging with the partition function in Eq. (24) is carried out by inserting, in sequence, the contribution for each chain bond so that it is adjacent to the statistical weight matrix $\underline{V}$ with the same serial index. The sums in Eq. (19) contain terms in which the bond indexes i, j, and k are not in sequential order from left to right. Those terms which do not conform to this ordering can be rearranged by means of the relationships,

$$
\underline{\mu}_i^T \hat{\underline{\alpha}}_j^0 \underline{\mu}_k = (\underline{\mu}_i^T \otimes \underline{\mu}_k^T) C(\hat{\underline{\alpha}}_j^0) = R(\hat{\underline{\alpha}}_j^0)(\underline{\mu}_i \otimes \underline{\mu}_k) \tag{26}
$$

where $R(\ )$ and $C(\ )$ indicate formation of a row or column vector from the enclosed square matrix. The direct matrix product is denoted by $\otimes$. Rearrangements, whose details are not given here, then permit direct calculation of the first average. The result [3,7,13,17] is

$$
\sum_{i,j,k} \langle \underline{\mu}_i^T \hat{\underline{\alpha}}_j^0 \underline{\mu}_k \rangle = 2Z_0^{-1} J^* P_1 P_2 \cdots P_N J \tag{27}
$$

where J^* is the row $(10\ldots0)$ and J is a column of zeros followed by ν 1's to conform in size to the arrays P_i. The central matrix is given by

$$
P_i = \begin{bmatrix}
\underline{V} & (\underline{V}\otimes\underline{\mu}^T)\|\underline{T}\| & [\underline{V}\otimes R(\hat{\underline{\alpha}}^0)]\|\underline{T}\otimes\underline{T}\| & \tfrac{1}{2}(\underline{V}\otimes\underline{\mu}^T\otimes\underline{\mu}^T)\|\underline{T}\otimes\underline{T}\| & \underline{V}\otimes[R(\hat{\underline{\alpha}}^0)(\underline{\mu}\otimes\underline{I}_3)]\|\underline{T}\| & \tfrac{1}{2}\underline{V}[R(\hat{\underline{\alpha}}^0)(\underline{\mu}\otimes\underline{\mu})] \\[4pt]
0 & (\underline{V}\otimes\underline{I}_3)\|\underline{T}\| & 0 & (\underline{V}\otimes\underline{I}_3\otimes\underline{\mu}^T)\|\underline{T}\otimes\underline{T}\| & (\underline{V}\otimes\hat{\underline{\alpha}}^0)\|\underline{T}\| & \underline{V}\otimes[(\underline{I}_3\otimes\underline{\mu}^T)C(\hat{\underline{\alpha}}^0)] \\[4pt]
0 & 0 & (\underline{V}\otimes\underline{I}_9)\|\underline{T}\otimes\underline{T}\| & 0 & (\underline{V}\otimes\underline{\mu}\otimes\underline{I}_3)\|\underline{T}\| & \tfrac{1}{2}(\underline{V}\otimes\underline{\mu}\otimes\underline{\mu}) \\[4pt]
0 & 0 & 0 & (\underline{V}\otimes\underline{I}_9)\|\underline{T}\otimes\underline{T}\| & 0 & \underline{V}\otimes C(\hat{\underline{\alpha}}^0) \\[4pt]
0 & 0 & 0 & 0 & (\underline{V}\otimes\underline{I}_3)\|\underline{T}\| & \underline{V}\otimes\underline{\mu} \\[4pt]
0 & 0 & 0 & 0 & 0 & \underline{V}
\end{bmatrix}
$$

(28)

The $\|\ \ \|$ indicates expansion of the enclosed array in a larger pseudodiagonal matrix. For example,

$$
\|T\| = \begin{bmatrix}
\underline{T}(\omega_1) & & & & \\
& \underline{T}(\omega_2) & & & \\
& & \cdot & & \\
& & & \cdot & \\
& & & & \cdot \\
& & & & \underline{T}(\omega_\nu)
\end{bmatrix}
$$

(29)

Terms on the diagonal are the matrices T evaluated for each of the ν rotational states. The other required term can be rewritten as [3,7]

$$
\langle \mathrm{Tr}(\hat{\underline{\alpha}}^0\hat{\underline{\alpha}}) \rangle = \langle R(\hat{\underline{\alpha}}^0)C(\hat{\underline{\alpha}}) \rangle
$$

(30)

The sum of this other required average is given by

$$\sum_{i,j} \langle R(\hat{\underline{\alpha}}_{-i}^0) C(\hat{\underline{\alpha}}_{-j}) \rangle = Z_0^{-1} J^* Q_1 Q_2 \cdots Q_N J \tag{31}$$

Thus both the sum and its average are performed by serial multiplication of these large matrices. The central array is

$$Q_i = \begin{bmatrix} \underline{V} & [\underline{V} \otimes R(\hat{\underline{\alpha}}^0)] \| \underline{T} \otimes \underline{T} \| & [\underline{V} \otimes R(\hat{\underline{\alpha}})] \| \underline{T} \otimes \underline{T} \| & R(\hat{\underline{\alpha}}^0) C(\hat{\underline{\alpha}}) \underline{V} \\ 0 & (\underline{V} \otimes \underline{I}_9) \| \underline{T} \otimes \underline{T} \| & & \underline{V} \otimes C(\hat{\underline{\alpha}}) \\ 0 & 0 & (\underline{V} \otimes \underline{I}_9) \| \underline{T} \otimes \underline{T} \| & \underline{V} \otimes C(\hat{\underline{\alpha}}^0) \\ 0 & 0 & 0 & \underline{V} \end{bmatrix} \tag{32}$$

A choice of bond polarizabilities is difficult. Let us consider the case of n-alkane chains. A suitable unit is the C - CH_2 group. The polarizability anisotropy for this group has a simple form if tetrahedral geometry prevails. If its anisotropy is considered to be derived from that of methane, we must subtract the amount for the two C - H bonds and add the amount for one C - C bond. Thus for this simple case the polarizability anisotropy is given by [18]

$$\Gamma^0 = \Delta\alpha_{CC}^0 - 2\Delta\alpha_{CH}^0 \tag{33}$$

Values of this quantity from several sources have been collected in Table 1. This is typical of the confusion that exists for the polarizabilities of most repeating units. Differences in solvent conditions and physical state account for the large discrepancies.

  Patterson and Flory [1] have measured and calculated the depolarized Rayleigh scattering for a homologous series of n-alkanes. By extrapolation to zero concentration and by correcting for collision-induced anisotropy, they obtained experimental results for molecules from C-5 to C-22 which agree with calculated values for a rotational isomeric state model. Failure to make these corrections to raw experimental data accounts for much previous lack of success in verifying the valence-optical

TABLE 1

Γ^0 [Å^3]	Investigators
0.54	Patterson and Flory [1]
0.72	Le Fèvre et al. [19]
1.60	Wang [20]
1.44	Denbigh [21]
0.283	Bunn and Daubeny [22]
1.25	Vuks [23]
0.44	Philippoff [24]
1.71	Zürcher [25]

principle. Their agreement lends strong support to this method of adding bond polarizabilities.

Polarizability tensors of some molecular units cannot be accurately estimated by simple addition of bond polarizabilities. An example of such a unit is the peptide monomer. The complicating feature in this unit is the resonance between two double-bonded forms. Recently, Ingwall and Flory [26] have evaluated the glycyl unit polarizability tensor. The anisotropic part of the polarizability tensor is identical to that for N-methylacetamide. They obtained the three distinct components of the anisotropic part of the polarizability tensor of N-methylacetamide from their measured values of the Rayleigh ratios for depolarized scattering of N-t-butylacetamide and N-methylacetamide. In addition, they utilized the Kerr constant for N-t-butylacetamide measured by Aroney et al. [27]. Relating the polarizabilities of the two amide molecules required the further specification of the polarizability of an alkane-type link. Their molecular coordinate system is: the x axis lies along the C - N bond, the y axis is perpendicular to it and forms an acute angle with the C=O bond, and the z axis is specified by the right-handness of the coordinate system. They obtained the following results for the components of the traceless polarizability tensor:

$$\hat{\alpha}_{xx} = 0.3\text{Å}^3, \quad \hat{\alpha}_{yy} = 0.9\text{Å}^3,$$

$$\hat{\alpha}_{zz} = -1.2\text{Å}^3, \quad \hat{\alpha}_{xy} = 0.1\text{Å}^3,$$

$$\hat{\alpha}_{xz} = 0, \quad \text{and} \quad \hat{\alpha}_{yz} = 0$$

Their preliminary calculations [28] of polypeptide Kerr constants using this tensor indicate a strong dependence on configuration. This method should be generally applicable for obtaining polarizability tensors of rigid repeating units in complicated polymers from data on model compounds.

Ishikawa and Nagai [29] used the rotational isomeric state model to calculate Kerr constants for n-alkanes and for polyoxyethylenes. In the alkanes, they assumed that the rotational states occur at $\omega = 0^\circ$ (trans), $\omega = 120^\circ$ (gauche$^+$), and $\omega = 240^\circ$ (gauche$^-$). Because it is a nonpolar molecule only the second term in Eq. (17) is required. Also, they took $\underline{V}_1 = \underline{I}_3$ since the terminal methyl group is symmetric. They chose $\Gamma = 1.1\Gamma^0$. By choosing $\Gamma^0 = 0.81$ Å^3 and the energy of a gauche above a trans state to be 800 cal mole^{-1}, they quantitatively reproduced the observed increase of K_m/N with increasing N. Adjacent rotational state pairs with gauche states of opposite signs were excluded. We have repeated their calculation, using slightly different energies and values of $\Gamma^0 = 0.54, 0.81,$ and 1.25 (see Fig. 2). Temperature chosen was $20^\circ C$. We likewise chose tetrahedral bond angles and $\Gamma = 1.1\Gamma^0$. Agreement was good for the choice of $\Gamma^0 = 0.81$. A better fit could have been obtained, but unfortunately the quality of the data by Stuart, Finck, and Kuss [29] does not warrant further fitting.

In the case of the polyoxyethylenes Nagai and Ishikawa [29] obtained only qualitative agreement with experiment. They were able to calculate the observed decrease in K_m/N with increasing chain length. The rotational isomeric state model, unlike the

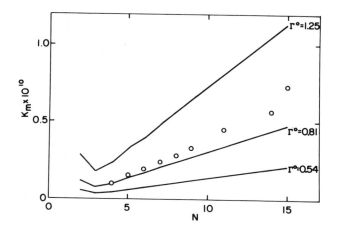

FIG. 2. Molar Kerr constants for n-alkanes. Calculated
results are solid lines; experimental points are circles [29]. The
energy of a gauche state was taken to be 500 cal mole^{-1} above that
of a trans state. The additional energy of a pair of gauche states
of opposite sign, $u_{g^+g^-}$, was 2500 cal mole^{-1} for these calculations.

freely jointed chain, yields various chain-length dependences,
which arise because of detailed molecular structure.

Le Fèvre and coworkers have observed that the ratio K_m/N increases with increasing N for polyvinylbromide [30] and
polyvinylchloride [31]. However, for polymethylmethacrylate [32],
polymethylacrylate [33], and polystyrene [34], this ratio decreases. Unfortunately, most of these data are for poorly
characterized polymers. The vinyl polymers are of unknown tactici-
ty. Experiments were not carried out at θ conditions. Further
calculations of Kerr constants by the present theory would be
justified only if more refined experimental results were available.

It would be possible to obtain an estimate of the effect of
excluded volume on the Kerr constant by calculating correlations
of polarizabilities and dipole moments with end-to-end distances.
This approach has been applied previously by Nagai and Ishikawa
[35] to treat the excluded volume of dipole moments.

VI. ELECTRIC DICHROISM

If group absorptivities are configurationally independent,
then it is possible to utilize directly all previous theory for
electric birefringence by simply replacing polarizability tensors
$\underline{\alpha}^0$ by absorptivity tensors $\underline{\varepsilon}$. Flory and Abe [36] treated strain
dichroism of flexible polymers in detail. For that property,
correlations between the transition dipoles and the end-to-end
vector are important. These same correlations would not necessarily be expected to assume importance for electric dichroism. Data
for flexible polymers are unavailable.

VII. DYNAMIC ELECTRO-OPTICS OF FLEXIBLE POLYMERS

The models discussed in the preceding section relate the
electro-optical properties of flexible polymers, at equilibrium in
an electric field, to the optical properties of the molecular
groups constituting the polymers and to the spatial configurations
of the polymers. The remaining sections will be devoted to the
corresponding nonequilibrium electro-optical properties. Since
studies of the nonequilibrium electro-optical properties are frequently directed toward gaining information about the motions of
polymers in dilute solutions, we shall focus on this part of the
topic. We assume that the polymer solutions are in a nonconducting
solvent, dilute enough, and at a suitable temperature to isolate
the individual polymer molecules from each other, thus minimizing
mutual polymer interactions. These conditions make the discussion
neater. The dynamic aspects of electric birefringence and electric
dichroism effects in solutions of flexible polymers are very similar; so these two subtopics will be treated in a parallel fashion.

No theoretical description of polymer motion covers the complete time scale of observable, motion-related (dynamic) properties. We have therefore bisected the time scale roughly according
to the number of chain repeat units that are moving coordinately,

to classify the present theoretical descriptions of motion. The
motion is rapid and local if only a few polymer chain units move
in a correlated way. If many, at least 50, polymer chain units
move coordinately, we call the motion long range and presume it to
be much slower than local motion would be in the same solvent. All
polymer motions are considered to be Brownian [37] or micro-
Brownian and are friction limited by solvent viscosity. Since we
are restricting our discussion to nonionized polymers in a spatially uniform but time-varying electric field, the movement of polymer
chains bearing dipoles is either rotational or dilational rather
than translational. To get some idea of the range of the time
scale for polymer motion, we shall outline a method of determining
the rotational correlation times for extreme cases of local and
long-range motion.

VIII. TIME SCALES FOR MOTION

Polymers can change from one rotational isomeric state to
another by intramolecular rotation. If the energy barrier hinder-
ing this rotation is negligible compared with thermal energy, the
polymer can be pictured as a chain of units of about monomer size
that can rotate easily to respond to the torques caused by an electric field acting on chain dipoles. In this extreme case of local
motion, the correlation time is about that expected for rotation of
a monomer molecule in the solvent. If the hindrance to rotation
is large, the chain units must rotate in unison to respond to the
electric field. In the extreme case of long-range motion, the
whole polymer chain rotates.

The methods for getting rotational correlation times in these
two cases differ in detail, but both are based on friction-limited
Brownian motion through a viscous solvent. Let us first consider
the case of local motion. Following the ideas of Debye [38], we
consider the time development of a conditional probability function
$f(\theta, \phi; t)$, which gives the probability that a spherical molecule

will have rotated to a position represented by angles θ and ϕ, at time t, after it started at the pole of a spherical coordinate system. Debye showed that the probability function f obeys a diffusion equation of the form

$$\frac{1}{\Theta} \frac{\partial f}{\partial t} = \nabla^2 f \tag{34}$$

whose solution is a series of normalized spherical harmonics [39] (or surface harmonics) $y_{\ell m}(\theta,\phi)$ damped exponentially, Θ is the rotational diffusion coefficient, and

$$f = \sum_{\ell,m} a_{\ell m} y_{\ell m}(\theta,\phi) \exp [-\Theta\ell(\ell + 1)t] \tag{35}$$

The coefficients $a_{\ell m}$ are found from the initial conditions by using the expansion theorem for spherical harmonics. The properties of spherical harmonics lead to the following useful result: The spatial average value of any function of angles θ and ϕ that is a spherical harmonic of order ℓ will decay exponentially to zero from its initial state, with a time constant given by

$$\tau_\ell = \frac{1}{[\Theta\ell(\ell + 1)]} \tag{36}$$

Thus for vector functions such as the molecular dipole moment, the relaxation of the component of the dipole moment resolved along the spatial direction of the electric field is given by [38]

$$<\cos \theta(t)>_{av} = <\cos \theta(0)>_{av} \exp \left(\frac{-t}{\tau_1}\right) \tag{37}$$

because $\cos \theta$ is a spherical harmonic of order unity, $(\ell = 1)$.

More apropos to the subject of this chapter is the time constant for the relaxation of tensor functions as used to describe birefringence and dichroism. These tensor averages, which involve spherical harmonics of second order $(\ell = 2)$, have time constants.

$$\tau_2 = \frac{1}{6\Theta} \tag{38}$$

If we are using electric birefringence relaxation to study the local motion of a polymer chain with freely rotating "spheres bearing dipoles" as elements, we expect the response to be 1/e of its original value in time τ_2. If the rotating chain elements are assumed to be spheres of radius r, we can use Stokes' [40] result in which the frictional coefficient ζ is

$$\zeta = 8\pi\eta r^3 \tag{39}$$

in a solvent of viscosity η. Combining this result with Einstein's [41] relationship for Brownian diffusion, we have

$$\Theta = \frac{kT}{\zeta} \tag{40}$$

where kT is the thermal energy per molecule. Using Eqs. (38) and (39) together with (40), we get

$$\tau_2 = \frac{4\pi\eta r^3}{3kT} \tag{41}$$

For typical organic solvents such as benzene or cyclohexane, $\eta \approx 0.007$ p at $25°C$ and τ_2, often called τ_c, the rotational correlation time, is 0.6×10^{-11} sec for $r = 2$ Å. This gives an estimate of the short-time (local motion) end of the time scale.

An expression for a flexible chain's rotational diffusion constant was derived originally by Kirkwood and Riseman [42] using their model that represents a flexible polymer chain by a sequence of beads that interact with the viscous solvent through Stokes' frictional coefficients and with each other, hydrodynamically, through an Oseen [43] tensor. Isihara [44] used the same model with a different mathematical treatment to get a similar expression for the rotational diffusion constant. The result is given below.

$$\Theta = \frac{RT}{4M\eta[\eta]} \tag{42}$$

where M is the polymer's molecular weight, $[\eta]$ is its intrinsic viscosity [52], η is the solvent viscosity, and RT is the thermal energy. The rotational correlation time is obtained from the combination of Eqs. (38) and (42). Figure 3 displays τ_2 thus

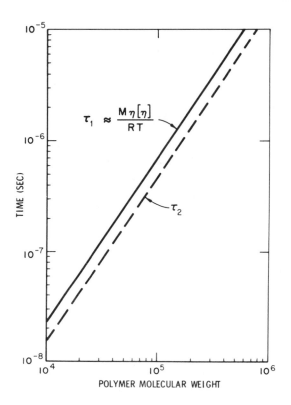

FIG. 3. Molecular weight dependence of relaxation times. The parameters used are: η = 0.007 p at 25°C. Equation (43) with k' = 0.08 and a = 0.50 was used to relate [η] to M.

calculated vs. M for a flexible polymer such as polystyrene in cyclohexane at 34°C [45] or polyisobutylene in benzene at 24°C [46]. In both these cases, the intrinsic viscosity and molecular weight are related by the same Mark-Houwink equation [47]

$$[\eta] = k'M^a \tag{43}$$

with k' = 0.08 and a = 0.50. This rotational correlation time gives an order-of-magnitude estimate of the time scale for long-range motion. More detailed models for dynamic electric birefringence and dichroism, which will be discussed later, treat cases which are more realistic for flexible polymers.

The arrangement of dipolar components along the polymer chain contour determines whether information on local motion or on long-range motion will be provided by dynamic electric birefringence or dichroism observations. If the dipoles are on long pendant side chains only, their electro-optical response will be insensitive to the motions of the polymer backbone. If the dipoles are rigidly attached to the polymer backbone, the dichroism and birefringence should be mainly sensitive to local motion. Models to describe this dipole situation will be discussed later.

IX. A DYNAMICAL MODEL FOR LONG-RANGE MOTION

The situation offering the most information on polymer motion is that in which the dipoles have components parallel to the chain contour. Parallel dipole components occur when the polymer chain is so constituted that a repeat unit cannot have a plane of symmetry perpendicular to the chain contour. The dipole vector for a sequence of such repeat units, without reversal of the directional sense, correlates completely with the displacement vector $\underline{r}_n$ of the sequence [48] (see Fig. 4). This means that the electric field can orient long sequences of the chain. The quantitative development of the parallel dipole situation has been successfully treated by Stockmayer and Baur [49] using the bead and spring model for polymer dynamics developed by Rouse [50] and refined by Zimm [51].

The Rouse-Zimm model represents a linear polymer of very high molecular weight in dilute solution by a chain of $N + 1$ beads that are joined sequentially by N Hookean springs of force constant $3kT/b^2$. The mean-square length of each spring is b^2, and kT is the thermal energy. Each "spring" represents a gaussian subchain made of axially symmetric "statistical chain elements." These elements bear the averaged optical characteristics of the molecule that will be needed for our development of birefringence and dichroism. Each bead is characterized hydrodynamically by a translational friction constant ζ. A $3N + 3$ dimensional

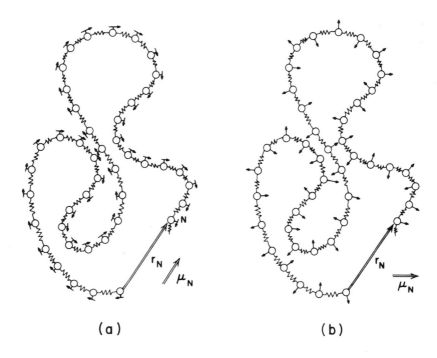

$$(a) \qquad\qquad (b)$$

FIG. 4.   Schematic representation of bead and spring chains
with (a) parallel dipoles and (b) perpendicular dipoles. Correlation of the dipole vector μ_N and the displacement vector r_N
occurs only in case (a). (Redrawn from Ref. 48 by permission of
Butterworths Publishers.)

configurational distribution function ψ gives the probability of
finding the j'th bead with coordinates between x_j and
$x_j + dx_j$, y_j and $y_j + dy_j$, z_j and $z_j + dz_j$. The relative
velocities of the beads and the fluid are modified by hydrodynamic
interaction between beads, but intramolecular friction caused by
local rotational barriers is neglected. The form of the hydrodynamic interaction is given by the Kirkwood-Riseman approximation to
the Oseen tensor [42,43]. This model focuses on long-range chain
motions and submerges the details of local rotameric transitions by
including their averaged effects in the behavior of the N springs.

Following Stockmayer and Baur [49], we imagine that the beads
bear electric charges, e_i on bead i. A spatially uniform but

time-varying electric field acts on the solution in the direction
of the x axis; thus a local field E acts on the polymers. The
field induces a polarization in the chains and causes an electrical
force to be applied to the charges located on the beads. The x
component of the electrical force on bead i is given by

$$F_{x(e_i)} = e_i E + \frac{3(\alpha_1 - \alpha_2)}{5b^2} E^2 (-x_{i-1} + 2x_i - x_{i+1}) \qquad (44)$$

The second term in Eq. (44) is based on the method used by Kuhn
and Grün [2] to calculate the polarizability of freely jointed
chains. Each of the N "springs" in the Rouse-Zimm model represents a gaussian subchain with a low-frequency polarizability component α_1 along the chain and a polarizability component α_2
perpendicular to the chain.

There are also mechanical forces and thermodynamic forces acting on the chain. When these forces are added to the electrical
force, the x component of force on the whole chain is, in matrix
notation,

$$\underline{F}_x = -kT \frac{\partial}{\partial x} \ln \psi + \underline{E} \cdot \underline{e} - \frac{3kT}{b^2} \underline{A} \cdot \underline{x} + \frac{3(\alpha_1 - \alpha_2)}{5b^2} E^2 \underline{A} \cdot \underline{x} \qquad (45)$$

The first term on the right side of Eq. (45) represents a Brownian
motion, thermodynamic force. It can be imagined to be analogous
to the driving force for diffusion. In this sense, the probability
density ψ in configurational space takes a role comparable to the
chemical potential in ordinary diffusion. The last two terms are
formed with the aid of matrix operators. The operator $(3kT/b^2)\underline{A}$
transforms a position vector into a mechanical force vector for the
"springs," while $[3(\alpha_1 - \alpha_2)E^2/5b^2]\underline{A}$ does the same for the electrical polarization.

Using Zimm's methods [51], we write the following equation for
the x part of the diffusion of configurational probability under
the influence of all the forces acting on the polymer chain:

$$\frac{\partial \psi}{\partial t} = \left(\frac{\partial}{\partial \underline{x}}\right)^T \cdot \left[D\underline{H} \cdot \left(\frac{\partial \psi}{\partial \underline{x}}\right) - D\beta E\psi\underline{H} \cdot \underline{e} + \sigma\left(1 - \beta^2 F^2 q^2\right)\underline{H} \cdot \underline{A} \cdot \underline{x}\psi\right]$$

(46)

The symbols are those used by Zimm [51] and by Stockmayer and Baur [49]. The diffusion coefficient for a bead is $D = kT/\zeta$; $\sigma = 3D/b^2$; $\beta = 1/kT$; $q = (\alpha_1 - \alpha_2)/5\beta$; and the superscript T signifies the transpose of a vector. The matrix operator $(1/\zeta)\underline{H}$ accounts for hydrodynamic interaction among the beads by transforming a force vector into a velocity vector. Both $\underline{A}$ and $\underline{H}$ are represented by square matrices of order $N + 1$. Expressions for their matrix elements are given by Zimm [51].

The solution of the diffusion equation is facilitated by choosing a set of normal coordinates that simultaneously diagonalizes the quadratic forms. The proper set of normal coordinates is constructed from the right-hand eigenvectors of the operator $\underline{H} \cdot \underline{A}$. The properties of this transformation have discussed in detail by Zimm [51]. Let $\underline{Q}$ be the transformation matrix that diagonalizes $\underline{H} \cdot \underline{A}$, and define the diagonal matrices $\underline{\Lambda}$, $\underline{M}$, and $\underline{N}$ with the matrix elements $\overline{\lambda}_k$, $\overline{\mu}_k$, and $\overline{\nu}_k$, respectively:

$$\underline{Q}^{-1} \cdot \underline{H} \cdot \underline{A} \cdot \underline{Q} = \underline{\Lambda} \tag{47}$$

$$\underline{Q}^T \cdot \underline{A} \cdot \underline{Q} = \underline{M} \tag{48}$$

$$\underline{Q}^{-1} \cdot \underline{H} \cdot \underline{Q}^{-1T} = \underline{N} \tag{49}$$

The matrix elements $\overline{\lambda}_k$, $\overline{\mu}_k$, and $\overline{\nu}_k$ are related as follows:

$$\overline{\nu}_k = \frac{\overline{\lambda}_k}{\overline{\mu}_k} \tag{50}$$

The set of normal coordinates $\{\overline{\xi}, \overline{\eta}, \overline{\zeta}\}$ is related to the set of Cartesian coordinates $\{x, y, z\}$ by the transformation $\underline{Q}$ in the following way:

$$\underline{x} = \underline{Q} \cdot \overline{\underline{\xi}} \tag{51}$$

$$\frac{\partial}{\partial \underline{x}} = \underline{Q}^{-1T} \cdot \frac{\partial}{\partial \overline{\underline{\xi}}} \tag{52}$$

The transformations connecting y and z with $\overline{\eta}$ and $\overline{\xi}$ and their partial derivatives are identical in form to the above relation for x and $\overline{\xi}$. The transformed diffusion equation becomes

$$\frac{\partial \psi}{\partial t} = \sum_{k=1}^{N} \left\{ D\overline{\nu}_k \left[\frac{\partial^2 \psi}{\partial \overline{\xi}_k^2} + \frac{\partial^2 \psi}{\partial \overline{\eta}_k^2} + \frac{\partial^2 \psi}{\partial \overline{\zeta}_k^2} \right] \right.$$

$$\left. + \sigma \overline{\lambda}_k \left[\gamma_1 \overline{\xi}_k \frac{\partial \psi}{\partial \overline{\xi}_k} + \overline{\eta}_k \frac{\partial \psi}{\partial \overline{\eta}_k} + \overline{\zeta}_k \frac{\partial \psi}{\partial \overline{\zeta}_k} \right] + 3\sigma\gamma_2 \overline{\lambda}_k \psi - D\overline{\nu}_k \beta E \overline{\varepsilon}_k \frac{\partial \psi}{\partial \overline{\xi}_k} \right\} \tag{53}$$

where

$$\gamma_1 = (1 - \beta^2 q^2 E^2); \tag{54}$$

$$\gamma_2 = (1 - \frac{1}{3}\beta^2 q^2 E^2); \tag{55}$$

and ε_k is the transformed charge distribution. The solution to Eq. (53) has been given by Stockmayer and Baur [49]. We do not need an explicit expression for ψ to compute the configuration averages required to treat dynamic electric birefringence and dichroism for a Rouse-Zimm chain; manipulation of the implicit Eq. (53) will yield the needed averages. We will move on to derive expressions for dynamic electric birefringence and dichroism before returning to compute these averages.

X.   OPTICAL ANISOTROPY OF A FLEXIBLE POLYMER CHAIN

The Kuhn and Grün method [2] for computing the optical anisotropy of freely jointed chains is used to calculate the birefringence and dichroism of Rouse-Zimm chains. If the electric field is along the x direction and the light beam used to measure the birefringence travels along the z direction, we find the difference in optical polarizability projected along the principal axes of the optical polarizability tensor to be

$$\Delta\Gamma = q'[<\underline{x}^T \cdot \underline{A} \cdot \underline{x}> - <\underline{y}^T \cdot \underline{A} \cdot \underline{y}>] \tag{56}$$

where

$$q' = \frac{3(\alpha_1^0 - \alpha_2^0)}{5b^2} \tag{57}$$

and α_1^0 and α_2^0 are, according to Kuhn and Grün, the optical polarizabilities along and perpendicular to the chain contour for a "statistical chain element." Here the symbol < > denotes an average weighted by the distribution function ψ. The birefringence, in excess of that of the solvent, is [49,8]

$$\Delta n = \frac{2\pi(n^2 + 2)^2 cN_A}{9nM} \Delta\Gamma \tag{58}$$

where n is the refractive index, c is the polymer concentration in mass per unit volume, N_A is Avogadro's number, and M is the polymer's molecular weight.

Callis and Davidson [4] have used the Kuhn and Grün model [2] to extend the Rouse-Zimm model to flow dichroism. Their extension follows Zimm's flow birefringence development [51] very closely. Using the same geometry for the electric field and the light beam as was used for the birefringence discussion, we find the difference in absorbance (called the dichroism in Chap. 6) for light vibrating along the two principal axes in the x, y plane to be

$$\Delta A = q''[<\underline{x}^T \cdot \underline{A} \cdot x> - <\underline{y}^T \cdot \underline{A} \cdot \underline{y}>] \tag{59}$$

where

$$q'' = \frac{3(\varepsilon_1 - \varepsilon_2)}{5b^2} \tag{60}$$

and ε_1 and ε_2 are absorption coefficients along and perpendicular to the chain contour for a "statistical chain element."

The expression for both birefringence and dichroism involves the same averages weighted by ψ. By using the normal coordinate

transformations mentioned for Eq. (46), we can write the averages
needed to compute them in terms of the corresponding averages of
the normal coordinates. Thus

$$<\underline{x}^T \cdot \underline{A} \cdot \underline{x}> = \iiint \psi(\overline{\underline{\xi}},\overline{\underline{\eta}},\overline{\underline{\zeta}})\overline{\underline{\xi}}^T \cdot \underline{Q}^T \cdot \underline{A} \cdot \underline{Q} \cdot \overline{\underline{\xi}} \, d\{\overline{\underline{\xi}},\overline{\underline{\eta}},\overline{\underline{\zeta}}\} \qquad (61)$$

with a similar equation for $<\underline{y}^T \cdot \underline{A} \cdot \underline{y}>$. Using Eq. (48) gives:

$$<\underline{x}^T \cdot \underline{A} \cdot \underline{x}> = \sum_{k=1}^{N} \overline{\mu}_k <\overline{\xi}_k^2> \qquad (62)$$

$$<\underline{y}^T \cdot \underline{A} \cdot \underline{y}> = \sum_{k=1}^{N} \overline{\mu}_k <\overline{\eta}_k^2> \qquad (63)$$

XI. DYNAMIC BIREFRINGENCE AND DICHROISM

At this point, we take a different route from that used by
Stockmayer and Baur [49] to compute averages. The normal coordinate averages can be calculated without solving the diffusion equation explicitly for the time-dependent distribution function ψ.
By multiplying both sides of Eq. (53) by $\overline{\xi}_k^2$, $\overline{\eta}_k^2$, and $\overline{\xi}_k$ in
order and integrating over all spatial coordinates, we get the
following set of ordinary differential equations:

$$\frac{d}{dt} <\overline{\xi}_k^2> = 2D\overline{\nu}_k - 2\gamma_1 \sigma\overline{\lambda}_k <\overline{\xi}_k^2> + 2D\overline{\nu}_k \beta E\overline{\varepsilon}_k <\overline{\xi}_k> \qquad (64)$$

$$\frac{d}{dt} <\overline{\eta}_k^2> = 2D\overline{\nu}_k - 2\sigma\overline{\lambda}_k <\overline{\eta}_k^2> \qquad (65)$$

$$\frac{d}{dt} <\overline{\xi}_k> = D\overline{\nu}_k \beta E\overline{\varepsilon}_k - \gamma_1 \sigma\overline{\lambda}_k <\overline{\xi}_k> \qquad (66)$$

Solutions for time-dependent birefringence and dichroism depend on
the solutions of this set of coupled ordinary differential equations when the appropriate initial conditions are imposed.

We shall now examine dynamic birefringence and dichroism in
response to a step-function in the applied electric field. The
polymer solution is originally in thermal equilibrium with E = 0;

then a small field E is applied stepwise. The polymer chains respond to the forces caused by the applied electric field and move toward a steady state with the field on. The expressions for normal coordinate averages at steady state are obtained by solving the set of differential equations (64), (65), and (66) with the following initial conditions. At equilibrium with $E = 0$, $(t < 0)$,

$$<\overline{\xi}_k^2>_E = \frac{D}{\overline{\mu}_k \sigma} \tag{67}$$

$$<\overline{\eta}_k^2>_E = \frac{D}{\overline{\mu}_k \sigma} \tag{68}$$

$$<\overline{\xi}_k>_E = 0 \tag{69}$$

The steady state is obtained as $t \to \infty$.

$$<\overline{\xi}_k^2>_s = \frac{D}{\overline{\mu}_k \sigma}\left[\frac{1}{\gamma_1} + \frac{D}{\overline{\mu}_k \sigma}\left(\frac{\beta E \overline{\epsilon}_k}{\gamma_1}\right)^2\right] \tag{70}$$

$$<\overline{\eta}_k^2>_s = \frac{D}{\overline{\mu}_k \sigma} \tag{71}$$

$$<\overline{\xi}>_s = \frac{D}{\overline{\mu}_k \sigma}\left(\frac{\beta E \overline{\epsilon}_k}{\gamma_1}\right) \tag{72}$$

The result for steady state can be used to compute the specific Kerr constant [49] defined as follows:

$$K'_{sp} = \left(\frac{2\pi N_A}{81 Mn^2}\right)(n^2 + 2)^2(\epsilon + 2)^2\left(\frac{\Delta\Gamma}{E^2}\right) \tag{73}$$

Using Eqs. (56), (62), and (63) and the steady-state averages we get

$$K'_{sp} = \left(\frac{2\pi N_A \beta^2}{1215\ Mn^2}\right)(n^2 + 2)^2(\epsilon + 2)^2(\alpha_1^0 - \alpha_2^0)\left[3Nq^2 + \frac{<\mu^2>_0}{\gamma_1^2}\right] \tag{74}$$

where $<\mu^2>_o$ is the mean-square permanent dipole moment of the chain. The term $3Nq^2$ represents the contribution from induced dipoles. The factor γ_1^{-2} is quite close to unity. The value of the constants in the induced moment term (proportional to q^2) is of questionable validity, as Stockmayer and Baur [49] point out.

For convenience in discussing relaxation experiments, we shift the origin of the time scale to have relaxation from the steady state at time $t = 0$. Thus, field-free relaxation is described by the solutions to Eqs. (64), (65), and (66), with the condition that E is zero after $t = 0$ and that the steady state described by Eqs. (7), (71), and (72) has been achieved prior to $t = 0$. With these conditions, the averages are:

$$<\overline{\xi}_k^2>_R = \frac{D}{\overline{\mu}_k \sigma}\left[\frac{1}{\gamma_1} - 1 + \frac{D}{\overline{\mu}_k \sigma}\left(\frac{\beta E \overline{\epsilon}_k}{\gamma_1}\right)^2\right] \exp\ (-2\sigma\overline{\lambda}_k t)\ +\ \frac{D}{\overline{\mu}_k \sigma} \tag{75}$$

$$<\overline{\eta}_k^2>_R = \frac{D\cdot}{\overline{\mu}_k \sigma} \tag{76}$$

$$<\overline{\xi}_k^2>_R = \frac{D}{\overline{\mu}_k \sigma}\left(\frac{\beta E \overline{\epsilon}_k}{\gamma_1}\right)\exp\ (-\sigma\overline{\lambda}_k t) \tag{77}$$

We have computed the normal coordinate averages needed to calculate the field-free relaxation of dichroism and birefringence from a steady state. The expression for field-free relaxation is especially interesting because most experimental designs for measuring dynamic electric birefringence or dichroism are set up to observe this relaxation. Using Eqs. (62) and (63) to transform the variables in Eqs. (56) and (59), we get

$$\Delta\Gamma = q'\left[\sum_{k=1}^{N} \overline{\mu}_k <\overline{\xi}_k^2> - \sum_{k=1}^{N} \overline{\mu}_k <\overline{\eta}_k^2>\right] \tag{78}$$

for the difference in optical polarizabilities, leading to birefringence, Eq. (58); and

$$\Delta A = q''\left[\sum_{k=1}^{N} \overline{\mu}_k <\overline{\xi}_k^2> - \sum_{k=1}^{N} \overline{\mu}_k <\overline{\eta}_k^2>\right] \tag{79}$$

for dichroism. Substituting the results from Eqs. (75) and (76)
into Eqs. (78) and (79) and keeping only terms to order $\beta^2 E^2$, we
have the following expressions, respectively, for field-free
relaxation of electric birefringence and dichroism.

$$\Delta n = C_1 \left\{ \frac{1}{3} \sum_{k=1}^{N} <\mu^2>_k \exp\left(\frac{-t}{\overline{\tau}_k}\right) + q^2 \sum_{k=1}^{N} \exp\left(\frac{-t}{\overline{\tau}_k}\right) \right\} \tag{80}$$

where

$$C_1 = \frac{2\pi (n^2 + 2)^2 c N_A b^2 \beta^2 E^2 q'}{27 n^2 M} \tag{81}$$

The relaxation time of the kth "normal mode" is τ_k,

$$\overline{\tau}_k = \frac{1}{2\sigma \overline{\lambda}_k} \tag{82}$$

Numerical values for $\overline{\tau}_k$ will be given later. The mean-squared
dipole moment associated with the kth normal mode is [49]

$$<\mu^2>_k = \frac{b^2 \overline{\epsilon}_k^2}{\overline{\mu}_k} \tag{83}$$

The actual value of C_1 is not usually needed, since the important
information on polymer motion gained from relaxation experiments is
contained in the leading term which here is called F:

$$F = \sum_{k=1}^{N} <\mu^2>_k \exp\left(\frac{-t}{\overline{\tau}_k}\right) \tag{84}$$

The second term of Eq. (80) corresponds to the time dependence of
the induced polarization. Stockmayer and Baur [49] point out that
this term cannot be taken too seriously, since it is dominated by
short relaxation times, which are not well accounted for by the
Rouse-Zimm model.

   An analogous expression for dichroism relaxation is

$$\Delta A = C_2 \left\{ \frac{1}{3} \sum_{k=1}^{N} <\mu^2>_k \exp\left(\frac{-t}{\bar{\tau}_k}\right) + q^2 \sum_{k=1}^{N} \exp\left(\frac{-t}{\bar{\tau}_k}\right) \right\} \qquad (85)$$

where

$$C_2 = \frac{b^2 \beta^2 E^2 q''}{3} \qquad (86)$$

The same comments on the validity of terms in Eq. (80) hold for Eq. (85). The time scale of observed polymer motions is determined, according to this model, by the function F. Observed relaxation is a sum of exponentially declining terms, each characteristic of the motion of a particular normal mode, and each weighted according to the mean-squared dipole moment for that mode. Let us now examine the values of the time constants $\bar{\tau}_k$.

The time constants for the normal modes of polymer motion can be calculated from the hydrodynamics of polymer chains, according to the Rouse-Zimm model. The most useful expression for the $\bar{\tau}_k$ is written in terms of the observable intrinsic viscosity of the polymer and the numerical eigenvalues of the operator $\underline{H} \cdot \underline{A}$ [51]:

$$\bar{\tau}_k = \frac{M\eta[\eta]}{RT\bar{\lambda}_k'} \sum_{k=1}^{N} \left[\left(\frac{1}{\bar{\lambda}_k'}\right) \right] \qquad (87)$$

where $\bar{\lambda}_k'$ is an eigenvalue of the operator $\underline{H} \cdot \underline{A}$ [53]. The factor $\bar{\lambda}_k'[\sum (1/\bar{\lambda}_k')]$ has been calculated by many authors who have treated embellished versions of the Rouse-Zimm model, incorporating variable hydrodynamic interaction [54,55], chain stiffness and ring formation [56], and polymer branching [57], to mention most.

In most experimental situations the first few relaxation times (especially $k = 1$) dominate the observed response. Thompson and Gill [58] measured $\bar{\tau}_1$ by using birefringence relaxation. They used Eq. (87) with good results, along with the measured intrinsic viscosity to determine the molecular weight of a high-molecular-weight polymer, T2 bacteriophage DNA. Their measured value of $\bar{\tau}_1$ was about 0.5 sec for $M = 1 \times 10^8$ daltons,

$[\eta] = 290 \pm 5$ dl/g and $\eta = 0.009$ p at 25°C. Stockmayer [48] has pointed out that expressions for $\bar{\tau}_1$ are nearly the same for several models of polymer motion. In all these cases, an approximation to $\bar{\tau}_1$ is given by

$$\bar{\tau}_1 \approx \frac{M[\eta]\eta}{RT} \tag{88}$$

Figure 3 displays $\bar{\tau}_1$ vs. M.

We can gain insight into the weighting of various terms in the relaxation response F by examining the factor $<\mu^2>_k$. The character of $<\mu^2>_k$, the mean-squared dipole moment associated with the kth normal mode of chain motion, is sensitive to the distribution of charge along the chain e_i.

The components of the electric charge vector in normal coordinate space $\bar{\epsilon}_k$ may be represented by the complete set of eigenfunctions of $\underline{H} \cdot \underline{A}$. These eigenfunctions $\underline{\alpha}$ are solutions of the eigenvalue problem

$$\underline{H} \cdot \underline{A}\underline{\alpha} = \lambda\underline{\alpha} \tag{89}$$

The eigenvalues of $\underline{H} \cdot \underline{A}$ are not very different from those of $\underline{A}$ (the so-called free draining case) [51]. Since the eigenfunctions of $\underline{A}$ are simple circular functions, they can be used to approximate $\underline{\alpha}$; hence we can use the concepts of Fourier analysis. Following Zimm [51], we convert the chain from a discrete sequence of beads and springs to a continuous ribbon. Then sums are replaced by integrals and the indexes j can be written as $j = N(s + 1)/2$ where s represents distance along the contour of the chain. Using this continuous Fourier representation, we get

$$<\mu^2>_k = \left[\frac{2b^2N}{\pi^2k^2}\right][\int_0^1 e(s) \cos (k\pi s) ds]^2 \tag{90}$$

where we have used the result [51]

$$\bar{\mu}_k = \frac{\pi^2k^2}{N^2} \tag{91}$$

The integral in brackets in Eq. (90) can be thought of as the
Fourier component of charge distribution along the polymer's contour. This value for each mode determines the way the chain responds to an electric field or, abstractly, the way the terms in the
function F are weighted. For example, if the chain is made up of
a sequence of unreversed, parallel dipoles as defined above, $\langle \mu^2 \rangle_k$
vanishes for the even k modes and becomes inversely proportional
to k^2 for the odd k modes. In this case, the first mode,
$k = 1$, dominates the relaxation, with the $k = 3$ mode being $1/9$
as heavily weighted in the relaxation spectrum. The symmetry of
the first few modes is shown in Fig. 5. As Stockmayer has observed, any specific arrangement of parallel dipoles responds to a
time-varying electric field by just those modes whose symmetry is
congruent to that of the charge distribution along the chain contour [48].

No experiments have been reported for measuring dynamic electric birefringence or dichroism of well-characterized, truly flexible polymers. The Rouse-Zimm model, however, has successfully
explained experiments on dynamic flow birefringence [58,59] and
flow dichroism [4]. This model has also been applied successfully

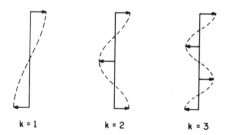

k = 1   k = 2   k = 3

FIG. 5. Schematic representation of the first few normal
coordinates of a linear polymer chain, according to the Rouse-Zimm
model. The $k = 0$ normal coordinate is translational. The solid
lines represent the chain contour, which would follow a kinked
space curve in reality. The dashed lines indicate relative magnitudes of the displacements. Arrows are at positions of maximum
amplitude. (Redrawn from Ref. 48 by permission of Butterworths
Publishers.)

to explain dynamic dielectric measurements of dilute parallel
dipole polymer solutions [48]. Its extension to the interpretation
of dynamic electric birefringence and dichroism experiments seems,
therefore, to be well founded for parallel dipole polymers with
very high molecular weights.

XII. DYNAMICAL MODELS FOR LOCAL POLYMER MOTION

Although the Rouse-Zimm model describes slower, long-range
polymer motions, it cannot be used reliably in the above form to
treat rapid, local polymer motion. Local motion would be expected
to be observable when the charge distribution along the polymer
chains leads to perpendicular dipoles. Good progress has been made
recently in developing models for describing local motion. Most of
these models have been aimed at describing dielectric loss experiments, but several of them can be extended to include dynamic electric birefringence and dichroism. As yet none of these extensions
has been made. We shall therefore be limited to giving general
descriptions of the pertinent models for local motion and to suggesting the form these extensions to dynamic birefringence and
dichroism might take.

Successful models for local motion should contain features
that correlate the motion of proximate chain repeat units. Models
that stipulate completely free and independent rotation, such as
that used above to get a rough estimate of the time scale for local
motion, are not adequate for describing real experimental situations. Various methods have been used to account for motion that
is local, but still correlated over a short range. The introduction of nearest-neighbor correlation is a step in this direction.

Work and Fujita [60] and Anderson [61] generalized Glauber's
[62] time-dependent, one-dimensional Ising model to treat local
polymer motion and dielectric loss. They made the following
assumption: The probability that the ith chain element will

change its orientation from one state to another is determined
by its orientation relative to its neighbors (i - 1) and
(i + 1). Correlations along the chain are propagated from neighbor
to neighbor. In a similar vein, Orwoll and Stockmayer [63] used a
stochastic model to treat polymer motion on a local scale.

Another approach has been taken by Helfand [70]. He catego-
rized internal chain motions on the basis of their effect on the
two attached chain ends:

1. Local configurational transitions in which only a few internal
   chain bonds move and no chain end motion occurs. A typical
   example is the so-called "crankshaft" motion.
2. Transitions in which one attached chain end is translated with
   respect to the other attached chain. A gauche migration is
   typical.
3. More general transitions in which one of the attached chain
   ends changes orientation with respect to the other chain end.

The two relevant configurations are the initial and activated
states rather than the initial and final states. He calculated the
rate of type 2 transitions by considering a simple double well potential. The equations of motion are the Langevin equations, and
the polymer distribution function satisfies a Fokker-Planck equation. It is assumed that parts of the chain attached to bonds
undergoing the central transition do not rigidly follow the motion.
Instead they move to asymptotically "heal" onto the original
configurations at distances along the chain away from the central
transition. His result for a high friction limit indicates that
the rates of type 2 transitions differ by no more than a factor of
10 from those of central chain portions without the encumbrances of
attached chain ends. Though Helfand's simple model does not specify in detail the perturbations of bond angles and bond rotations
necessary to achieve this effect, his model does provide an indica-
tion that local motions may occur under less stringent conditions
than had previously been assumed.

The various approaches, cited above, do give insight into the possible nature of local motion in polymers, but none is sufficiently developed to be used quantitatively for interpreting experimental electrical responses, whether from dielectric studies or electro-optical studies.

Clark and Zimm [64] have taken another route to developing a theoretical description of the electric response of polymer chains with sequences of perpendicular dipoles. The Clark-Zimm model represents a polymer with perpendicular dipoles by a chain of rotational Maxwell elements [65] (Fig. 6). Each element has a torsional spring to account for small-amplitude oscillations of bond angles near the bottoms of their potential energy "wells"; and a "dashpot" to account for friction in large-amplitude oscillations that lead to transitions, over an energy barrier, to other rotational potential wells. Another dashpot is included to account for viscous drag exerted by the solvent when the chain elements move. The magnitudes of the aforementioned parameters have been estimated

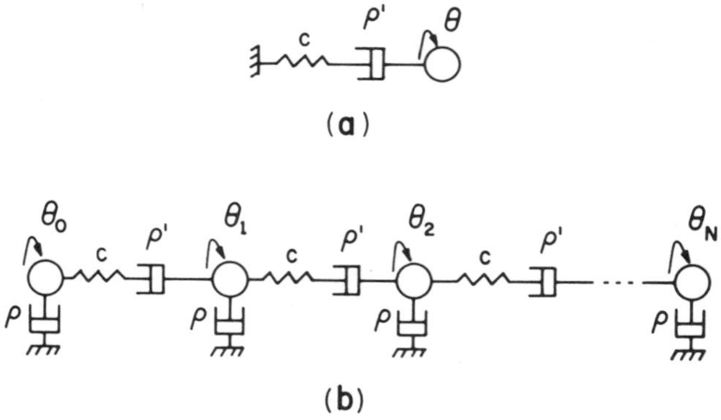

FIG. 6. Schematic representation of the Clark-Zimm model. (a) Maxwell element. Rotational dashpot of frictional coefficient ρ' represents barriers to internal rotation. Torsional spring of spring constant C represents parabolic potential minimum. (b) Chain of Maxwell elements. Rotational dashpot of frictional coefficient ρ represents viscous drag exerted by fluid. (Reproduced from Ref. 64, p. 50, by permission of Plenum Publishing Corp.).

by Clark [66]. The ratio of torsional spring constant to thermal energy is $C/kT \approx 1.8$ for a polymer such as polyvinyl chloride with a rotational barrier of about 4 kcal/mole. The ratio of the frictional coefficients for the dashpots is $\rho/\rho' \cong 1$. The relaxation time for a monomer is about what we calculated above for the correlation time for extreme local motion.

Clark and Zimm [64] wrote the equations of motion in the presence of an oscillating electric field and derived an equation for the diffusion of the configurational distribution function ψ -- analogous to ψ in the Rouse-Zimm model. Clark and Zimm carry the solution for ψ only to linearity in $\beta E\mu$, where β and E are the same as defined previously, and μ is the permanent moment of a dipole-bearing Maxwell-element bead. A distribution function of this order is sufficient for calculating the average of the cosine of the angle between a dipole and the field; so dielectric loss can be computed for this one-dimensional model. This first-order function ψ cannot be used to calculate dynamic electric birefringence or dichroism, since terms to the order $(\beta E\mu)^2$ are needed for this task. Zimm and Clark [67] have shown that it is easy to compute ψ to order $(\beta E\mu)^2$. Once this is done, the Clark-Zimm model could be extended to treat a chain of Maxwell elements with beads that bear optical polarizability tensors as well as permanent dipoles. With this extended model, averages could be computed for the squares of direction cosine relating the electric field to the dipole moments along the chain, and expressions for dynamic electric birefringence and dichroism would be available. We expect that this extension will soon be made by Zimm and Clark.

Application of linear response theory [69] to treat the time-dependent polarizability anisotropy leads to a general expression in two different time correlation functions. The decay of polarizability anisotropy can be described in terms of

$$\Phi_1 = <\mu(0)\Delta\alpha^0(t)>$$

and

$$\Phi_2 = <\alpha(0)\Delta\alpha^0(t)> \tag{92}$$

These functions represent the correlations between the dipole moment and the optical polarizability anisotropy, and between the electric polarizability and the optical polarizability anisotropy. The effect of the electric field is manifested in the energy terms in Eq. (3). These correlation functions describe only a first perturbation away from equilibrium.

Previously [6], a calculation of the dipolar time autocorrelation function for a family of α,ω-dibromo-n-alkanes was presented. Here we outline application of the same method to calculate the correlation functions in Eq. (92). This method describes the transitions between internal configurations in terms of passages over energy barrier heights between rotational isomers. In addition, for smaller molecules, the possibility of relaxation by rotation of the entire molecule must be included. We make the approximation that internal and external relaxation occur independently,

$$\Phi_2 = \Phi_{2,int} \exp\left(\frac{-t}{\tau_2}\right) \tag{93}$$

where τ_2 is the rotational relaxation time in Eq. (38). (Hereafter only expressions for Φ_2 will be given; there are analogous expressions for Φ_1.) Values of τ_2 can be estimated from Eqs. (38) and (42) or obtained from experimental results [6].

Each bond rotation will be considered to be independent of all others. For simplicity, a method appropriate only for short chains will be presented. Neighbor dependences of the transition rates could be introduced; the chain length limitation also could be overcome if no more than nearest-neighbor dependences were included.

For an example, we return to the n-alkane chain treated above and in Ref. 6. Three rotational isomers for an internal carbon-

carbon bond in such molecules are located at the energy minima at $0°$ (tr), $120°$ (g^+), and $240°$ (g^-). The two latter states have equal energies of about 500 cal mole^{-1} above the trans minimum. Energy barriers of about 3 kcal mole^{-1} occur near $60°$ and $300°$. A larger barrier of 9 to 10 kcal mole^{-1} appears at $180°$. Internal rotational state transitions are described by the following scheme:

$$(g^-) \underset{r_1}{\overset{r_2}{\rightleftharpoons}} (tr) \underset{r_2}{\overset{r_1}{\rightleftharpoons}} (g^+) \tag{94}$$

The direct transition $g^+ \rightleftharpoons g^-$ can be ignored at normal temperatures. By means of Eyring specific rate theory [63], the two transition rates can be expressed in terms of the barrier heights as $r_1 = (kT/h) \exp(-3.0/RT)$ and $r_2 = (kT/h) \exp(-2.5/RT)$, where h is Planck's constant. If $\underline{p}_j(t)$ is the probability vector for the three rotational states of bond j at time t, then the differential equation for this bond's probabilities is

$$\frac{d\underline{p}_j(t)}{dt} = \underline{S}_j \underline{p}_j(t) \tag{95}$$

where the transition rate array is given by

$$\underline{S}_j = \begin{bmatrix} -2r_1 & r_2 & r_2 \\ r_1 & -r_2 & 0 \\ r_1 & 0 & -r_2 \end{bmatrix} \tag{96}$$

Indexes on the row and column of $\underline{S}_j$ are in the order tr, g^+, and g^-. A probability vector for the entire N bond chain is given by

$$\underline{p}^{(N)}(t) = \underline{p}_2(t) \otimes \underline{p}_3(t) \otimes \ldots \otimes \underline{p}_{N-2}(t) \otimes \underline{p}_{N-1}(t) \tag{97}$$

For independent rotations the time dependence of this vector is described by

$$\frac{d\underline{p}^{(N)}}{dt} = [(\underline{S}_2 \otimes \underline{I}_3 \otimes \underline{I}_3 \cdots \otimes \underline{I}_3) + (\underline{I}_3 \otimes \underline{S}_3 \otimes \underline{I}_3 \cdots \otimes \underline{I}_3)$$

$$+ (\underline{I}_3 \otimes \underline{I}_3 \otimes \underline{S}_4 \cdots \otimes \underline{I}_3)$$

$$+ \cdots \tag{98}$$

$$+ (\underline{I}_3 \otimes \underline{I}_3 \otimes \underline{I}_3 \cdots \otimes \underline{S}_{N-1})]\underline{p}^{(N)}$$

$$= \underline{S}p^{(N)}$$

A solution is given in terms of $\underline{L}$, the diagonal array of eigen-values of $\underline{S}$, and $\underline{B}$ the array of eigenvectors:

$$\underline{p}^{(N)} = \underline{B} \exp (\underline{L}t)\underline{B}^{-1}\underline{p}^{(N)} (t = 0)$$

$$= \underline{C}\ \underline{p}^{(N)} (t = 0) \tag{99}$$

At this point it is convenient to reexpress $\underline{p}^{(N)} (t = 0)$ in terms of the usual equilibrium nearest-neighbor dependent statistics. Note that $\underline{p}^{(N)}$ will be a column of 3^{N-2} probabilities in which each element corresponds to one permitted rotational isomer. If $\underline{k}$ is a vector of the values of $\Delta\alpha^0$ for each configuration and $\underline{k}^T$ is a pseudorow vector of the values of α for each configuration, then the internal electro-optical polarizability anisotropy time correlation function is given by

$$\Phi_{2,int} = \underline{k}^T(\underline{C} \otimes \underline{I}_3)[diag(\underline{p}^{(N)} (t = 0)) \otimes \underline{I}_3](k \otimes \underline{I}_3) \tag{100}$$

Although calculations by this method are feasible, they have not been performed. Here the left end of the chain has been assumed to be fixed. A better model would permit simultaneous movement of both chain ends. Such modification should not change the order of magnitude of the results. Any desired sequence of bond types could be treated by using the appropriate transition matrices in Eq. (98). This model is designed to account for short range motion corresponding to very high-frequency motions. Somewhat longer range motions could be included by considering neighbor-dependent transition rates.

XIII.   CONCLUSION

A large fraction of current research activity in the field of polymer electro-optics is theoretical. The basic experimental data for testing theories and for directing their development are scarce. More experimental work is needed in this field.

ACKNOWLEDGMENT

Dr. K. Kajiwara read the manuscript carefully and offered us many valuable suggestions. We are very grateful to him for this.

REFERENCES

1. G. D. Patterson and P. J. Flory, J. Chem. Soc., J. Faraday Trans. II, 68, 1098 (1972).

2. W. Kuhn and F. Grün, Kolloid-Z., 101, 248 (1942).

3. P. J. Flory, Statistical Mechanics of Chain Molecules, Interscience, New York (1969).

4. P. R. Callis and N. Davidson, Biopolymers, 7, 355 (1969).

5. K. Nagai and T. Ishikawa, J. Chem. Phys., 43, 4508 (1965).

6. R. L. Jernigan, Dielectric Properties of Polymers, (F. E. Karasz, ed.), Plenum, New York (1972).

7. P. J. Flory and R. L. Jernigan, J. Chem. Phys., 48, 3823 (1968).

8. M. V. Volkenstein, Configurational Statistics of Polymeric Chains, Interscience, New York (1963).

9. A. D. Buckingham, Proc. Phys. Soc., 68A, 910 (1955); D. A. Dows, J. Chem. Phys., 41, 2656 (1964).

10. R. L. Jernigan and P. J. Flory, J. Chem. Phys., 50, 4178 (1969).

11. H. A. Stuart and A. Peterlin, J. Polymer Sci., 5, 551 (1950).

12. Section 47 of Ref. 8.

13. K. Jagai, J. Chem. Phys., 51, 1091 (1972).

14. C. V. Goebel, W. L. Dimpfl, and D. A. Brant, Macromolecules, 3, 644 (1970); D. A. Brant and W. L. Dimpfl, Macromolecules, 3, 655 (1970).

15. W. K. Oldon and P. J. Flory, Biopolymers, 11, 1 (1972).

16. P. J. Flory, Principles of Polymer Chemistry, Cornell University Press, Ithaca, N. Y. (1953).

17. P. J. Flory, R. L. Jernigan, and A. E. Tonelli, J. Chem. Phys., 48, 3822 (1968).

18. R. P. Smith and E. M. Mortensen, J. Chem. Phys., 32, 502 (1960).

19. R. J. W. Le Fèvre, B. J. Orr, and G. L. D. Ritchie, J. Chem. Soc. (B), 1966, 273.

20. S. N. Wang, J. Chem. Phys., 7, 1012 (1939).

21. K. G. Denbigh, Trans. Faraday Soc., 36, 936 (1940).

22. C. W. Bunn and R. P. Daubeny, Trans. Faraday Soc., 50, 1173 (1954).

23. M. F. Vuks, Opt. i Spektroskopiya, 2, 494 (1957); Chem. Abstr., 51, 13485e (1957).

24. W. Philippoff, J. Appl. Phys., 31, 1899 (1960).

25. R. F. Zürcher, J. Chem. Phys., 37, 2421 (1962).

26. R. T. Ingwall and P. J. Flory, Biopolymers, 11, 1527 (1972).

27. M. J. Aroney, R. J. W. Le Fèvre, and A. N. Singh, J. Chem. Soc. (B), 1965, 3179.

28. R. T. Ingwall, E. A. Czurylo, and P. J. Flory, Biopolymers, 12, 1137 (1973).

29. T. Ishikawa and K. Nagai, Polymer J., 2, 263 (1971).

30. R. J. W. Le Fèvre and K. M. S. Sundaram, J. Chem. Soc., 1962, 4003.

31. R. J. W. Le Fèvre and K. M. S. Sundaram, J. Chem. Soc., 1962, 1494.

32. R. J. W. Le Fèvre and K. M. S. Sundaram, J. Chem. Soc., 1963, 1880.

33. R. J. W. Le Fèvre and K. M. S. Sundaram, J. Chem. Soc., 1963, 3188.

34. C. G. Le Fèvre, R. J. W. Le Fèvre, and G. M. Parkins, J. Chem. Soc., 1959, 1468.

35. K. Nagai and T. Ishikawa, Polymer J., 2, 416 (1971).

36. P. J. Flory and Y. Abe, Macromolecules, 2, 335 (1969).

37. E. Nelson, Dynamic Theories of Brownian Motion, Princeton University Press, Princeton, N. J. (1967).

38. P. Debye, Polar Molecules, Dover, New York (1929).

39. M. Abramowitz and I. A. Stegun (eds.), Handbook of Mathematical Functions, Dover, New York (1965).

40. H. Lamb, Hydrodynamics, Dover, New York (1932).

41. A. Einstein, Investigations on the Theory of Brownian Motion (edited with notes by R. Fürth; translated by A. D. Cowper), Dover, New York (1956).

42. J. G. Kirkwood and J. Riseman, J. Chem. Phys., 16, 565 (1948); J. Riseman and J. G. Kirkwood, J. Chem. Phys., 17, 442 (1949).

43. J. M. Burgers, Second Report on Viscosity and Plasticity of the Amsterdam Academy of Sciences, Nordemann, New York (1938), Chap. III.

44. A. Isihara, J. Chem. Phys., 47, 3821 (1967).

45. A. R. Shultz and P. J. Flory, J. Polymer Sci., 15, 231 (1955).

46. T. G. Fox, Jr. and P. J. Flory, J. Phys. Colloid Chem., 53, 197 (1949); J. Am. Chem. Soc., 73, 1909 (1951).

47. H. Mark, Z. Elektrochem., 40, 449 (1934); R. Houwink, J. Prakt. Chem., 157, 15 (1940).

48. W. H. Stockmayer, Pure Appl. Chem., 15, 539 (1967).

49. W. H. Stockmayer and M. E. Baur, J. Am. Chem. Soc., 86, 3485, (1964).

50. P. E. Rouse, J. Chem. Phys., 21, 1272 (1953).

51. B. H. Zimm, J. Chem. Phys., 23, 269 (1956).

52. E. O. Kraemer, Ind. Eng. Chem., 30, 1200 (1938).

53. B. H. Zimm, G. M. Roe, and L. F. Epstein, J. Chem. Phys., 24, 279 (1956).

54. J. E. Hearst, J. Chem. Phys., 37, 2547 (1962).

55. N. W. Tschoegl, J. Chem. Phys., 39, 149 (1963); 40, 473 (1964); 44, 4615 (1966).

56. V. Bloomfield and B. H. Zimm, J. Chem. Phys., 44, 315 (1966).

57. B. H. Zimm and R. W. Kilb, J. Polymer Sci., 37, 19 (1959).

58. D. S. Thompson and S. J. Gill, J. Chem. Phys., 47, 5008 (1967).

59. G. B. Thurston and J. L. Schrag, J. Chem. Phys., 45, 3373 (1966).

60. R. N. Work and S. Fujita, J. Chem. Phys., 45, 3779 (1966).

61. J. E. Anderson, J. Chem. Phys., 52, 2821 (1970).

62. R. J. Glauber, J. Math. Phys., 4, 294 (1963).

63. R. A. Orwoll and W. H. Stockmayer, Adv. Chem. Phys., 15, 305 (1969).

64. M. B. Clark and B. H. Zimm, in Dielectric Properties of
    Polymers (F. E. Karasz, ed.), Plenum, New York (1972), p. 45.

65. J. D. Ferry, Viscoelastic Properties of Polymers, 2nd ed.,
    Wiley, New York (1970).

66. M. B. Clark, Dielectric Properties of Polymers (F. E. Karasz,
    ed.), Plenum, New York (1972), p. 73.

67. B. H. Zimm and M. B. Clark, Private communication (1972).

68. S. Glasstone, K. J. Laidler, and H. Eyring, The Theory of
    Rate Processes, McGraw-Hill, New York (1941).

69. R. Kubo, J. Phys. Soc. Japan, 12, 570 (1957).

70. E. Helfand, J. Chem. Phys., 54, 4651 (1971).

Chapter 6

OPTICAL ABSORPTION IN AN ELECTRIC FIELD

Wolfgang Liptay

Institute for Physical Chemistry
University of Mainz
Mainz, Germany

NOTATION

The symbols and units used for physical quantities conform to
the recommendations adopted by the IUPAC Council on July 7, 1969
(SI units). Vectors are represented by underlined letters, tensors
by double-underlined letters. Thus, for example, $\underline{p}$ is a column
vector, $\overline{\underline{p}}$ the corresponding row vector, and $\underline{\underline{\alpha}}$ a tensor; hence,

$\underline{p} \, \underline{p}$ is the scalar product and $\underline{p} \, \overline{\underline{p}}$ the second-order tensor
product.

I.   INTRODUCTION

The most simple electronic spectra are those of isolated
atoms. Therefore, before starting to discuss the influence of an
electric field on the optical absorption of molecules, I will remind the reader of a few facts rather well known in atomic spectra
[1].

An external electric field influences the state functions of
an atom and, consequently, the energy as well as other physical
quantities of this system. Since the components of the total
angular momentum $\underline{J} [|\underline{J}|^2 = \hbar^2 \, J(J + 1)]$ can take only the values
$M\hbar$ (M = J, J - 1, ..., -J) with respect to any preferred direc-
tion, where J and M are the respective quantum numbers of the
state considered and $h = 2\pi\hbar$ is the Planck constant, space
quantization takes place in an electric field. The shift of the
energy of states with different magnitudes $|M|$ is usually differ-
ent. Hence the energy level of a state with quantum number J,
which is at least (2J + 1)-fold degenerate in the absence of a
field, is split into J + 1 levels in the presence of an electric
field. Consequently, a line, which may be observed in optical
absorption or emission and which results from the totality of
transitions between all unperturbed states of the two levels, will
be split into several components in an electric field. This
phenomenon was discovered by Stark in 1913 and is known as the
Stark effect.

In an applied electric field, there are not only the electrostatic interactions between the nucleus and the electrons and
between the electrons themselves, but also between all charges of
the atom and the external field. These interactions cause an
additional term to the potential energy which is in a uniform

electric field E_z parallel to the z axis of a space-fixed Cartesian coordinate system given by $-E_z \sum_i e_i z_i = -E_z P_z$, where z_i is the z coordinate of the particle i with charge e_i and P_z is the z component of the dipole moment. Consequently, in an applied field, the Hamiltonian H of the unperturbed atom has to be supplemented by the corresponding operator and becomes

$$H^E = H - E_z P_z = H - E_z \sum_i e_i z_i \tag{1}$$

For large positive values of z_i the potential energy tends to $-\infty$ and consequently the solution of the Schrödinger equation leads to unbound states (where the electrons are separated from the nucleus) with a continuous energy spectrum. Hence, strictly, there are no stationary bound states of an atom in an applied electric field; every bound state has a finite lifetime in respect to transitions to an unbound state. But, as long as the field E_z is not too large, for many bound states this lifetime is very large, at least in comparison with lifetimes in respect to transitions to other bound states. Then such states may be considered as quasi-stationary states with quasi-discrete values of the energy and the influences of the applied field can be investigated by treating $-E_z P_z$ as a perturbation in the usual way.

An unperturbed state is characterized by the state function $|k\rangle$, its energy is consequently

$$W_k = \langle k|H|k\rangle \tag{2}$$

A first-order effect, which is proportional to E_z, may occur only if the matrix element

$$(P_z)_{kk} = \langle k|P_z|k\rangle = \sum_i e_i \langle k|z_i|k\rangle \tag{3}$$

is different from zero. This matrix element representing the z component (in the space-fixed coordinate system) of the permanent dipole moment of the system in the unperturbed state $|k\rangle$ is for

isotropic systems necessarily zero. Hence, in such systems there
are usually no first-order effects. A different situation is given
if there are two or more states whose unperturbed energy levels are
close spaced in comparison with the shifts caused on those levels
by an applied field. In such cases the first-order perturbation
method for nearly degenerate states has to be applied, and all
pairs of individual states of the set of initially nearly degenerate states (which combine according to the optical selection rules
for an electric dipole transition) lead to finite matrix elements
$(p_z)_{kk'}$, and hence to linear field effects. The linear splittings
of energy levels are well known for the H atoms and also for H-
like ions; they are caused by the fact that states with the same
value of the quantum number n and different orbital angular
momentum $\underline{L}$ are degenerate with one another, except for relativity
effects. According to Eqs. (1) and (3) the order of magnitude of
linear shifts is $E_z(p_z)_{kk} \approx E e a_o \approx 8.5 \cdot 10^{-30} E_z$ C m, where
$a_o = 5.292 \cdot 10^{-11}$ m is the Bohr radius. Hence, in a rather
strong field $E_z = 10^7$ V m^{-1}, the first-order splitting of the
lines, i.e., the linear Stark effect, is of the order
$\Delta \tilde{\nu}^E \approx E e a_o/hca \approx 4$ cm^{-1}, where c is the speed of light in
vacuum and $a = 10^2$ m^{-1} cm is a conversion constant (thus the unit
of the wavenumber $\tilde{\nu}$ is cm^{-1} also in the SI system). The line
pattern caused by the first-order splitting will be symmetric
around the position of the unperturbed line. The second-order
effects lead to the square Stark effect dependent on E_z^2. Those
shifts and splittings are of the order of magnitude
$\overline{\Delta\nu}^E \approx 10^{-15} E_z^2$ V^{-2}m^{2}cm^{-1}; in an applied field $E_z = 10^7$ V m^{-1}
they will be $\overline{\Delta\nu}^E \approx 0.1$ cm^{-1}. Hence, if there is a first-order
effect, the second-order effect may be neglected up to quite large
field strength. But usually there is only the rather small second-
order effect causing line patterns not symmetrical about the original line.

The patterns observed in a transverse direction (x
direction, for example) will show a parallel polarization (z

direction of the electric field vector of the light wave) for
$\Delta M = 0$ transitions and a perpendicular polarization (y direction) for $\Delta M = \pm 1$ transitions. The components in pattern differ
in intensity because (a) individual transition probabilities
differ, (b) some components are the sum of several transitions,
and (c) the experimental conditions may put unequal numbers of
atoms into the various states of the initial level. For atoms,
it is usually possible to apply a sufficiently large field causing
a splitting larger than the width of the lines. Then the Stark
shifts $\Delta \overline{\nu}^E$ may be determined, and from such data information
about the matrix elements describing the first- and second-order
effects may be obtained. A further evaluation of these data would
lead to the changes of the polarizability of the atoms during the
considered optical transition process.

For not too large molecules in the gas phase, a resolved
rotational fine structure may be observed. In such cases an
applied electric field may cause quite similar effects as are described for atoms above, that is, a broadening, a shift and a
splitting of certain lines. The effects are dependent on the
electric dipole moments and the polarizabilities of the ground
state and the excited electronic state and therefore are used
extensively for the determination of these quantities [2-5]. The
methods and results have been reviewed recently [6]. In such research, the field strength which may be applied is usually limited
to a few 10^6 V m^{-1}. The maximum values even of linear Stark splitting are of the order 0.1 cm^{-1} and hence comparable to the Doppler
width of the lines in an electronic band with resolved rotational
fine structure. Hence the individual Stark components arising, in
the presence of an applied field, from a rotational line are
actually not fully resolved into individual transitions, but rather
an intensity-weighted blending of them is obtained. What may be
observed in optical absorption experiments, for example, is actually not the Stark shift and splitting of individual lines, but a
field-dependent change of the molar absorption coefficient of the
molecule, a phenomenon usually called electrochromism.

In dense media the rotational motions are inhibited by
intermolecular interactions, hence rotational states are usually
not well defined. As a consequence, the orientational distribution
of molecules in space may be described according to the Maxwell-
Boltzmann statistic. In homogeneous and isotropic systems as in
liquid or solid solutions for example, the molecules are randomly
distributed -- each orientation has the same probability. In an
applied electric field the orientational distribution becomes
anisotropic if the molecules have a permanent electric dipole moment or an anisotropic polarizability. This orientational effect,
which leads to a dichroism of absorbing solutions in an electric
field, was described by Kuhn et al. [7] and could be used to
determine the transition moments of electronic bands. However, at
least two other effects are superimposed on the orientational
effect. First, if there is a change of the dipole moment or of the
polarizability during the electronic excitation process, the absorption band will be shifted, similarly to the Stark effect.
Second, the transition moment may be dependent on the electric
field, then an applied field causes a change of the intensity also.
The shifts of the bands are of the order 0.1 to 1 cm^{-1}, the widths
of vibronic or electronic bands in liquid or solid solutions are of
the order 100 to a few 1000 cm^{-1}, which prohibits the observation
of any shift caused by an external electric field in such system.
Another situation may happen for molecules in crystals at very low
temperatures, where lines with a width comparable to the shifts,
and therefore the shifts induced by an external field also may be
observed [8]. The field effect, which may be measured in solutions
as well as in other dense systems, is the change of the absorption
coefficient, that is, the electrochromism. The evaluation of such
data leads to information on the direction and the electric field
dependence of the transition moment and on dipole moments and
polarizabilities in the ground state as well as in electronically
excited states [9,10]. The methods and results have been reviewed
in several papers [6,11-13].

At the location of a molecule in a dense medium there may be an electric field even in absence of an applied electric field. This field is caused by the charge distributions of the surrounding molecules; the linear average of it is known as the reaction field introduced by Onsager [14]. The electric field due to the surrounding molecules will influence the absorption coefficient as well as an applied field. This is the most important cause of the solvent dependence of absorption and emission bands of polar molecules usually called solvatochromism. Therefore, the solvatochromism, that is, the solvent dependence of the wavenumber of electronic bands [15-24] and the solvent dependence of the in intensity [25], is closely related to electrochromism [12,26,27].

II. ELECTROCHROMISM IN THE GASEOUS PHASE

An isolated molecule is characterized by its Hamiltonian H and a complete set of eigenstates $|k>$ satisfying the Schrödinger equation

$$H|k> = W_k|k> \tag{4}$$

where W_k is the energy of the molecule. The state function $|k>$ may be approximated by a product function $|v>\,|r>\,|t>$ and the energy W_k by a sum $W_v + W_r + W_t$, where $|v>$ describes the electronic-vibrational (vibronic) state, $|r>$ the rotational state, and $|t>$ the translational state with energies W_v, W_r, and W_t, respectively.

The wavenumber ν_{lk}, corresponding to a transition between the states $|l>$ and $|k>$ $(W_l > W_k)$ in the absence of an applied electric or magnetic field besides the usually weak radiation field, is given by the Einstein relation

$$hca\,\bar{\nu}_{lk} = W_l - W_k \tag{5}$$

The molar absorption coefficient $\varepsilon_{1k}(\bar{\nu})$ for a line corresponding to an individual unperturbed electric dipole transition $|1\rangle \leftarrow |k\rangle$ has been obtained by Weisskopf and Wigner [28],

$$\frac{\varepsilon_{1k}(\bar{\nu})}{\bar{\nu}} = S \left| \underline{e} \, \underline{\mu}_{1k} \right|^2 s_{1k}(\nu) \tag{6}$$

where $\underline{\mu}_{1k}$ is the electric transition dipole moment, that is, with the operator $\underline{p}$ of the electric dipole moment of the molecule

$$\underline{\mu}_{1k} = \langle 1 | \underline{p} | k \rangle \tag{7}$$

and where $\underline{e}$ is a unit vector parallel to the electric field vector of the incident light wave and S is a universal constant,

$$S = \frac{2\pi^2 bN_A \log e}{hca\varepsilon_o} = 29.356 \cdot 10^{60} \, c^{-2} \, m^{-2} \, mol^{-1} \, dm^3 \, cm^{-1}$$

where N_A is the Avogadro constant, ε_o is the permittivity of vacuum, $\log e = 0.43429$, and $b = 10^3 \, m^{-3} \, dm^3$ is another conversion constant (thus the unit of ε_{1k} is $mol^{-1} \, dm^3 \, cm^{-1}$ in the SI system). The quantity $s_{1k}(\bar{\nu})$ in Eq. (6) is a normalized line-shape function, that is

$$\int_0^\infty s_{1k}(\bar{\nu}) \, d\nu = 1 \tag{8}$$

which for an individual transition, where the width γ_{1k} of the line at half-maximum is much smaller than ν_{1k}, may be represented by

$$s_{1k}(\bar{\nu}) = \frac{2\gamma_{1k}\bar{\nu}_{1k}}{\pi} \frac{\bar{\nu}}{(\bar{\nu}^2 - \bar{\nu}_{1k}^2)^2 + \bar{\nu}^2\gamma_{1k}^2} \tag{9}$$

Even a resolved rotational line of an electronic transition of a molecule is usually (and in case of rotationally degenerate states necessarily) a superposition of some individual lines caused by transitions between some individual states $|k\rangle$ of the lower electronic state and some individual states $|1\rangle$ of the upper electronic state of the molecule. Hence the molar absorption

coefficient which may be observed on a sufficiently dilute gaseous
system is due to a superposition and can be represented by

$$\frac{\varepsilon(\bar{\nu})}{\bar{\nu}} = S \sum_k \sum_\ell w_k |\bar{e} \ \underline{\mu}_{1k}|^2 s_{1k}(\bar{\nu}) \tag{10}$$

where w_k is the probability that a molecule will be in the
individual initial state $|k>$.

If a uniform electric field E is applied to a dilute gaseous
system, the Hamiltonian H^E of each molecule becomes

$$H^E = H - \bar{E} \ \underline{p} \tag{11}$$

The state function $|k(E)>$ and the energy W_k^E of the molecule in
the applied field may be obtained by the Schrödinger perturbation
method, which yields

$$|k(E)> = |k> + \sum_{j \neq k} \frac{\bar{E} \ \mu_{jk}}{W_j - W_k} \ |j> + \dots \tag{12}$$

and

$$W_k^E = W_k - \bar{E} \ \underline{\mu}_{kk} - \frac{1}{2} \bar{E} \ \underline{\underline{\alpha}}_{kk} \ \underline{E} - \dots \tag{13}$$

where the matrix elements $\underline{\mu}_{jk}$ are given by Eq. (7) and where

$$\underline{\underline{\alpha}}_{kk} = 2 \sum_{j \neq k} \frac{\underline{\mu}_{kj} \underline{\mu}_{jk}}{W_j - W_k} \tag{14}$$

The quantities $\underline{\mu}_{kk}$ and $\underline{\underline{\alpha}}_{kk}$ are the expectation values of the
permanent dipole moment and the polarizability of an isolated
molecule in a space-fixed coordinate system. Both these quantities
depend on the permanent dipole moment of the molecule in a molecule-fixed coordinate system and on the rotational quantum numbers
J, K, and M, and the latter quantity on the polarizability of
the molecule in a molecule-fixed coordinate system also [6].

In the presence of an applied electric field the wavenumber
$\bar{\nu}_{1k}^E$ of the distinct transition $|1> \leftarrow |k>$ becomes

$$\overline{\nu}_{lk}^E = \overline{\nu}_{lk} + \Delta\overline{\nu}_{lk}^E = \frac{W_l^E - W_k^E}{hca} \tag{15}$$

where according to Eq. (13) the wavenumber shift $\Delta\overline{\nu}_{lk}^E$ due to the external field is given by

$$hca \; \Delta\overline{\nu}_{lk}^E = -\underline{E}(\underline{\mu}_{11} - \underline{\mu}_{kk}) - \frac{1}{2}\underline{E}(\underline{\alpha}_{11} - \underline{\alpha}_{kk})\underline{E} - \cdots \tag{16}$$

Higher order terms in $\underline{E}$ usually may be neglected. If in this case $\underline{\mu}_{11} \neq \underline{\mu}_{kk}$ and $\underline{\alpha}_{11} = \underline{\alpha}_{kk}$ the shift $\Delta\overline{\nu}_{lk}^E$ depends linearly on the field, and if $\underline{\mu}_{11} = \underline{\mu}_{kk}$ and $\underline{\alpha}_{11} \neq \underline{\alpha}_{kk}$, the shift depends on the square of the field. These shifts are referred to as first-order (linear) Stark shift and second-order (square) Stark shift, respectively. At practicable field strength, $E = 2 \cdot 10^6 \; V \; m^{-1}$, the first-order shifts due to a change of the dipole moment $|\underline{\mu}_{11} - \underline{\mu}_{kk}| = 3.3 \cdot 10^{-30} \; C \; m \; (= 1 \; D)$ will be 0.33 cm^{-1}, the second-order shifts usually will be even smaller. Only in an electronic transition band when there are very sharp lines (as in the resolvable rotational fine structure of vaporized molecules) is a direct observation of a shift possible in the visible or near ultraviolet range, as has been demonstrated in absorption spectra by Klemperer et al. [3,4] and in emission spectra by Phelps and Dalby [2]. But even in these favorable cases one cannot observe in an applied electric field a shift of lines corresponding to transitions between individual levels, but rather a broadening and splitting of a rotational line due to blending of different transitions between different close spaced levels. Hence electro-optical absorption measurements do not lead to the Stark shifts directly, but to the change of the absorbance due to an applied electric field (electrochromism).

The light absorption of a homogeneous and isotropic medium is isotropic in absence of an applied field. In the presence of an applied uniform electric field the medium becomes anisotropic and the light propagation under this situation is similar to the light propagation in a uniaxial crystal. Hence, not only the absorbance

is influenced by an applied field, but the refractive index $n(\bar{\nu})$
and therefore the dispersion of the Kerr constant also. There is
a close relationship between electrochromism and the dispersion of
the Kerr constant, given by a Kramers-Kronig integral transformation, as was recognized by Kuball [29]. He has shown [30] that
some, but not all, of the information which can be obtained from
electrochromic measurements is also obtainable from dispersion
measurements of the Kerr effect in the vicinity of absorption
bands.

In the presence of an applied uniform electric field E the
attenuation of the light wave will follow Lambert's law only in
special cases [6], namely, if either the electric field vector e
of the incident plane-polarized light wave is parallel to E or if
e is perpendicular to E. One may therefore define two absorption
coefficients $a_p^E(\bar{\nu})$ (e parallel to E) and $a_s^E(\bar{\nu})$ (e perpendicular to E) and two molar absorption coefficients $\varepsilon_p^E(\bar{\nu})$ and
$\varepsilon_s^E(\bar{\nu})$ using Beer's law

$$a_\alpha^E(\bar{\nu}) = c_B^E \varepsilon_\alpha^E(\bar{\nu}) \qquad (\alpha = p, s) \tag{17}$$

where c_B^E is the concentration of the absorbing species (in unit
mol dm^{-3}) in the presence of an applied field. In common systems
electrostriction and other field effects on the concentration
usually may be neglected, thus in satisfactory approximation
$c_B^E = c_B$, where c_B is the concentration in absence of an external
field. The molar absorption coefficients $\varepsilon_\alpha^E(\bar{\nu})$ in the presence
of an applied electric field can be represented analogous to Eq.
(10) by

$$\frac{\varepsilon_\alpha^E(\bar{\nu})}{\bar{\nu}} = S \sum_k \sum_\ell w_k^E \, | (p^E)_{1k} |^2 \, s_{1k}^E(\bar{\nu}) \qquad (\alpha = p, s) \tag{18}$$

where $(p^E)_{1k}$ is the component of the transition dipole moment
$\underline{p}_{1k}^E$ parallel to $\underline{e}$. The comparison of Eq. (18) with Eq. (10) implies that, under the assumption $c_B^E = c_B$, there are three

reasons which may cause an electric field dependence of $\varepsilon_\alpha^E(\bar{\nu})$:

1. Changes of the line-shape functions $s_{1k}^E(\bar{\nu})$
2. Changes of the probabilities w_k^E of the initial states of the absorption process
3. Changes of the transition dipole moments $\underline{p}_{1k}^E$

The line-shape function, Eq. (9), depends on $\bar{\nu}_{1k}$, which determines the origin of the line, and on γ_{1k}, which determines the width of the line. One may expect that usually the field dependence of γ_{1k} can be ignored, at least as long as the transition dipole moment is not too strongly field-dependent and as long as preionization does not play a role. In this approximation it is assumed that the form of the line-shape function is independent of an applied field and that only the origin will be shifted due to the Stark shift $\Delta\bar{\nu}_{1k}^E$, Eq. (16), of the transition $|1> \leftarrow |k>$. Hence the value of the line-shape function for this definite transition at the wavenumber $\bar{\nu}$ in the presence of an applied electric field is the same as at the wavenumber $(\bar{\nu} - \Delta\bar{\nu}_{1k}^E)$ in the absence of the field or

$$s_{1k}^E(\bar{\nu}) = s_{1k}(\bar{\nu} - \Delta\bar{\nu}_{1k}^E) \tag{19}$$

The field dependence of the probability w_k^E in a system in thermal equilibrium at a temperature T may be represented by [6]

$$w_k^E = w_k \left\{ 1 + \frac{1}{kT}\,\bar{E}\,\underline{\mu}_{kk} \right.$$
$$\left. + \frac{1}{6kT}\left[\frac{1}{kT}\left\{3(\bar{E}\,\underline{\mu}_{kk})^2 - E^2\mu_{vv}^2\right\} + 3\bar{E}\underline{\alpha}_{kk}\bar{E} - E^2\,\mathrm{tr}\,\underline{\alpha}_{vv}\right] \right\} \tag{20}$$

where k is the Boltzmann constant, w_k is the probability of the state $|k>$ in absence of the field and $\underline{\mu}_{vv}$ is the permanent electric dipole moment and $\mathrm{tr}\,\underline{\alpha}_{vv}$ the trace of the polarizability of the molecule in its vibronic state $|v>$.

The field dependence of the transition moment $\underline{p}_{1k}^E$ results by evaluating the matrix element given by Eq. (7) but using perturbed state functions as given by Eq. (12); this yields

$$\underline{p}_{1k}^{E} = \underline{\mu}_{1k} + \underline{\underline{\alpha}}_{1k}\underline{E} + \cdots \tag{21}$$

where

$$\underline{\underline{\alpha}}_{1k} = \sum_{j \neq k} \frac{\underline{\mu}_{ij}\underline{\mu}_{jk}}{W_j - W_k} + \sum_{j \neq \ell} \frac{\underline{\mu}_{jk}\underline{\mu}_{1j}}{W_j - W_1} \tag{22}$$

Equation (21) is obviously similar to the equation for the electric dipole moment in an electric field [see Eq. (29)]. Consequently, $\underline{\mu}_{1k}$ may be referred to as permanent electric transition dipole moment and the second-order tensor $\underline{\underline{\alpha}}_{1k}$ as transition polarizability.

Equations (19) to (21) introduced into Eq. (18) lead to

$$\frac{\varepsilon_{\alpha}^{E}(\overline{\nu})}{\overline{\nu}} = S \sum_{k} \sum_{\ell} w_k |(\mu_{\alpha})_{1k}|^2 s_{1k}(\overline{\nu} - \Delta\overline{\nu}_{1k}^{E})\left\{1 + \frac{1}{kT}\,\overline{E}\,\underline{\mu}_{kk} + \overline{E}\,R_{1k}^{(\alpha)}\right.$$
$$+ \frac{1}{6k^2T^2}\left[3(\overline{E}\,\underline{\mu}_{kk})^2 - E^2\mu_{vv}^2\right] \tag{23}$$
$$+ \frac{1}{6kT}\left[3\overline{E}\,\underline{\underline{\alpha}}_{kk}\underline{E} - E^2\,\mathrm{tr}\,\underline{\underline{\alpha}}_{vv} + \Pi\,6(\overline{ER}_{1k}^{(\alpha)})(\overline{E}\,\underline{\mu}_{kk})\right]\left.\right\}$$
$$(\alpha = p,\ s)$$

where $\underline{R}_{1k}^{(\alpha)}$ is a vector with the component $(R_{\beta}^{(\alpha)})_{1k}$,

$$(R_{\beta}^{(\alpha)})_{1k} = \frac{(\mu_{\alpha})_{1k}(\alpha_{\alpha\beta})_{k1} + (\mu_{\alpha})_{k1}(\alpha_{\alpha\beta})_{1k}}{(\mu_{\alpha})_{1k}(\mu_{\alpha})_{k1}} \tag{24}$$

which represents the first-order field effect on the transition dipole moment.

Equation (23) is developed to terms square in $\underline{E}$ and describes the electrochromism of vaporized molecules at sufficient dilution (thus intermolecular interactions may be neglected) in third-order in $\underline{E}$ because if for any system, which is isotropic in absence of an applied field, $\varepsilon_{\alpha}^{E}(\overline{\nu})/\overline{\nu}$ is expanded in powers of $\underline{E}$, all terms with odd powers of $\underline{E}$ have to vanish. For small rotational lines in a sufficiently strong electric field the Stark shift $\Delta\overline{\nu}_{1k}^{E}$ is comparable to the breadth of the lines or even larger, then individual values for the line-shape functions $s_{1k}(\bar{\nu} - \Delta\bar{\nu}_{1k}^{E})$ have to be introduced into Eq. (23). For given values of the Stark shift $\Delta\bar{\nu}_{1k}^{E}$ they may be obtained from experimental data or one may use Lorentzian line-shape functions, as given by Eq. (9), or similar Gaussian line-shape functions taking into account the Doppler effect [6]. The values of $\Delta\bar{\nu}_{1k}^{E}$ can be calculated according to Eq. (16), where one should remember that the quantities $\underline{\mu}_{11}$, $\underline{\mu}_{kk}$, $\underline{\alpha}_{11}$ and $\underline{\alpha}_{kk}$ are expectation values in absence of an applied field and in relation to a space-fixed coordinate system. Therefore the further evaluation depends on the initial and final rotational states of the electronic transition being considered; the rather involved treatment has been reviewed previously [6]. The way for the evaluation of the experimental data obtained from electrochromic measurements is now to use a computer to generate the expected contours of $\varepsilon_{\alpha}^{E}(\bar{\nu})/\bar{\nu}$ with an appropriate known value for the permanent dipole moment of the molecule in the ground state (in a molecule-fixed coordinate system) and suitable trial values for this quantity in the excited state and fitting these contours to the experimental contour of $\varepsilon_{\alpha}^{E}(\bar{\nu})/\bar{\nu}$. This procedure, which neglects polarizability effects, applied to a set of reasonably selected rotational lines of a favorable molecule may lead to a unique value of the magnitude of the excited-state dipole moment and to its direction relative to the ground-state dipole moment (in a molecule-fixed coordinate system).

For sufficiently broad lines and small electric fields the Stark shifts $\Delta\bar{\nu}_{1k}^{E}$ may be much less than the widths of the lines at half-maximum. Then the line-shape functions may be expanded in powers of the shift [6] similar to the procedure which will be used in the next section. This method leads to an equation similar to Eq. (65) below.

III. ELECTROCHROMISM IN LIQUID SOLUTIONS

A. Excitation from an Equilibrium State to a Franck-Condon State

The system we will consider now is a solution of molecules, whose electrochromism we want to investigate, in a solvent not absorbing in the wavenumber range in question. The system shall be at a definite temperature T and volume V. We assume a sufficiently small number density N_B of solute molecules, thus intermolecular interactions between a solute molecule and any other solute molecule can be neglected. In spite of the general interactions between the solute molecules and the surrounding solvent molecules, the solute molecules usually keep their individuality as is well known. Otherwise, if there are specific interactions between a solute molecule and the solvent molecules, new identities will be formed, as for example electron donor-acceptor complexes or complexes due to hydrogen bonding. In such a case the new identities have to be treated as new species of solute molecules. Due to the individuality of the molecules the vibronic states of a molecule are maintained in a solution but more or less perturbed by interactions with surrounding solvent molecules. The intermolecular interactions inhibit the translational and rotational motions peculiar to free molecules and cause vibrational and torsional motions called librations, i.e., displacements of the molecules as wholes from momentary equilibrium positions. The momentary equilibrium positions slowly drift in physical space causing translational diffusion and rotational diffusion of the molecules and related phenomena. The total energy of the system may approximately be represented as the sum of the energy of the system with fixed molecules and the librational energy.

The energies and eigenfunctions of the system with fixed molecules can be obtained by a perturbation treatment. Even for the most simplified model -- neglect of overlap of electron distri-

butions on different molecules and approximation of the intermolecular interaction by a point dipole interaction operator -- the
calculations lead to rather lengthy equations, which cannot easily
be applied to investigated systems [21].

The system may also be considered from a classical point of
view. The energy of an isolated solute molecule in the vibronic
state $|k>$ shall be W_k, the Helmholtz energy of the pure solvent
shall be A_{solv}. If a solute molecule is brought into the solvent,
there is a change of the Helmholtz energy of the system, which may
be divided into three parts:

1. The work W_{Ck} required to create a cavity in the solvent into
   which the solute molecule shall be introduced.

2. The work W_{Pk} required to polarize the medium, equivalent to
   the polarization due to the total dipole moment of the solute
   molecule in the solution. This polarization of the medium
   causes a contribution to the electric field inside the cavity.

3. The work to place the molecule with the permanent electric
   dipole moment $\underline{\mu}_k$ and the polarizability $\underline{\underline{\alpha}}_k$ into the effective electric field at the location of the solute molecule,
   which will be assumed to be a homogeneous field
   $$\underline{E}_k : -\underline{\bar{\mu}}_k \underline{E}_k - \frac{1}{2} \underline{\bar{E}}_k \underline{\underline{\alpha}}_k \underline{E}_k.$$

The comparison of the results of the above-mentioned perturbation
treatment with these classical changes of the Helmholtz energy
shows that all but one of the terms of the perturbation treatment
can be identified with classical terms. The additional term
represents, as may be expected:

4. The energy W_{Dk} corresponding to the dispersion interactions
   between the solute molecule and the surrounding solvent molecules.

Consequently, the Helmholtz energy A_S of the system in the equilibrium ground state becomes

$$A_S = A_{solv} + \sum_{p=1}^{VN_S} \left\{ W_k + W_{Lk\kappa} + W_{Ck} + W_{Dk} + W_{Pk} \right.$$
$$\left. - \bar{\mu}_k \bar{E}_k - \frac{1}{2} \bar{E}_k \underline{\alpha}_k \bar{E}_k \right\}_p \tag{25}$$

where $W_{Lk\kappa}$ is the change of the Helmholtz energy due to librational interactions and the subscript p to the side of the expression in braces designates that the quantities in this expression are those corresponding to the solute molecule p in the solution.

Next we consider an optical absorption process, where one of the solute molecules in the vibronic state $|k>$ together with its surrounding in the librational state $|\kappa>$ is excited into the vibronic state $|l>$ and librational state $|\lambda>$, and where all other molecules are still in the same states as they were in the initial state of the system. During this optical excitation process there will be a change of the electronic distributions of the one solute molecule and its neighboring solvent molecules but there will be no change of the nuclear configuration according to the Franck-Condon principle. The initial state of the excitation process will be called equilibrium ground state and the final state, which is not a thermodynamical equilibrium state, will be called Franck-Condon excited state. For the Franck-Condon excited state an equation similar to Eq. (25) will hold, only for one of the solute molecules all quantities in the expression in curly brackets should be substituted by the appropriate quantities for the Franck-Condon excited state, which we will designate by a superscript FC. The wavenumber $\bar{\nu}^{SE}_{l\lambda,k\kappa}$ corresponding to the transition in question of a solute molecule in presence of an applied electric field will be

$$\overline{\nu}^{SE}_{\ell\lambda,k\kappa} = \overline{\nu}_{1k} + \Delta\overline{\nu}^{SE}_{\ell\lambda,k\kappa} = \frac{A^{FC}_S(1\lambda) - A_S(k\kappa)}{hca} \tag{26}$$

where the wavenumber shift $\Delta\overline{\nu}^{SE}_{\ell\lambda,k\kappa}$ due to the solvent effect and due to an applied field is according to Eq. (25) given by

$$hca\Delta\overline{\nu}^{SE}_{1\lambda,k\kappa} = W^{FC}_{L1\lambda} - W_{Lk\kappa} + W^{FC}_{C1} - W_{Ck} + W^{FC}_{D1} - W_{Dk} + W^{FC}_{P1} - W_{Pk}$$
$$- \underline{E}^{FC}_1 (\underline{\mu}^{FC}_1 + \frac{1}{2} \underline{\alpha}^{FC}_1 \underline{E}^{FC}_1) + \overline{E}_k (\underline{\mu}_k + \frac{1}{2} \underline{\alpha}_k \underline{E}_k) \tag{27}$$

This formula is similar to Eq. (16), only the effects caused by the solvent molecules are also included. The changes of the libration-al energy $(W^{FC}_{L1} - W_{Lk\kappa})$, of the work required to create a cavity $(W^{FC}_{C1} - W_{Ck})$ and of the dispersion interactions $(W^{FC}_{D1} - W_{Dk})$ are independent of an applied field in a good approximation. Hence the field dependence of the wavenumber shift is due to the last two terms of Eq. (27), which is obvious because of the dependence on $\underline{E}_k$ and $\underline{E}^{FC}_1$, respectively, and also due to the change of the polarization energy $(W^{FC}_{P1} - W_{Pk})$.

B.   The Electric Fields at the Location of a Solute Molecule

For the further evaluation of Eq. (27) we need the electric fields $\underline{E}_k$ and $\underline{E}^{FC}_1$ at the location of the solute molecule in the equilibrium ground state and in the Franck-Condon excited state, respectively. For the derivation of these quantities we use the same model as before. In the empty cavities, into which the solute molecules shall be introduced, there is an electric field caused by external charges and by the charge distribution of the surrounding solvent molecules. The charge distribution of the solvent molecules will be modified by the solute molecule and consequently its contribution to the field at the location of the solute molecule also. In the following, we consider first cavities with a definite size, shape, and orientation in physical space and

call these equivalent cavities. The average value of a quantity
with respect to an assembly of equivalent cavities will be designated by $< >_{avC}$. We also have to consider solute molecules in
the vibronic state $|k>$ with a definite orientation, and hence we
may assume inside of equivalent cavities, and with a definite per-
manent electric dipole moment $\underline{\mu}_k$ and a polarizability $\underline{\underline{\alpha}}_k$; we
call such molecules equivalent ones. The average of a quantity
with respect to an assembly of equivalent molecules will be designated by $< >_{avM}$.

The field $\underline{E}_k$ in the thermodynamical equilibrium state of a
system may be represented as a superposition of three contributions:

1. The electric multipole moments of the solvent molecules
   surrounding the empty cavity cause an electric field $\underline{E}_f$
   inside of the cavity. In a homogeneous and isotropic system
   the average $<\underline{E}_f>_{avC}$ will vanish, but the average of even
   powers $<E_f^{2n}>$ $(n = 1, 2, ...)$ will be finite. The field $\underline{E}_f$
   which depends on the random distribution of solvent molecules,
   hence on their nuclear configuration, will be called fluctuation field.

2. If an electric field is applied to the system due to external
   charges there will be a contribution $\underline{E}_{hk}$, which depends on
   the external field strength $\underline{E}_a$ and on the configuration and
   the dielectric properties of the solvent molecules. The average value $<\underline{E}_{hk}>_{avC}$ may be represented by

   $$<\underline{E}_{hk}>_{avC} = \underline{\underline{f}}_e \underline{E}_a \tag{28}$$

   where $\underline{\underline{f}}_e$ is a tensor, which may be calculated on the basis of
   suitably chosen models [see Eq. (81)].

3. The total electric dipole moment $\underline{p}_k$ of the solute molecule
   causes a polarization of the medium surrounding it. This
   polarization is the cause of a further contribution to the
   field at the location of the solute molecule called reaction
   field. The total electric dipole moment $\underline{p}_k$ of a solute

molecule is, if hyperpolarizabilities are neglected,

$$\underline{p}_k = \underline{\mu}_k + \underline{\underline{\alpha}}_k \underline{E}_k \tag{29}$$

The electric field $\underline{E}_k$ may be represented by

$$\underline{E}_k = <\underline{E}_{dip}>_{avM} + <\underline{E}_{ext}>_{avM} + \underline{E}_{fluc} \tag{30}$$

where $\underline{E}_{dip}$, $\underline{E}_{ext}$, and $\underline{E}_{fluc}$ are the contributions caused by the permanent dipole moment $\underline{\mu}_k$, by external charges and by the fluctuation field, respectively. Hence $<\underline{E}_{dip}>_{avM}$ is the reaction field, as introduced by Onsager [14], which may be represented by

$$<\underline{E}_{dip}>_{avM} = \underline{\underline{f}} \, \underline{p}_{dip} \tag{31}$$

where $\underline{\underline{f}}$ is also a tensor describing the effects due to the total polarization (orientational, atomic and electronic polarization) of the solvent (see Eq. (82)) and $\underline{p}_{dip}$ is a contribution to the total dipole moment of the solute molecule given by

$$\underline{p}_{dip} = \underline{\mu}_k + \underline{\underline{\alpha}}_k <\underline{E}_{dip}>_{avM} \tag{32}$$

The field $<\underline{E}_{ext}>_{avM}$ due to external charges may be represented as a sum of the cavity field $\underline{\underline{f}}_e \underline{E}_a$ and an additional contribution to the reaction field, which is caused by the induced dipole moment $\underline{\underline{\alpha}}_k <\underline{E}_{ext}>_{avM}$ due to the field $<\underline{E}_{ext}>_{avM}$ and hence this additional contribution is $\underline{\underline{f}} \, \underline{\underline{\alpha}}_k <\underline{E}_{ext}>_{avM}$. Thus it is

$$<\underline{E}_{ext}>_{avM} = \underline{\underline{f}}_e \underline{E}_a + \underline{\underline{f}} \, \underline{\underline{\alpha}}_k <\underline{E}_{ext}>_{avM} \tag{33}$$

Similarly, the quantity $\underline{E}_{fluc}$, which represents the fluctuation of the field, is the sum of $\underline{E}_f$ and a further contribution to the reaction field caused by the induced dipole moment $\underline{\underline{\alpha}}_k \underline{E}_f$. In case of sufficiently slow changes of the fluctuation this contribution to the reaction field will be $\underline{\underline{f}} \, \underline{\underline{\alpha}}_k \underline{E}_f$, in case of very fast changes $\underline{f}$ should be substituted by $\underline{f}'$, a tensor describing the effects due to electronic polarization of the solvent only, but not due to orientational and atomic polarization [see Eq. (83)]. Hence it holds generally

$$\underline{E}_{fluc} = \underline{E}_f + \underline{f}\, \overset{*}{\underline{\alpha}}_k \underline{E}_{fluc} \tag{34}$$

where

$$\underline{f} \geq \overset{*}{\underline{f}} \geq \underline{f}' \tag{35}$$

Using Eqs. (31) to (34) the electric field $\underline{E}_k$, Eq. (30), at the location of the solute molecule becomes

$$\underline{E}_k = (\underline{1} - \underline{f}\, \underline{\alpha}_k)^{-1}(\underline{f}\, \mu_k + \underline{f}\, \underline{E}_a) + (\underline{1} - \overset{*}{\underline{f}}\, \underline{\alpha}_k)^{-1}\underline{E}_f \tag{36}$$

For the evaluation of Eq. (27) we also need the electric field $\underline{E}_1^{FC}$ at the location of the solute molecule in the Franck-Condon excited state. The reaction field of the solute molecule is due to the electronic polarization and due to the nuclear polarization (atomic and orientational polarization) of the solvent molecules surrounding the solute molecule. Since in the Franck-Condon excited state the nuclear configuration is the same as in the ground state, the value of the reaction field due to nuclear polarization in the Franck-Condon excited state is equal to the value in the ground state. Only the contribution to the reaction field due to the electronic polarization will be changed in the Franck-Condon excited state. The contribution $\underline{E}_{EPk}$ to the reaction field of the solute molecule in the ground state caused by electronic polarization is given by the formula

$$\underline{E}_{EPk} = \underline{f}'\underline{p}_k = \underline{f}'(\mu_k + \underline{\alpha}_k\underline{E}_k) \tag{37}$$

where $\underline{f}'$ is a tensor, which may be calculated on basis of appropriate models [see Eq. (83)]. The contribution to the total field $\underline{E}_k$ at the location of the solute molecule caused by the external charges, the nuclear polarization of the solvent molecules

and the fluctuation, whose value in the Franck-Condon excited state
remains equal to the value in the equilibrium ground state, is
consequently $\underline{E}_k - \underline{E}_{EPk}$. Hence the total field $\underline{E}_1^{FC}$ in the
Franck-Condon excited state becomes

$$\underline{E}_1^{FC} = (\underline{E}_k - \underline{E}_{EPk}) + \underline{E}_{EP1}^{FC} \tag{38}$$

where $\underline{E}_{EP1}^{FC}$ is the contribution to the reaction field in the
Franck-Condon excited state caused by the electronic polarization
of the solvent molecules, which is given by

$$\underline{E}_{P1}^{FC} = \underline{f}'\underline{p}_1^{FC} \tag{39}$$

where

$$\underline{p}_1^{FC} = \underline{\mu}_1 + \underline{\alpha}_1 \underline{E}_1^{FC} \tag{40}$$

and where $\underline{\mu}_1 = \underline{\mu}_1^{FC}$ is the permanent electric dipole moment and
$\underline{\alpha}_1 = \underline{\alpha}_1^{FC}$ is the polarizability of the solute molecule in the
Franck-Condon excited state. Using Eqs. (37) to (40) yields the
total field $\underline{E}_1^{FC}$,

$$\underline{E}_1^{FC} = (\underline{1} - \underline{f}'\underline{\alpha}_1)^{-1}\left[\underline{f}'\underline{\mu}_1 + (\underline{1} - \underline{f}\,\underline{\alpha}_k)^{-1}(\underline{f} - \underline{f}')\underline{\mu}_k + (\underline{1} - \underline{f}\,\underline{\alpha}_k)^{-1}\right.$$

$$(\underline{1} - \underline{f}'\underline{\alpha}_k)\underline{f}_e\underline{E}_a \tag{41}$$

$$\left. + (\underline{1} - \underline{f}\,\underline{\alpha}_k^*)^{-1}(\underline{1} - \underline{f}'\underline{\alpha}_k)\underline{E}_f\right]$$

C.   The Wavenumber Shift Caused by
Solvent Effects and an Applied Electric Field

For the calculation of the wavenumber shift $\Delta\bar{\nu}_{1\lambda,kk}^{SE}$
according to Eq. (27) we furthermore need the difference of the

polarization energies $W_{P1}^{FC} - W_{Pk}$. Quite similarly to the separation of the reaction field into two parts (see Sec. IIIC), the
polarization energy W_{Pk} can be divided into a contribution W_{PEk},
corresponding to the electronic polarization of the solvent molecules, and into a contribution W_{POk}, corresponding to the nuclear
polarization, where the latter contribution remains unchanged during the optical excitation process. The electronic parts of the
polarization energies in the ground state and Franck-Condon excited state are given by the formulas

$$W_{PEk} = \frac{1}{2} \, \underline{p}_k \underline{\underline{f}}' \underline{p}_k \tag{42}$$

and

$$W_{PE1}^{FC} = \frac{1}{2} \, \underline{p}_1^{FC} \underline{\underline{f}}' \underline{p}_1^{FC} \tag{43}$$

Hence

$$W_{P1}^{FC} - W_{Pk} = W_{PE1}^{FC} - W_{PEk} = \frac{1}{2} \, \underline{p}_1^{FC} \underline{\underline{f}}' \underline{p}_1^{FC} - \frac{1}{2} \, \underline{p}_k \underline{\underline{f}}' \underline{p}_k \tag{44}$$

Provided that all tensors have identical principal axes, as is
necessarily the case for molecules with sufficient symmetry and as
will be assumed generally in the following equations, the introduction of Eqs. (29), (36), (40), (41), and (44) into Eq. (27)
leads to the wavenumber shift

$$\Delta\overline{\nu}_{1\lambda,\,k\kappa}^{SE} = \Delta\overline{\nu}_{1\lambda,k\kappa}^{S} + \Delta\overline{\nu}_{1\lambda,k\kappa}^{E} \tag{45}$$

where

$$hca\Delta\bar{\nu}^S_{1\lambda,k\kappa} = W^{FC}_{L1\lambda} - W_{Lk\kappa} + W^{FC}_{C1} - W_{Ck} + W^{FC}_{D1} - W_{Dk}$$

$$- \frac{1}{2}(\bar{\underline{\mu}}_1 - \bar{\underline{\mu}}_k)(\underline{1} - \underline{f}'\underline{\alpha}_1)^{-1}\underline{f}'(\underline{\mu}_1 - \underline{\mu}_k)$$

$$- (\bar{\underline{\mu}}_1 - \bar{\underline{\mu}}_k)(\underline{1} - \underline{f\alpha}_k)^{-1}\underline{f}\,\underline{\mu}_k$$

$$- \bar{\underline{\mu}}_k(\underline{1} - \underline{f}'\underline{\alpha}_1)^{-1}(\underline{1} - \underline{f\alpha}_k)^{-2}f(\underline{\alpha}_1 - \underline{\alpha}_k)$$

$$\left[\frac{1}{2}(\underline{1} - \underline{f}'\underline{\alpha}_k)\underline{f}\,\underline{\mu}_k + (\underline{1} - \underline{f\alpha}_k)f'(\underline{\mu}_1 - \underline{\mu}_k)\right]$$

$$- \bar{\underline{E}}_f(\underline{1} - \underline{f}'\underline{\alpha}_1)^{-1}(\underline{1} - \underline{f}\,\overset{*}{\underline{\alpha}}_k)^{-1}(\underline{1} - \underline{f}'\underline{\alpha}_k) \qquad (46)$$

$$\left[\underline{\mu}_1 - (\underline{1} - \underline{f\alpha}_k)^{-1}(\underline{1} - \underline{f}'\underline{\alpha}_1)\underline{\mu}_k\right]$$

$$- \frac{1}{2}\bar{\underline{E}}_f(\underline{1} - \underline{f}'\underline{\alpha}_1)^{-1}(\underline{1} - \underline{f}\,\overset{*}{\underline{\alpha}}_k)^{-2}(\underline{1} - \underline{f}'\underline{\alpha}_k)(\underline{\alpha}_1 - \underline{\alpha}_k)\underline{E}_f$$

and

$$hca\Delta\bar{\nu}^E_{1\lambda,k\kappa} = -\bar{\underline{E}}_a\underline{f}_e\Delta\overset{*}{\underline{\mu}} - \frac{1}{2}\bar{\underline{E}}_a\underline{f}_e\Delta\underline{\alpha f}_e\underline{E}_a \qquad (47)$$

where

$$\Delta\overset{*}{\underline{\mu}} = \Delta\underline{\mu} + (\underline{1} - \underline{f}'\underline{\alpha}_1)^{-1}(\underline{1} - \underline{f\alpha}_k)^{-1}(\underline{1} - \underline{f}\,\overset{*}{\underline{\alpha}}_k)^{-1}$$

$$(\underline{1} - \underline{f}'\underline{\alpha}_k)(\underline{\alpha}_1 - \underline{\alpha}_k)\underline{E}_f \qquad (48)$$

$$\Delta\underline{\mu} = (\underline{1} - \underline{f}'\underline{\alpha}_1)^{-1}(\underline{1} - \underline{f\alpha}_k)^{-1}(\underline{1} - \underline{f}'\underline{\alpha}_k)(\underline{\mu}_1 - \underline{\mu}_k) + f\Delta\underline{\alpha\mu}_k \qquad (49)$$

$$\Delta\underline{\alpha} = (\underline{1} - \underline{f}'\underline{\alpha}_1)^{-1}(\underline{1} - \underline{f\alpha}_k)^{-2}(\underline{1} - \underline{f}'\underline{\alpha}_k)(\underline{\alpha}_1 - \underline{\alpha}_k) \qquad (50)$$

$\Delta\bar{\nu}^{SE}_{1\lambda,k\kappa}$ is the wavenumber shift for the transition $|1\lambda> \leftarrow |k\kappa>$ of an equivalent solute molecule (that is with a definite orientation in space) at a given fluctuation field $\underline{E}_f$. In Eq. (45) this shift is split into a part $\Delta\bar{\nu}^S_{1\lambda,k\kappa}$ independent of an applied field, given in Eq. (46) and describing the solvent dependence of the wavenumber shift, and into a part $\Delta\bar{\nu}^E_{1\lambda,k\kappa}$, given in Eq. (47) and describing the dependence on an applied electric field. Equation (47) shows, if the abbreviation introduced in Eqs. (48) to (50) are used, that the field dependence of the shift in solutions is quite similar to the field dependence in a gaseous phase as represented in Eq. (16). $\Delta\overset{*}{\underline{\mu}}$ as well as $\Delta\underline{\mu}$ is essentially the change of the permanent dipole moment during the considered excitation process. Since the quantities $(1 - f\alpha_k)$, etc. are of the order 1, the additional factors and terms, dependent on polarizabilities, lead to corrections which are usually small but not negligible. Similarly, $\Delta\underline{\underline{\alpha}}$ is essentially the change of the polarizability.

D. The Field Dependence of the Transition Moment

The field dependence of the transition moment $\underline{p}_{lk}^{E}$ can be described by Eq. (21) using the transition polarizability $\underline{\underline{\alpha}}_{kl}$, Eq. (22), higher developments have been given before [26,31]; in solutions only the field $\underline{E}$ has to be substituted by the average of the fields in the ground state and the Franck-Condon excited state [31]. Hence

$$\underline{p}_{lk}^{E} = \underline{\mu}_{lk} + \underline{\underline{\alpha}}_{lk}\underline{E}_{M} + \dots \tag{51}$$

where

$$\underline{E}_{M} = \frac{1}{2} (\underline{E}_{k} + \underline{E}_{l}^{FC})$$

With Eqs. (36) and (41) and the abbreviations

$$\underline{E}_{M}^{S} = (\underline{1} - \underline{\underline{f\alpha}}_{k})^{-1}\left\{\underline{\underline{f}}\,\underline{\mu}_{k} + \frac{1}{2}(\underline{1} - \underline{\underline{f'\alpha}}_{l})^{-1}\underline{f}'\left[(\underline{1} - \underline{\underline{f\alpha}}_{k})\underline{\mu}_{l}\right.\right.$$
$$\left.\left. - (\underline{1} - \underline{\underline{f\alpha}}_{l})\underline{\mu}_{k}\right]\right\} \tag{52}$$

$$\phi = \left[\underline{1} + \frac{1}{2}(\underline{1} - \underline{\underline{f'\alpha}}_{l})^{-1}\underline{f}'(\underline{\underline{\alpha}}_{l} - \underline{\underline{\alpha}}_{k})\right](\underline{1} - \underline{\underline{f\alpha}}_{k})^{-1} \tag{53}$$

$$\overset{*}{\phi} = \left[\underline{1} + \frac{1}{2}(\underline{1} - \underline{\underline{f'\alpha}}_{l})^{-1}\underline{f}'(\underline{\underline{\alpha}}_{l} - \underline{\underline{\alpha}}_{k})\right](\underline{1} - \underline{\underline{f}}\,\overset{*}{\underline{\underline{\alpha}}}_{k})^{-1} \tag{54}$$

the effective field $\underline{E}_{M}$ becomes

$$\underline{E}_M = \underline{E}_M^S + \overset{*}{\underline{\phi}\underline{E}}_f + \underline{\phi f}\,\underline{E}_a \tag{55}$$

and the transition dipole moment

$$\underline{p}_{1k}^E = \underline{p}_{1k}^S + \underline{\alpha}_{1k}\underline{\phi f}\,\underline{E}_a + \ldots \tag{56}$$

where in $\underline{p}_{1k}^S$ all terms independent of the external field are collected, thus this quantity is the transition dipole moment of a solute molecule in absence of an applied field. The development to next higher-order has been given recently [31].

E. The Field Dependence of the Molar Absorption Coefficient

As well as in a gaseous phase, discussed in Sec. II, the light absorption of a solution becomes anisotropic if an electric field is applied. Hence there are only well defined molar absorption coefficients $\varepsilon_\alpha^{SE}(\bar{\nu})$ for the case that either the electric field vector $\underline{e}$ of the incident plane-polarized light wave is parallel to $\underline{E}_a (\alpha = p)$ or $\underline{e}$ is perpendicular to $\underline{E}_a (\alpha = s)$. For these cases the molar absorption coefficients $\varepsilon_\alpha^{SE}(\bar{\nu})$ of the solute molecules in the presence of an applied electric field, at wavenumbers where the solvent does not absorb, can be represented analogous to Eq. (18) by

$$\frac{\varepsilon_\alpha^{SE}(\bar{\nu})}{\bar{\nu}} = S \sum_{k\kappa} \sum_{\ell\lambda} \iint w_{k\kappa}^E w_{ok}^E w_f \left| (p_\alpha)_{1\lambda,k\kappa}^E \right|^2 s_{1\lambda k\kappa f}^{SE}(\bar{\nu})\, d\tau dE_f \tag{57}$$

$$(\alpha = p,s)$$

In the above equations of this section we considered the excitation of only one equivalent solute molecule at a location with a definite fluctuation field. In order to obtain the absorption coefficient one has to take the average with respect to the assembly of solute molecules or what is equivalent to this to take the average with respect to (1) all vibronic states $|k>$ and $|l>$ and (2) all

librational states and $|\kappa>$ and $|\lambda>$ of the ground state and
the Franck-Condon excited state, respectively, and (3) all
fluctuations of the electric field and (4) all orientations of the
solute molecules. $w^E_{k\kappa}$ is the probability of a solute molecule
being in the vibronic state $|k>$ together with its surrounding in
the librational state $|\kappa>$. $w^E_{ok} = w^E_{ok}(\theta,\phi,\psi)$ is the probability
for a solute molecule in the vibronic state $|k>$ to have in pre-
sence of an applied electric field a definite orientation described by Eulerian angles θ, ϕ, and ψ. $d\tau$ is a volume element
in the orientational configuration space $(d\tau = \sin\theta\ d\theta\ d\phi\ d\psi)$;
the integral in Eq. (57) has to be extended over $0 \le \theta \le \pi$,
$0 \le \phi \le 2\pi$, $0 \le \psi \le 2\pi$. $w_f = w_f(E_f^2)$ is the probability density
for the fluctuation field; according to the model considered it is
independent of the states of the solute molecules and from the
applied field. Expressing the total probability as a product is
possible if the considered solute molecule is sufficiently rigid,
that is if the molecule does not have two or more different
configurations with energy differences comparable to or smaller
than kT and with different dipole moments or polarizabilities.
Under these circumstances w^E_k is independent of $\underline{E}_a$ and w^E_{ok} is
equal for all low lying vibronic states of the ground state in a
sufficient approximation. Then taking the average with respect to
the orientational distribution can be performed independently from
the other variables. If the molecule is not sufficiently rigid,
that is if there are two or more different configurations with
different dipole moments or polarizabilities, the averaging process
becomes more complicated and therefore we restrict the following
discussions to sufficiently rigid molecules. The orientational
distribution density $w^E_{ok} = w^E_o$, which within this restriction is
independent of the vibrational state, is given by the Maxwell-
Boltzmann formula

$$w^E_{ok} = w^E_o = \left[\exp\left\{ -\frac{W^E_k}{kT} \right\} d\tau \right]^{-1} \exp\left\{ -\frac{W^E_k}{kT} \right\} \tag{58}$$

where W_k^E represents the sum of terms of the expression in the brackets of Eq. (25) dependent on the external field. Since $W_k^E \ll kT$ the exponential function may be expanded in powers of W_k^E/kT. Using Eqs. (25) and (36) and the abbreviations

$$\overset{*}{\underline{\mu}} = \underline{\mu} + (\underline{1} - \underline{f\alpha}_k)^{-1} \underline{\alpha}_k \underline{E}_f \tag{59}$$

$$\underline{\mu} = (\underline{1} - \underline{f\alpha}_k)^{-1} \underline{\mu}_k \tag{60}$$

$$\underline{\underline{\alpha}} = (\underline{1} - \underline{f\alpha}_k)^{-1} \underline{\alpha}_k \tag{61}$$

we obtain similarly to Eq. (20)

$$w_o^E = (8\pi^2)^{-1} \left\{ 1 + \frac{1}{kT} \underline{\overline{E}}_a \underline{f}_e \overset{*}{\underline{\mu}} \right.$$
$$+ \frac{\underline{\overline{E}}_a}{6kT} \left[\frac{1}{kT} 3\underline{f}_e \overset{*}{\underline{\mu}} \underline{\overline{\mu}f}_e - \overline{\underline{\mu}f}_e^{2*} \underline{\mu}) + 3\underline{f}_e^2 \underline{\overline{\alpha}} - \operatorname{tr} \underline{f}_e^2 \underline{\alpha} \right] \left. \underline{E}_a \right\} \tag{62}$$

As in section II we may assume that the form of the line-shape function $s_{l\lambda\kappa\kappa f}^{SE}(\overline{\nu})$ for an individual transition is independent of the applied electric field and only the origin of the line is shifted by an amount $\Delta\overline{\nu}_{l\lambda, \kappa\kappa}^E = \Delta\overline{\nu}_{lk}^E$ independent of the libration-al states as given by Eq. (47). Hence in this approximation it is similar to Eq. (19)

$$s_{l\lambda\kappa\kappa f}^{SE}(\overline{\nu}) = s_{l\lambda\kappa\kappa f}^S(\overline{\nu} - \Delta\overline{\nu}_{lk}^E) \tag{63}$$

where $s_{l\lambda\kappa\kappa f}^S$ is the line-shape function of this transition of the molecule in the solution in the absence of an applied field. For sufficiently broad individual lines and a not too large applied electric field the shift will be small compared to the width of the line at half-maximum. Then the line-shape function may be expanded in powers of the shift,

$$
s_{1\lambda k\kappa f}^{SE}(\bar{\nu}) = s_{1\lambda k\kappa f}^{S}(\bar{\nu}) - \left(\frac{ds_{1\lambda k\kappa f}^{S}(\bar{\nu}')}{d\bar{\nu}'}\right)_{\bar{\nu}'=\bar{\nu}} \Delta\bar{\nu}_{1k}^{E}
$$

$$
+ \frac{1}{2}\left(\frac{d^2 s_{1\lambda k\kappa f}^{S}(\bar{\nu}')}{d\bar{\nu}'^2}\right)_{\bar{\nu}'=\bar{\nu}} (\Delta\bar{\nu}_{1k}^{E})^2 + \dots
$$

(64)

Introducing Eqs. (62) and (64) into Eq. (57) and averaging with respect to the orientational distribution leads generally to a rather complicated formula [6]. For further simplifications we have to restrict the considerations to wavenumber ranges with isolated absorption bands. A wavenumber range with an isolated band is one for which the following conditions are fulfilled:

1. For all transitions $|1\lambda> \leftarrow |k\kappa>$ contributing to the absorption in this wavenumber range the transition dipole moment $\underline{p}_{1k}^{S}$ has the same direction $\underline{m}^{S}$ in the molecule-fixed coordinate system and the same relative dependence on the external applied field [according to Eq. (51)].

2. All initial states have the same permanent electric dipole moment $\underline{\mu}_g$ and the same polarizability $\underline{\underline{\alpha}}_g$.

3. All final states have the same permanent electric dipole moment $\underline{\mu}_a$ and the same polarizability $\underline{\underline{\alpha}}_a$.

Under these restrictions the averaging process leads to

$$
\varepsilon_{\alpha}^{SE}(\bar{\nu}) = \varepsilon^{S}(\bar{\nu})\left[1 + L_{\alpha}(\bar{\nu})E_a^2 + O(E_a^4)\right] \quad (\alpha = p,s)
$$

(65)

where $\varepsilon^{S}(\bar{\nu})$ is the molar absorption coefficient of the solute molecule in the absence of an applied field. The expression Eq. (65) is correct up to third-order in $\underline{E}_a$ because all terms with odd powers of $\underline{E}_a$ have to vanish as long as the system in absence of an applied field is isotropic. Since $L_{\alpha}(\bar{\nu})E_a^2$ is usually of the order 10^{-4} to 10^{-6} and hence very small compared to 1 a molar absorption coefficient $\varepsilon^{SE}(\chi,\bar{\nu})$ can be defined for any angle χ between the electric field vector $\underline{e}$ of the incident light wave and the applied electric field $\underline{E}_a$, namely [6],

$$\varepsilon^{SE}(\chi,\bar{\nu}) = \varepsilon^{S}(\bar{\nu}) \left[1 + L(\chi,\bar{\nu})E_a^2 + O(E_a^4) \right] \tag{66}$$

where

$$L(\chi,\bar{\nu}) = L_s(\bar{\nu}) \sin^2 \chi + L_p(\bar{\nu}) \cos^2 \chi \tag{67}$$

or expressing it by the results of the averaging process

$$\begin{aligned} L(\chi,\bar{\nu}) = A(\chi) &+ \frac{1}{15hca} \left(\frac{d \ln [\varepsilon^{S}(\bar{\nu}')/\bar{\nu}']}{d\bar{\nu}'} \right)_{\bar{\nu}'=\bar{\nu}} B(\chi) \\ &+ \frac{1}{30h^2c^2a^2} \left[\left(\frac{d \ln (\varepsilon^{S}(\bar{\nu}')/\bar{\nu}')}{d\bar{\nu}'} \right)^2_{\bar{\nu}'=\bar{\nu}} \\ &+ \left(\frac{d^2 \ln (\varepsilon^{S}(\bar{\nu}')/\bar{\nu}')}{d\bar{\nu}'^2} \right)_{\bar{\nu}'=\bar{\nu}} \right] C(\chi) \end{aligned} \tag{68}$$

where

$$A(\chi) = \frac{1}{3} D + \frac{1}{30} (3 \cos^2 \chi - 1)E \tag{69}$$

$$B(\chi) = 5F + (3 \cos^2 \chi - 1)G \tag{70}$$

$$C(\chi) = 5H + (3 \cos^2 \chi - 1)I \tag{71}$$

The quantity

$$L(\chi,\bar{\nu}) = \frac{\varepsilon^{SE}(\chi,\bar{\nu}) - \varepsilon^{S}(\bar{\nu})}{\varepsilon^{S}(\bar{\nu})E_a^2} \tag{72}$$

determines the electrochromism of solute molecules and may be obtained from such measurements in dependence on $\bar{\nu}$ and χ. The first and second derivatives of $\ln(\varepsilon^{S}(\bar{\nu}')/\bar{\nu}')$ with respect to the wavenumber $\bar{\nu}'$ at $\bar{\nu}' = \bar{\nu}$ may be determined experimentally from the molar absorption coefficient of the solute molecules in their

solution in absence of an applied field. This quantities are
dependent on $\bar{\nu}$ and therefore the wavenumber dependence of
$L(\chi,\bar{\nu})$ can lead to the quantities $A(\chi)$, $B(\chi)$ and $C(\chi)$ according to Eq. (68). From the dependence of these quantities on χ
the six quantities D to I can be obtained. These quantities
are dependent on properties of the solute molecule and are reported for the most general case recently [31]. For many molecules
the field dependence of the transition moment is relatively small.
Furthermore, in suitable solvents as in aliphatic hydrocarbons, for
example (but not in benzene or dioxane where the molecules own
relatively large quadrupole moments), the fluctuation field is
nearly zero. Then in a sufficient approximation it is

$$D = \frac{f_e^2}{kT} \, \underline{\bar{R}}^{(1)} \underline{\mu} \tag{73}$$

$$E = \frac{1}{k^2T^2} \left[3 (\underline{\bar{m}}^S \underline{f}_e \underline{\mu})^2 - \underline{\bar{\mu}f}_e^2\underline{\mu} \right]$$
$$+ \frac{1}{kT} \left[3\underline{\bar{m}}^S \underline{f}_e^2 \underline{\underline{\alpha}} \underline{m}^S - \mathrm{tr}\, (\underline{f}_e^2\underline{\underline{\alpha}}) + f_e^2 (3\underline{\bar{R}}^{(2)} - 2\underline{\bar{R}}^{(1)})\underline{\mu} \right] \tag{74}$$

$$F = \frac{1}{kT} \, \underline{\bar{\mu}f}_e^2\Delta\underline{\mu} + \frac{1}{2} \, \mathrm{tr}\, (\underline{f}_e^2\Delta\underline{\underline{\alpha}}) + f_e^2\underline{\bar{R}}^{(1)} \Delta\underline{\mu} \tag{75}$$

$$F + G = \frac{3}{kT} \, (\underline{\bar{m}}^S \underline{f}_e \underline{\mu}) (\underline{\bar{m}}^S \underline{f}_e \Delta\underline{\mu}) + \frac{3}{2}\underline{\bar{m}}^S \underline{f}_e^2 \Delta\underline{\underline{\alpha}}\underline{\bar{m}}^S + \frac{3}{2} \, f_e^2\underline{\bar{R}}^{(2)} \Delta\underline{\mu} \tag{76}$$

$$H = \Delta\underline{\bar{\mu}f}_e^2\Delta\underline{\mu} \tag{77}$$

$$H + I = 3 (\underline{\bar{m}}^S \underline{f}_3 \Delta\underline{\mu})^2 \tag{78}$$

The quantities $\underline{\mu}$, $\Delta\underline{\mu}$, $\underline{\underline{\alpha}}$, and $\Delta\underline{\underline{\alpha}}$ have been defined in Eqs.
(60), (49), (61) and (50), respectively ($k = g$, $l = a$). The
vectors $\underline{R}^{(1)}$ and $\underline{R}^{(2)}$, which are related to the transition
polarizability $\underline{\underline{\alpha}}_{ag}$ of the considered transition, describe the
effects due to the electric field dependence of the transition
moment and are given explicitly by Liptay et al. [26] (where transition hyperpolarizabilities have been considered also). The permanent dipole moment and the average polarizability of the molecules
in the ground state may be obtained from dielectric and

refractometric measurements. With knowledge of such data $\underline{R}^{(1)}$ and $\underline{R}^{(2)}$ may be obtained from D and E in favorable cases [26,27] and these quantities lead to a few components of the transition polarizability $\underline{\alpha}_{ag}$. From E furthermore the direction of the transition dipole moment m^S can be obtained and in case of molecules without a permanent dipole moment the component of the polarizability $\underline{\alpha}_g$ in direction of m^S [32]. The quantities F to I are dependent on $\Delta\mu$ and $\Delta\underline{\alpha}$, and hence the change of the dipole moment and of the polarizability during an electronic excitation process can be obtained from these quantities; the results of such measurements have been reviewed recently [6].

For the evaluation of the quantities D to I the values of the tensors $\underline{f}_e$, $\underline{f}$, and $\underline{f}'$ have to be known. They have to be calculated on the basis of an appropriate model. Usually it is assumed that the solvent can be approximated by a homogeneous and isotropic dielectric where the solute molecules are localized in cavities with a definite shape. In the most simple approximation the shape of the cavities is assumed to be spherical, then the tensor $\underline{f}_e$ is reduced to the scalar f_e,

$$\underline{\underline{f}}_e = f_e\underline{\underline{1}} = \frac{3\varepsilon_r}{2\varepsilon_r + 1}\underline{\underline{1}} \tag{79}$$

where ε_r is the relative permittivity of the solution. If, furthermore, the electric dipole moment of the solute molecule is approximated by a point dipole localized in the center of the sphere, then

$$\underline{\underline{f}} = f\underline{\underline{1}} = \frac{2(\varepsilon_r - 1)}{4\pi\varepsilon_o a_w^3(2\varepsilon_r + 1)}\underline{\underline{1}} \tag{80}$$

and

$$\underline{\underline{f}}' = f'\underline{\underline{1}} = \frac{2(n^3 - 1)}{4\pi\varepsilon_o a_w^3(2n^2 + 1)}\underline{\underline{1}} \tag{81}$$

where n is the refractive index of the solution (for $\bar{\nu} \to 0$) and a_w is the radius of the sphere (interaction radius). If the shape of the cavity is assumed to be ellipsoidal with axes 2a, 2b, and 2c, then the principal components of the tensors become

$$f_{e\lambda} = \frac{\varepsilon_r}{\varepsilon_r - \kappa_\lambda(\varepsilon_r - 1)} \tag{82}$$

$$f_\lambda = \frac{3}{4\pi\varepsilon_o abc} \frac{\kappa_\lambda(1 - \kappa_\lambda)(\varepsilon_r - 1)}{(1 - \kappa_\lambda)\varepsilon_r + \kappa_\lambda} \tag{83}$$

$$f'_\lambda = \frac{3}{4\pi\varepsilon_o abc} \frac{\kappa_\lambda(1 - \kappa_\lambda)(n^2 - 1)}{(1 - \kappa_\lambda)n^2 + \kappa_\lambda} \tag{84}$$

where

$$\kappa_\lambda = \frac{abc}{2} \int_0^\infty \frac{ds}{(s + x_\lambda^2)\{(s + a^2)(s + b^2)(s + c^2)\}^{1/2}} \tag{85}$$

$$(x_1 = a, \ x_2 = b, \ x_3 = c)$$

The equations above refer to the case of absorption measurements. They may be modified in order to describe electrochromic emission measurements. Due to the finite viscosity of solvents, allowance has to be made for the fact that the equilibrium distribution of orientation in an applied electric field may not be reached during the lifetime of the excited states [33,34].

IV. ELECTROCHROMISM IN OTHER DENSE MEDIA

In other dense media but liquid solvents the mobility of molecules is usually strongly restricted. Therefore the equilibrium state of the orientational distribution, as described by Eq. (62), will not be reached during reasonable times. If, furthermore, the system is already anisotropic in absence of an applied field, as is usually the case in crystals, for example, the

light propagation in the absorbing medium influenced by an applied
electric field is described by rather complex equations for arbi-
trary angles of the applied field $\underline{E}_a$ and of the field vector $\underline{e}$
of the incident plane light wave relative to the principal axes
of the system. If the system is isotropic in absence of an applied
field or if the anisotropic permittivity is neglected for
simplification, then similar arguments as have been used in Sec.
III lead to a formula similar to Eq. (57). For the further treatment one only has to introduce the appropriate function for the
orientational distribution density w_{ok}^E of the absorbing molecules. The averaging process then leads to similar but modified
Eqs. (65) to (78). In case of systems anisotropic in absence of an
applied field the terms with odd powers in $\underline{E}_a$ may be finite also
and a separation of linear and square electrochromic effects is
possible. A further problem may be the evaluation of the terms
$\underline{f}_e$, $\underline{f}$, and $\underline{f}'$ since often Eqs. (79) to (85) cannot be applied
for such systems.

A rather simple example of such systems are rigid solutions of
appropriate molecules in suitable glasses. If solutions are solid-
ified in absence of an applied field and an electric field is
applied afterwards the orientational distribution will usually remain isotropic, at least if the viscosity of the rigid solvent is
large enough and the temperature is low enough. This orientational
distribution is equal to the distribution for $T \rightarrow \infty$, and therefore all terms dependent on T vanish in Eqs. (73) to (78).
Consequently, measurements on rigid solutions allow the independent
determination of the remaining terms in Eqs. (73) to (78).

In crystals or in systems as monolayer assemblies or membranes
the orientational distribution is determined by the structure of
the system and the field dependence of the distribution function
will often be negligible. Electrochromic measurements on such
systems also lead to information on dipole moments and
polarizabilities in excited states [8,35-37]. In less rigid system
the relaxation time for the orientational distribution may be

reasonably short, then a time dependence of the electrochromic
effects may be observed from which information on the relaxation
processes and therefore on the rigidity of the system can be obtained [38].

REFERENCES

1. E. U. Condon and G. H. Shortley, The Theory of Atomic Spectra,
    University Press, London (1959).

2. D. H. Phelps and F. W. Dalby, Can. J. Phys., 43, 144 (1965).

3. D. E. Freeman and W. Klemperer, J. Chem. Phys., 45, 52 (1966).

4. J. R. Lombardi, D. E. Freeman, and W. Klemperer, J. Chem.
    Phys., 46, 2746 (1967).

5. A. D. Buckingham, D. A. Ramsay, and J. Tyrell, Can. J. Phys.,
    48, 1242 (1970).

6. W. Liptay, Advances in Electronic Excitation and Relaxation,
    (E. C. Lim, ed.), Academic, New York (1974).

7. W. Kuhn, H. Dührkop, and M. Martin, Z. physik. Chem. Abs., B45,
    121 (1940)

8. R. M. Hochstrasser and L. J. Noe, J. Chem. Phys., 48, 514
    (1968).

9. W. Liptay and J. Czekalla, Z. Naturforsch., 15a, 1072 (1960).

10. H. Labhart, Chimia (Aarau), 15, 20 (1961).

11. W. Liptay, Modern Quantum Chemistry, Part 3, Academic Press,
    New York (1965), p. 45.

12. W. Liptay, Angew. Chem., 81, 195 (1969); Int. Ed., 8, 177
    (1969).

13. H. Labhart, Adv. Chem. Phys., 13, 179-204 (1967).

14. L. Onsager, J. Am. Chem. Soc., 58, 1486 (1936).

15. Y. Ooshika, J. Phys. Soc. Japan, 9, 594 (1954).

16. E. Lippert, Z. Naturforsch., 10a, 541 (1955), Ber. Bunsenges.
    Physik. Chem., 61, 962 (1957).

17. N. Mataga, Y. Kaifu, and M. Koizumi, Bull. Chem. Soc. Japan,
    29, 465 (1956).

18. E. G. McRae, J. Phys. Chem., 61, 562 (1957).

19. N. G. Bakshiev, Opt. Spectry. (USSR) (Engl. Transl.), 10, 379
    (1961).

20. L. Bilot and A. Kawski, Z. Naturforsch., 17a, 621 (1962).

21. W. Liptay, Z. Naturforsch., 20a, 1441 (1965).

22. W. Liptay, Modern Quantum Chemistry, Part 2, Academic, New York (1965), p. 173.

23. R. A. Marcus, J. Chem. Phys., 43, 1261 (1965).

24. W. Liptay, H.-J. Schlosser, and R. Weber, unpublished.

25. W. Liptay, Z. Naturforsch., 21a, 1605 (1966).

26. W. Liptay, B. Dumbacher, and H. Weisenberger, Z. Naturforsch., 23a, 1601, 1613 (1968).

27. W. Liptay, H. Weisenberger, F. Tiemann, W. Eberlein, and G. Konopka, Z. Naturforsch., 23a, 377 (1968).

28. V. Weisskopf and E. Wigner, Z. Physik, 63, 54 (1930).

29. H.-G. Kuball, Z. Naturforsch., 22a, 1407 (1967).

30. H.-G. Kuball and D. Singer, Ber. Bunsenges. Physik. Chem., 73, 403 (1969).

31. W. Liptay and G. Walz, Z. Naturforsch., 26a, 2007 (1971).

32. W. Liptay and H.-J. Schlosser, Z. Naturforsch., 27a, 336 (1972).

33. W. Liptay, Z. Naturforsch., 18a, 705 (1963).

34. G. Weber, J. Chem. Phys., 43, 521 (1965).

35. R. M. Hochstrasser and J. W. Michaluk, J. Chem. Phys., 55, 4668 (1971).

36. R. M. Hochstrasser and D. A. Wiersma, J. Chem. Phys., 55, 5339 (1971).

37. H. Bücher and H. Kuhn, Z. Naturforsch., 25b, 1323 (1970).

38. C. T. O'Konski and K. Bergmann, J. Chem. Phys., 37, 1573 (1962).

Chapter 7

ELECTRIC DICHROISM OF MACROMOLECULES

Charles M. Paulson, Jr.

Engineering Physics Laboratory
E. I. du Pont de Nemours & Co., Inc.
Wilmington, Delaware

I. INTRODUCTION

Since about 1959 various techniques of electric dichroism
have been increasingly used to study the electro-optic properties
of macromolecules in dilute solution. Some 40 papers in the
literature deal with the subject. The term "electric dichroism"
as used here refers to the ability of a medium to absorb light to
different extents, depending on the direction of polarization of
the light incident on the medium under the influence of an external electric field.* Throughout this discussion only linear
dichroism, as opposed to the widely used circular dichroism, will
be considered. Moreover, we shall consider only dichroism arising
from orientation-induced anisotropy without field-induced changes
in the frequency or overall intensity of the transition. A general treatment of optical absorption in electric fields including
such effects (electrochromism) has been given by Czekalla,
Labhart, and Liptay and is reviewed in the preceding chapter and
elsewhere [3,4]. Several recent reviews on applications of linear
electric dichroism with emphasis on proteins [5], nucleic acids
[6], and colloids [7] have also appeared.

Electric dichroism techniques are usually employed either to
measure hydrodynamic and electrical properties of absorbing molecules, or to determine in solution the orientation of specific
transition dipole moments within the macromolecule. Apparatus and
techniques utilizing pulsed electric fields, which were first
developed for studies of electric birefringence of macromolecules
[36,16], have been readily extended to electric dichroism measurements [57,65]. From relaxation and field dependence experiments,
a measure of the rotational diffusion coefficient, dipole moment,

---

*An alternate definition of dichroism, the ability of some
materials to change color as the thickness of the material is
varied, is often used in the dye and pigment industry [1]. That
phenomenon does not depend on an inherent anisotropy in the medium
for the change in color but is due to the shape of the absorption
band [2].

and polarizability of the macromolecule can be obtained. In addition, if the polarization direction of a transition is known relative to a particular group within the macromolecule, the orientation of that group to the principal electrical axis can be determined from the magnitude of the dichroism. Alternatively, if the structure of the macromolecule is known from x-ray or other data, the polarization of the transition moment can be established. The use of electric dichroism techniques will no doubt continue to grow, particularly for the study of macromolecules of biological interest, where the technique offers speed, sensitivity, and the opportunity for study of the macromolecule in dilute solution.

II. THEORY

A. Basic Definitions

In analogy with the definition of birefringence, dichroism can be defined as [8]

$$\Delta A = A_{\parallel} - A_{\perp} \tag{1}$$

where $A_{\parallel}$ and $A_{\perp}$ are the optical absorbances of the solution for light polarized parallel and perpendicular to the applied electric field, with the direction of light propagation perpendicular to the field. A more useful concentration-independent quantity, the reduced dichroism, is given by

$$\frac{\Delta A}{A} = \frac{A_{\parallel} - A_{\perp}}{A} \tag{2}$$

where A is the absorbance of the isotropic solution. The reduced dichroism can be expressed in terms of the molar absorptivities of the solute

$$\frac{\Delta A}{A} = \frac{\Delta \varepsilon}{\varepsilon} = \frac{\varepsilon_{\parallel} - \varepsilon_{\perp}}{\varepsilon} \tag{3}$$

where, assuming the Beer-Lambert law is followed, $A = \varepsilon c \ell$,
$A_\| = \varepsilon_\| c \ell$, $A_\perp = \varepsilon_\| c \ell$, c is the molar concentration, and ℓ is
the optical pathlength. Another quantity, the dichroic ratio,
$R = A_\| / A_\perp$, is frequently encountered in studies of infrared
dichroism of oriented films [9]. For many of the optical arrange-
ments, either $A_\|$ or $A_\perp$ is measured at any instant. In those
cases, the specific parallel dichroism

$$D_\| = \frac{A_\| - A}{A} \tag{4}$$

and specific perpendicular dichroism

$$D_\perp = \frac{A_\perp - A}{A} \tag{5}$$

are often the most convenient quantities to calculate [10]. For
pure absorption without any field-induced frequency or overall
intensity changes, the relation between the isotropic absorbance
and the polarized components is [41]

$$A = \frac{A_\| + 2A_\perp}{3} \tag{6}$$

Introducing Eq. (6) into the definitions of $D_\|$ and $D_\perp$, we have

$$D_\| = - 2D_\perp \tag{7}$$

Dichroism for light propagated in a direction parallel to the
field $\underline{E}$ has been measured in several cases [11,12]. Since the
absorption tensor is symmetric about the field direction, absorption is independent of polarization for this mode of propagation.
Defining the absorbance in this direction (longitudinal) as A_L
gives an expression for the longitudinal dichroism as

$$\frac{\Delta A_L}{A} = \frac{A_L - A}{A} \tag{8}$$

which is related to the reduced dichroism by [11]

$$\frac{\Delta A_L}{A} = \frac{\varepsilon_L - \varepsilon}{\varepsilon} = -\frac{1}{3}(\frac{\Delta\varepsilon}{\varepsilon}) \tag{9}$$

since $A_L \equiv A_\perp$

B. Relation of
Dichroism to Orientation Function

The theory of the orientation of macromolecules in electric fields has been extensively developed, mainly in terms of electric birefringence (see Chap. 3). These expressions can be readily extended to electric dichroism by the substitution of molecular absorptivities for optical polarizabilities [14]. A representation of a general model given by Holcomb and Tinoco [14] is shown in Fig. 1 where ε_j, α_j, and μ_j are the molar absorptivity, electrical polarizability, and dipole moment components, respectively, along the molecular axis j. A condition of the model is that the principal axes of the electrical and optical polarizabilities must coincide. Orientation of the molecule is expressed in terms of the three Eulerian angles, θ, ϕ, and ψ, where θ is the angle between the electric field E and axis a. Using Holcomb's and Tinoco's development, the dichroism is given by by

$$\Delta\varepsilon = \left(\varepsilon_a - \frac{\varepsilon_b + \varepsilon_c}{2}\right)\Phi_1 - \frac{3}{4}(\varepsilon_b - \varepsilon_c)\Phi_2 \tag{10}$$

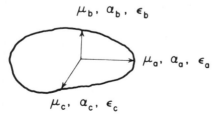

FIG. 1. Representation of the model for calculation of electric dichroism.

where the following definitions apply:

$$\Phi_1 = \int_0^{2\pi} \int_0^{\pi} \frac{3 \cos^2 \theta - 1}{2} f(\theta,\psi) \sin \theta \, d\theta \, d\psi \tag{11}$$

$$\Phi_2 = \int_0^{2\pi} \int_0^{\pi} (1 - \cos^2 \theta)(2 \cos^2 \psi - 1) f(\theta,\psi) \sin \theta \, d\theta \, d\psi \tag{12}$$

and

$$f(\theta,\psi) = \frac{\exp (-U/kT)}{\int_0^{2\pi} \int_0^{\pi} \exp (-U/kT) \sin \theta \, d\theta \, d\psi} \tag{13}$$

and U, the electric interaction energy, is a function of $\underline{\mu}$, $\underline{\alpha}$, and $\underline{E}$. In Eq. (10), Φ_1 is identical to $\Phi(\beta,\gamma)$ given by O'Konski et al. [15]. Expressions for Φ_1 and Φ_2 for various electrical parameters have been given elsewhere [14,15]. For limiting low fields the dichroism becomes proportional to E^2, as expected for Kerr law behavior. For limiting high fields as the orientation approaches saturation, $\Phi_2 = 0$ with $\Phi_1 = 1.0$ if the orientation is such that axis a is parallel to the field, and $\Phi_1 = -0.5$ if axes b or c are parallel to the field. For axially symmetric particles, the right-hand term in Eq. (10) is always zero.

Analysis of the dichroism as a function of E to obtain μ and α as well as field saturation values of $\Delta\epsilon$ can best be carried out by graphical or computer analysis based on the above theories [10]. Since the dichroism is related to the orientation distribution function in the same manner as birefringence, theories for field-free relaxation [16] and transient behavior in reversing [17] and alternating [18] fields should apply. For polydisperse systems, the type of average expressed by the measured parameter is generally a complicated function of the field strength as well as the electrical and optical properties of the macromolecules. The averages have been worked out in some detail for a rodlike model [10,19].

It has been demonstrated that many rodlike macromolecules, which are rigid at low molecular weight, become flexible with increasing degree of polymerization [18-20]. The above theories are, of course, not strictly valid for such systems, but can be applied as a function of molecular weight to demonstrate departures from rigidity. A theory of electric dichroism for highly flexible polymers is given in Chap. 4.

C. Orientation of Transition Dipole Moment

The magnitude of the dichroism can be related to orientation of the transition dipole moment of the chromophore at some fixed angle χ to the molecular symmetry axis. Since absorption is proportional to the square of the transition moment, it follows that [21,22]

$$\varepsilon_a = \varepsilon_o \cos^2 \chi \tag{14}$$

and

$$\varepsilon_b = \varepsilon_c = \frac{\varepsilon_o}{2} \sin^2 \chi \tag{15}$$

where ε_a and ε_b are molar absorptivities defined above, and $\varepsilon_o = 3\varepsilon$ is the molar absorptivity with the polarization vector parallel to the transition moment. Substitution of Eqs. (14) and (15) into Eq. (10) and simplifying gives

$$\frac{\Delta\varepsilon}{\varepsilon} = \frac{3}{2} (3 \cos^2 \chi - 1)\Phi_1 \tag{16}$$

Limiting field saturation values of the specific dichroism, reduced dichroism, and dichroic ratio are given in Table 1 for orientation of the symmetry axis a parallel and perpendicular to the field. For this calculation, orientation of the transition moment is taken as either parallel or perpendicular to the symmetry axis.

TABLE 1

Limiting Values of the Specific Dichroism,
Reduced Dichroism, and Dichroic Ratio at Saturating Fields

Orientation of symmetry axis	Orientation of transition moment to symmetry axis	$D_{\parallel}$	$D_{\perp}$	$\dfrac{\Delta A}{A}$	R
$\parallel$	$\parallel$	2.0	-1.0	3.0	∞
	$\perp$	-1.0	0.5	-1.5	0
$\perp$	$\parallel$	-1.0	0.5	-1.5	0
	$\perp$	0.5	-0.25	0.75	2.0

D. Form Dichroism

In direct analogy with form birefringence, an oriented system of isotropic absorbing particles of nonspherical shape imbedded in a medium of different refractive index will exhibit dichroic behavior. Such behavior is caused by the variation in complex refractive index for light propagated in different directions through the medium. This effect has been largely ignored in interpretation of electric dichroism data. Using dielectric theory, Wiener has treated the two limiting cases of infinitely long absorbing rods and lamellae oriented parallel in a nonabsorbing medium [23]. His expressions for the real and complex parts of the refractive indexes for rays propagated parallel and perpendicular to the orientation direction can be used to calculate birefringence and dichroism. Mayfield [24,25] has shown that, for low concentrations of weakly absorbing rods, Wiener's expressions for the dichroic ratio of the completely oriented system simplify to

$$R = \frac{(n^2 + 1)^2}{4} \tag{17}$$

where n is the relative refractive index of the particles to the medium. This equation was applied to tobacco mosaic virus (TMV), where it was found that form dichroism did not make an important contribution to the experimentally observed ultraviolet dichroism of the flow-oriented macromolecule.

E. Anisotropic Light Scattering

In the preceding discussions, the attenuation of light in the medium has been attributed only to absorption. In general, however, the particles will possess extinction cross sections containing both absorption and scattering terms, both of which contribute to the experimentally observed electric dichroism. Heller [26] has designated intrinsic and form dichroism as "consumptive" dichroism and the change in light attenuation from scattering as a consequence of orientation as "conservative" dichroism. Obviously, dichroism in a nonabsorbing region can only be conservative. Tolstoi and coworkers [27-32] have used techniques of conservative electric dichroism to study a number of colloidal particles.

Several workers have attempted to correct for the effects of scattering to the total dichroism [33,34]. In a study on TMV [33], the effect of scattering to the dichroism in the absorption region was subtracted by using an inverse fourth power extrapolation of the measured dichroism in a nonabsorbing region at higher wavelengths, similar to the correction of absorption spectra [35]. In another case, the measured dichroism on ribosomes, which was assumed to be due to scattering, was subtracted from the dichroism observed with the proflavine ribosome complex to obtain the contribution due to the anisotropy of the dye alone [34].

III.   MEASUREMENT OF ELECTRIC DICHROISM

A.   Pulsed-Field Apparatus

Much of the experimental approach to pulsed-field electric
dichroism has evolved from techniques and apparatus developed by
Benoit [36] and O'Konski et al. [16,37] for electric birefringence
measurements. These techniques use pulsed electric fields as the
orienting waveform to measure rotational diffusion coefficients
and to minimize artifacts due to electrophoresis and Joule heating. An electric dichroism arrangement employing pulsed fields is
shown in Fig. 2. This apparatus is typical of that used by many
workers [8,38-40]. Light from a monochromator at a wavelength λ
is passed through a polarizer and focused between the electrodes
in the center of the Kerr cell. The beam then passes through a
slit assembly to eliminate stray light and is detected at the
photomultiplier. An oscilloscope is used to monitor the transient
signal. The absorbance in the presence of the field for parallel
polarized light is given by [10,13]

$$A_{\parallel} = \log \frac{I_o}{I + \Delta I_{\parallel}} \tag{18}$$

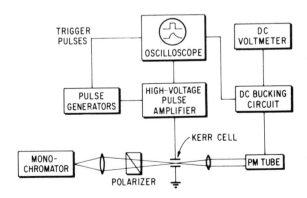

FIG. 2. Typical pulsed electric dichroism apparatus.

where I_o is the light intensity impinging on the cell, I is the light intensity following the cell in the absence of the field, and $\Delta I_\parallel$ is the change in light intensity produced by the field. The specific dichroism is then

$$D_\parallel = \frac{\log[I_o/(I + \Delta I_\parallel)] - A}{A} \tag{19}$$

which reduces to

$$D = -\frac{\Delta I_\parallel}{2.303AI} \tag{20}$$

for sufficiently small values of $\Delta I_\parallel/I$ [10,40]. A similar expression is obtained for $D_\perp$, which, by use of Eq. (7), gives

$$\Delta I_\parallel = -2\Delta I_\perp \tag{21}$$

when $\Delta I_{\parallel,\perp}/I \ll 1$. An analysis of the effect of concentration of the absorbing species on the signal-to-noise ratio for the dichroism signal has been given [13].

A useful variant of the above optics uses a Wollaston prism to split the light beam into two orthogonally polarized beams that are detected at separate matched photomultipliers [41,42], as shown in Fig. 3. When the dc components from each photomultiplier are combined with a differential amplifier and balanced, the transient signal is proportional to $\Delta\varepsilon/\varepsilon$ for small values of $\Delta\varepsilon\ell c/3$ [42].

Measurement of longitudinal dichroism, which is discussed in some detail by Baily and Jennings [11,53] offers advantages over measurements in the transverse direction. The apparatus is similar to that in Fig. 1, except that the polarizer is eliminated and a cell with conducting windows is substituted for the usual cell. Glass or quartz cell windows are rendered conducting by deposition of a thin stannic oxide film on the surface [43,44]. With care, coatings can be made thin enough to permit measurements in the ultraviolet to at least 220 nm [11,12]. The primary advantage of longitudinal dichroism is that it can reduce the electrode gap

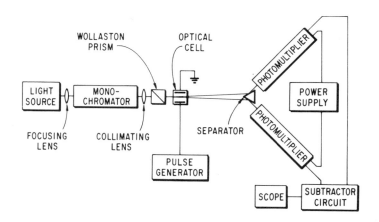

FIG. 3. Electric dichroism apparatus of Allen and Van Holde. (Reprinted in part from Ref. 42, p. 212, by courtesy of the authors and the American Institute of Physics.)

to an extremely small value, permitting the use of smaller voltages than would otherwise be necessary to produce a comparable degree of orientation. This property may be important for measurements on systems that are highly conducting. The smaller spacing also allows measurement of more concentrated solutions than could be measured with the longer optical pathlengths used in the transverse measurements. Another advantage is that conventional spectrophotometers can easily be adapted to make dichroism measurements by the placement of such cells in the cell compartment [11,53].

B. Static-Field Apparatus

If the transient portion of the dichroism signal is not of interest, the signal/noise ratio can be greatly increased by introduction of narrow bandwidth detecting electronics. Several instruments have been designed which use phase-sensitive detectors to lock-in on the dichroism signal produced by the application of a sinusoidal orienting field [4]. These instruments use high

fields (10^5 V/cm) and are primarily designed for the measurement
of electrochromism of small molecules.

Recent modifications to commercial spectropolarimeters permit
the continuous recording of linear dichroism as a function of
wavelength [45,46]. A modification to a Cary Model 60
spectropolarimeter, reported by Troxell and Scheraga [45], is
shown schematically in Fig. 4. Monochromatic, linearly polarized
light from the Cary instrument is passed through the electric
dichroism cell and into the Faraday modulator of the Cary, which
introduces a 60-Hz component to the major axis of the polarized
light. The beam then passes through a fixed analyzer to the
photomultiplier. The 60-Hz component, which is detected with a
lock-in amplifier, rotates the polarizer until the combined polarizer and the dichroism cell are crossed with the analyzer. The
orientation of the polarizer is simultaneously recorded. Troxell
and Scheraga have analyzed the response of this optical system
[45] with the Mueller matrix treatment following a procedure by
$\overline{\text{Go}}$ [47]. For the change in rotation induced by the field, α_E,
they obtain

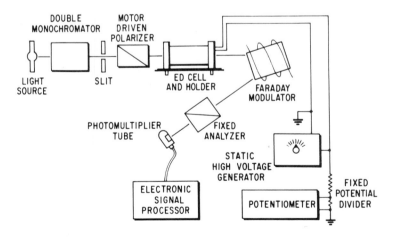

FIG. 4. Schematic diagram of dichroism apparatus of Troxell
and Scheraga. (Reprinted from Ref. 45, p. 525, by courtesy of
American Chemical Society.)

$$\alpha_E = 180\left\{\left[\frac{\ln 10}{4\pi}\right](\varepsilon_{\parallel} - \varepsilon_{\perp})\ell c \sin 2\theta'\right.$$

$$\left. - \left(\frac{\pi}{2}\right)\left[(n_{\parallel} - n_{\perp})\left(\frac{\ell}{\lambda}\right)\right]^2 \sin 4\theta' + \Delta(n_L - n_R)\left(\frac{\ell}{\lambda}\right)\right\} \quad (22)$$

where θ' is the angle between the applied field and the incident polarization, $n_{\parallel} - n_{\perp}$ is the birefringence, and $(n_L$ and $n_R)$ are refractive indexes for left- and right-handed circularly polarized light. By setting $\theta' = 45°$,

$$(\alpha_E)_{45°} = 33(\varepsilon_{\parallel} - \varepsilon_{\perp})\ell c \quad (23)$$

when the change in optical rotation term $(n_L - n_R)$ is negligible, as is the usual case within an absorption band. The limiting sensitivity of this instrument is very high, about 3×10^{-5} absorbance units. A spectrum taken with this instrument is shown in Fig. 8

Another modification to a commercial spectropolarimeter which permits continuous recording of linear dichroism with wavelength has been described by Mandel and Holzworth [46]. Although apparently used only for the measurement of film dichroism, it appears eminently suited for measurement of electric dichroism. The modifications consist only of adding an electronic power supply and a quartz depolarizer to commercial instruments that measure circular dichroism by the electro-optic Grosjean-Legrand method [48]. A diagram of the optics is shown in Fig. 5. A

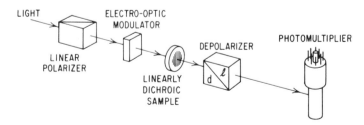

FIG. 5. Diagram of the optical system of the dichroism apparatus of Mandel and Holzwarth. (Reprinted from Ref. 46, p. 755, by courtesy of the authors and the American Institute of Physics.)

monochromatic beam of light linearly polarized at 90° to the field
direction is phase-modulated to 60 Hz by a Pockels cell with its
fast axis at 45° to the field. The the linear dichroism modification, an additional dc voltage is applied to the Pockels cell to
give it an additional quarter-wave bias. The light beam then
passes through the dichroic medium and onto the photomultiplier.
A depolarizer is inserted before the photomultiplier to reduce its
sensitivity to linearly polarized light. For the measurement of
linear dichroism, the ac part of the signal is separated from the
dc part, synchronously rectified, and compared with the dc part to
give a ratio R'. Mandel and Holzworth [ ] have analyzed the
optic system by the Jones matrix method. For small values of the
dichroism, they obtain

$$R' = 3.16\ell c (\varepsilon_{\|} - \varepsilon_{\perp}) \tag{24}$$

A limiting sensitivity of less than 10^{-4} absorbance units is
claimed for their modifications to the Cary Model 60 spectropolarimeter with a Cary Model 6001 CD attachment.

IV. APPLICATIONS

In this section, the application of electric dichroism
techniques to the study of macromolecules is surveyed. For com-
pleteness and easy referral, the available references to electric
dichroism studies on specific macromolecules are tabulated in
Table 2.

A. Determination of the Orientation of Specific Groups

By far the largest single use of electric dichroism has been
to establish the orientation of functional groups with respect to
some macromolecular axis. One of the earliest studies of electric

TABLE 2

Summary of References to Electric Dichroism of Macromolecules

Macromolecule	References
Poly(γ-benzyl-L-glutamate) (PBLG)	12, 49, 52-54
Poly(p-chloro-γ-benzyl-L-glutamate)	45
Poly(p-chloro-β-benzyl-L-aspartate)	45
Poly(β-benzyl-L-aspartate)	52
Poly-L-tyrozine	55
Poly(n-butylisocyanate)	19
Deoxyribonucleic acid (DNA)	57-63
Polyadenylic acid	6, 64
Ribonucleic acid (RNA)	65
Deoxyribonucleohistone (DNH)	38, 60, 61, 67
Ribosome	34, 39
Tobacco mosaic virus (TMV)	33, 53, 59
Quantasomes	40
Amylose triiodide	8, 10, 11, 53, 70
DNA/proflavine	60
DNH/proflavine	60, 61, 76, 77
rRNA/proflavine	34
DNA/crystal violet	62, 78
DNA/methylene blue	63
DNA/acridine orange	63
Polyuridylic acid/acridine orange	63
Polyadenylic acid/acridine orange	63
Polyglutamic acid/acridine orange	63
Polystyrene sulfonate/crystal violet	62
Hemoglobin	79
Kx brown and Congo red hydrosols	80, 81
Montmorillonite/methylene blue	13, 82
Colloidal particles	32

dichroism on macromolecules was reported by Spach in 1959 on the
infrared dichroism of poly(γ-benzyl-L-glutamate) (PBLG) oriented
in an electric field [49]. PBLG is an ideal molecule for electro-
optic studies, since it is largely rodlike and possesses a large
dipole moment along the helix axis [50]. The spectrum obtained
for a 2-3% solution of the polymer in chloroform is shown in Fig.
6. In the presence of a 15,000 V/cm field, an increase in intensity was found for the NH stretching band at 330 cm^{-1} and the
amide I band at 1655 cm^{-1}, while a decrease in intensity was
found for the amide II band at 1550 cm^{-1}. Solid-state studies of
PBLG have shown the 3300 cm^{-1} and 1655 cm^{-1} bands to be polarized
at 28° and 39° to the symmetry axis of the helical molecule [51].
Orientation of the molecule occurs with the symmetry axis parallel
to the electric field [50], thus increasing the intensity of the
bands. Conversely, intensity is decreased for the amide II band,

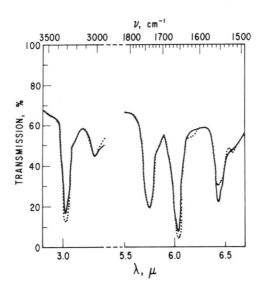

FIG. 6. Infrared absorption spectrum of PBLG in chloroform.
Solid line is in the absence of the field. Dotted line is in the
presence of the field with the vector of the incident radiation
parallel to the field direction. (Reprinted from Ref. 49, p. 668,
by courtesy of L'Académie des Sciences (Paris).

which is polarized at about 75° to the axis [51]. These results tend to confirm the preservation of the helical structure found in the solid state in solution.

A more recent study by Charney et al. [52] has been directed toward determination of the orientation of the benzyl ester side chains in PBLG. The dichroism spectrum for PBLG in the region of the 258 nm absorption band is shown in Fig. 7. One of the more interesting features of the spectrum is the higher resolution of the vibronic bands found in the reduced dichroism spectrum compared with the absorption spectrum. The higher resolution, which results from different polarization states of the vibronic bands, suggests that a dichroism spectrum is useful for resolving a complicated absorption spectrum. Since the dichroism spectrum was obtained at saturating fields, the magnitude of the reduced dichroism can be used to calculate the orientation of the transition moment by Eq. (16), giving a value of $\chi = 53.5^{\circ}$ for the 258 nm transition. This wavelength has been assigned to a B_2

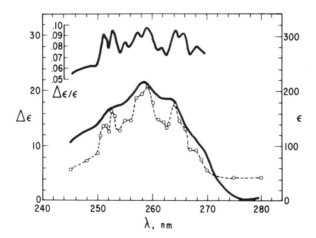

FIG. 7. Absorption spectrum (———) and wavelength dependence of electric dichroism (— — —) for PBLG in 1,2-dichloroethane (EDC) in the region of the 258 nm absorption band. The insert shows the dispersion of the reduced dichroism. (Reprinted from Ref. 52, p. 2659, by courtesy of American Chemical Society.)

electronic transition, with dipole moment located in the plane of
the phenyl group at right angles to the C_{2v} axis [52]. The
orientation of the phenyl ring can be compared with a value of
$\chi = 57^{\circ}$ obtained by Tsuboi from infrared film dichroism studies
[51]. However, the magnitude of the film dichroism is negative,
since the sign of the dichroism changes at $\chi = 54.7^{\circ}$.

Shirai [54] has found that PBLG solutions exhibit a small
positive dichroism at 263 nm constant in magnitude at temperatures
of 18°C and above. Below 18°C the dichroism decreases with decreasing temperature, becoming negative at 5°C. These results
indicate that the side chains change orientation slightly upon
transition from the solid state to solution. The possibility
exists that the measured dichroism in solution is decreased somewhat from the value expected for a rigid helix by the introduction
of flexibility at high molecular weights [20]. The electric
dichroism measurements provided some evidence for flexibility in
their linear rather than quadratic dependence on field strength at
the lowest field strengths measured.

The longitudinal dichroism has also been measured on PBLG in
DCE at 226 nm and 234 nm, where the dichroism was found to be
positive at the lower wavelength and negative at the higher wavelength [12]. Measurements have also been reported on p-Cl-PBLG in
dioxane [45], where the dichroism was found to be positive from
210 to 235 nm.

Electric dichroism measurements have been reported on other
synthetic polypeptides, including poly(β-benzyl-L-aspartate)
(PBLA) [52] and poly-L-tyrosine (PT) [55]. No dichroism signal
was observed from PBLA in the region of absorption of the benzyl
side chains, indicating either that the side chains are randomly
oriented, or that the dipole moment of the polymer is too small
to permit sufficient molecular orientation at the fields strengths
employed [52]. Dichroism spectra on PT in dioxane indicated
positive dichroism over the side-chain band at 278 nm and negative
dichroism over the 227 nm band [55].

Similar to PBLG, poly(n-butylisocyanate) PBIC has a fairly
rigid rodlike structure with a large dipole moment parallel to the
rod axis [56]. The electric dichroism spectra as measured by
Troxell and Scheraga [19] is shown in Fig. 8. Positive dichroism
is observed throughout the near ultraviolet bands. By an analysis
of the magnitude of the dichroism for low molecular weight PBIC
where the molecule is expected to be rigid, values of $\chi = 44^\circ$ and
$\chi \approx 53^\circ$ were assigned for the 249 nm and ~203 nm transitions.
These values correspond to polarizations in the plane of the
N-C=O group, but perpendicular to the C=O band and along the C=O
band, respectively [19]. This analysis uses values of the dipole
moment obtained from dielectric measurements.

Dvorkin's early electric dichroism measurements on solutions
of DNA showed the presence of negative dichroism in the 260 nm
band [57,58]. Absorption in this region is due to π-π* transitions in the nucleotide bases, and the helical DNA is known to
orient in an electric field with its long axis parallel to the
field. The results were thus interpreted as evidence that the
nucleotide bases primarily were oriented perpendicular to the

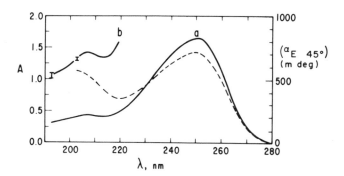

FIG. 8. Absorption spectrum of PBIC in cyclohexane in the
absence of electric field (— — —) and electric dichroism spectrum
(———) at (a) 45 and (b) 90 statvolts/cm, respectively.
(Reprinted from Ref. 19, p. 534, by courtesy of American Chemical
Society.)

helix axis, in accord with the Watson-Crick structure.
Subsequently, numerous electric dichroism measurements on DNA have
been reported [59-63]. Some variation in the magnitude of the reduced dichroism was observed, probably reflecting varying experimental conditions and effects of flexibility [6]. A synthetic
polynucleotide, polyadenylic acid (PolyA), has also been examined
and found to yield perpendicular dichroism in the 252-nm band [6,
64].

In contrast to DNA, measurements on ribonucleic acid (RNA)
indicated positive dichroism at 260 nm which decreased with
temperature, becoming negative above $40^{\circ}C$ [65]. However, recent
measurements have shown negative birefringence at room temperature
(implying perpendicular orientation of the nucleotide bases) for
RNA preparations with high helical content, and positive birefringence for completely unfolded molecules [66]. These findings
suggest that the earlier dichroism measurements were made on unfolded preparations and should be reexamined in terms of orientation of compact coiled conformations of RNA with the net orientation of the bases parallel to the long axis.

One nucleoprotein, deoxyribonucleohistone (DNH), has been
studied in some detail [38,60,61]. Similar to DNA, DNH was found
to exhibit negative dichroism at 260 nm; however, at comparable
ionic strength and applied field, the magnitude of the dichroism
from DNH was smaller than that from DNA [38]. Dichroism and birefringence measurements on the nucleoprotein as a function of histone increases continuously toward the pure DNA value [61]. The
results have been interpreted as evidence for presence of aggregates in the high histone content preparations. The UV dichroism
results have also been interpreted as indicating the presence of a
n-π* transition at about 280-290 nm [38].

An attempt was made to determine the orientation of the
tyrosyl ring at 275 nm in the histone component of DNH [67].
Since the UV absorption band of tyrosine is masked by the nucleotide base absorption in DNA, the band was shifted toward the

310 nm peak of 3,5-diiodotyrosine by iodination. However, no dichroism was observed in the shifted band, suggesting that either the group is oriented at 54° to the helix axis [67] or the group is randomly oriented [61].

The electric dichroism spectrum of TMV has been measured from 240 to 320 nm at saturation fields [33]. The dichroism is positive at $\lambda < 260$ nm and becomes negative at about 295 nm. The negative portion has been interpreted as evidence of a $n-\pi^*$ transition in the tryptophan spectrum while the positive portion at 280 nm is attributed to $\pi-\pi^*$ transitions of the aromatic amino acids oriented parallel to the helix axis. The stronger dichroism at $\lambda < 260$ nm is a combination of both nucleotide and amino acid transitions.

A small positive dichroism signal is observed in the 260 nm band of yeast ribosomes [34,39]. Initial interpretation was that the dichroism sign was due to orientation of the nucleotide bases predominantly parallel to the electric axis of the ribosome [39]. More recent studies, however, suggest that most of the observed dichroism arises from scattering and that a symmetrical arrangement of the double helical segments about the folding axis of the ribosome reduces the orientational dichroism to zero [34].

An interesting application of electric dichroism to the study of a particle of biological importance, spinach quantasomes, was reported by Sauer and Calvin [40]. They observed weak positive dichroism from 330 to 730 nm. A pronounced peak in the dichroic ratio spectrum at 695 nm was interpreted as resulting from a small fraction ($\sim5\%$) of chlorophyll, which is strongly oriented at a particular site (the quantatrope) in the quantasome.

B. Measurement of Rotational Diffusion Coefficients

The techniques of pulsed-field electric birefringence relaxation to measure rotational diffusion coefficients of macromolecules are readily extended to electric dichroism

relaxation. TMV, which has been studied by both methods [33,68],
is an ideal system for relaxation experiments, since it can be
obtained nearly monodisperse and is well represented by a rigid
cylinder model. Analysis of the dichroism relaxation curves [33]
has yielded values of the rotational diffusion coefficients of
$\theta_M = 340$ sec^{-1} for the monomer and $\theta_D = 77$ sec^{-1} for the dimer,
which compare fairly well with the values of $\theta_M = 333$ sec^{-1} and
$\theta_D = 56$ sec^{-1} measured by birefringence relaxation [68]. With an
assumed rod diameter of 180 Å, substitution of the values of θ
determined by dichroism relaxation into the Burgers-Broersma equation [69] for rotational diffusion of a cylinder yielded rodlength
values of 2990 Å and 5300 Å. The monomer length is in excellent
agreement with the electron microscopy value of about 3000 Å; however, the dimer length is somewhat less than expected for the end-
to-end dimer.

Other examples of relaxation measurements include studies on
ribosomes [39], quantasomes [40], and the amylose-triiodide complex [8,70].

C. Determination of Electrical Properties

Information on the electrical properties of macromolecules
has been gained, in several instances, by use of electric dichroism field-dependent behavior or transient behavior in reversing or
alternating fields [8,10,33,39,52,71-73]. From field-dependence
studies at low fields, Allen and Van Holde [33] have determined
values of the electrical polarizability of TMV ranging from
5.4×10^{-14} cm^3 at pH = 5.0 to 0.5×10^{-14} cm^3 at pH = 10.9.
The higher polarizability at lower pH is attributed to the pre-
sence of end-to-end dimers with approximately double the
polarizability of the monomer. The tendency of TMV to dimerize at
low pH is shown by the development of longer field-free relaxation
curves at low pH.

From reversing- and alternating-field experiments, the amylose-triiode complex was shown to possess permanent dipole moment behavior at degrees of triiode binding insufficient to completely saturate the complex [10]. At saturation levels, however, an induced polarization behavior was observed. A dispersion of the polarization was found at about 180 kHz, which was interpreted as evidence of ion atmosphere polarization [8]. Polarizabilities for the complex at various molecular weights were calculated from field-dependence data. The results were interpreted by use of the O'Konski and Krause theory [74].

D.   Studies of Conformation

Several workers have noted the effects of flexibility upon electric dichroism for rodlike macromolecules at high molecular weights [8,10,19,52,73]. Lengths of the amylose-triiodide helix calculated from field-free relaxation of the dichroism were found to be less than those calculated for the rigid helix using structural parameters from x-ray data [8]. The observed lengths were also found to reach a limiting value with increased molecular weight of the amylose [8,10]. These phenomena were interpreted as evidence of increased flexibility with increasing molecular length. The Hearst theory for rotational diffusion of stiff coils [75] was used to obtain a persistence length value of about 200 Å for the complex [8]. Despite the fairly high flexibility of the amylose-triiodide chain, large values of specific dichroism ($D_{\parallel}$ = 1.2 to 1.4) were found at saturating fields for the highest amylose molecular weights studied (the transition moment in the triiodide complex is directed along the helix axis). These findings provided evidence for the extension of the macromolecular coil at high fields [8,70].

Departure from rigidity has also been observed for rodlike PBIC [19,73]. From dichroism data at low fields (apparently in the Kerr law region), values of $\mu_2 \, g^{1/2}$, where μ_2 is the dipole moment along the symmetry axis and $g = (3 \cos^2 \chi - 1)/2$, were determined for a series of samples of varying molecular weights. A plot of this factor against μ_2 determined by dielectric measurements revealed a linear curve that passed through the origin with $g = 0.25$. However, a break in the curve occurred at about 1000 D (equivalent to degree of polymerization of about 600), suggesting the onset of flexibility at this point.

Electric dichroism measurements have been used to study the effects of small amounts of trifluoroacetic acid (TFA) on the conformation of PBLG [72]. The change in birefringence and reduced dichroism for PBLG in EDC with the addition of TFA is compared in Fig. 9. The reduced dichroism drops off rapidly with increasing TFA, while the birefringence reaches a limiting plateau. With the support of nuclear magnetic resonance (NMR) and infrared data, the difference between the dichroism and

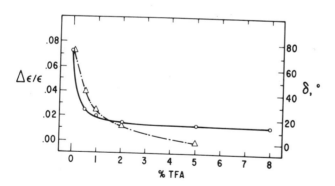

FIG. 9. Comparison of the Reduced Dichroism, $\frac{\Delta\epsilon}{\epsilon}$ $(- \cdot -)$ at 258 nm, to the electric birefringence (———) at 366 nm, of PBLG on the addition of TFA. $E^2 = 8.00 \times 10^8$ $(\text{V/cm})^2$. (Reprinted from Ref. 72, p. 995, by courtesy of Interscience Publishers.)

birefringence behavior has been interpreted as evidence of a
loosening of the orientation of the benzyl ester side chains by
interaction of the acid with the ester carbonyl. Since the NMR
data indicate that the phenyl group is not freely rotating, the
reduction of the dichroism must be due to a change in χ from
53.5° to 54.7° or, more likely, to a randomization effect resulting from an increase in the number of conformations available to
the phenyl group.

        E.   Studies of the Interaction of Dyes with Macromolecules

        The interaction of certain dyes with macromolecules
(primarily polynucleotides) has been a subject of considerable interest and is especially amenable to study by electric dichroism.
Houssier and Fredericq have investigated the binding of proflavine
to deoxyribonucleohistone (DNH) by this technique [60,61,76,77].
Similar to DNA [60], the complex exhibits negative dichroism in
the band of the bound proflavine at 459 nm. The absolute value of
the dichroic ratio was found to decrease with increasing ratios of
dye molecules bound per atom of phosphorous that were greater than
0.1. These results were interpreted as evidence that the proflavine molecules are intercalated between the stacked nucleotide
bases, with at most one dye per five nucleotide pairs. Excess
proflavine was thought to bind in a random manner on the outside
of the helix.

        Dichroism measurements in combination with spectroscopic
binding and fluorescence quenching studies, have shown the existence of two types of intercalation complexes of proflavine with
rRNA [34]. A strong binding region (0.02 sites per rRNA phosphorous) is believed to be located in the (A + U)-rich single
strands and gives a low-magnitude dichroism. A weaker binding region ($\sim$0.08 sites per phosphorus) is located in the (G + C)-
rich, double strands where the dye molecules are more oriented, as

shown by the higher dichroism. Somewhat similar results were
obtained on proflavine complexes with unfolded ribosomes, which
were used to support a model of the arrangement of RNA with the
ribosome.

In other studies, the electric dichroism of crystal violet/
DNA has been measured [62,78]. The specific dichroism was found
to be negative and passed through a maximum when the ratio of
polymer to dye was about 10. Negative dichroism has also been
reported for the methylene blue/DNA complex and the complexes of
acridine orange with DNA and polyuridylic acid [63]. A small
positive dichroism was noted for the acridine orange/polyadenylic
acid complex [63].

F. Assignment of Transitions

With some knowledge of the structure of the macromolecule,
one may use dichroism information obtained by electrical orientation to establish polarization of transitions or assignment of the
transition to a particular symmetry notation. As an example of
the latter, the rich vibronic substructure found in the dichroism
spectrum of PBLG shown in Fig. 7 has been completely assigned to
states of symmetry $^1A_1$ and $^1B_1$ within the $^1B_2$ electronic
transition in the substituted phenyl group [52].

REFERENCES

1. R. M. Evans, An Introduction to Color, Wiley, New York (1948),
    p. 73.

2. T. G. Webber, SPE J., 24, 29 (1968).

3. W. Liptay, Modern Quantum Chemistry, Part III, (O. Sinanoǧlu,
    ed.) Academic, New York (1965), p. 45.

4. H. Labhart, Adv. Chem. Phys., 13, 179 (1967).

5. K. Yoshioka and H. Watanabe, Physical Principles and
    Techniques of Protein Chemistry Part A (S. J. Leach, ed.),
    Academic Press, New York (1969), p. 335.

6. E. Charney, Procedures in Nucleic Acid Research, Vol. 2
    (G. L. Cantoni and D. R. Davies, eds.), Harper & Row, New
    York (1971), p. 176.

7. S. P. Stoylov, Adv. Colloid Interface Sci., 3, 45 (1971).

8. C. M. Paulson, Jr. and C. T. O'Konski, Polymer Preprints, 7,
    1175 (1966).

9. C. Y. Liang, Newer Methods of Polymer Characterization
    (Bacon Ke, ed.), Interscience, New York (1964), p. 33.

10. C. M. Paulson, Jr., Ph.D. thesis, University of California,
    Berkeley (1965).

11. E. D. Baily and B. R. Jennings, Appl. Opt., 11, 527 (1972).

12. S. J. Hoffman and R. Ullman, J. Polymer Sci., C31, 205 (1970).

13. K. Bergmann, unpublished laboratory report, University of
    California, Berkeley (1960); K. Bergmann and C. T. O'Konski,
    in preparation (1976).

14. D. N. Holcomb and I. Tinoco, J. Phys. Chem., 67, 2691 (1963).

15. C. T. O'Konski, K. Yoshioka, and W. H. Orttung, J. Phys.
    Chem., 63, 1558 (1959).

16. C. T. O'Konski and B. H. Zimm, Science, 111, 113 (1950).

17. I. Tinoco and K. Yamaoka, J. Phys. Chem., 63, 423 (1959).

18. A. Peterlin and H. A. Stuart, Hand- und Jahrbuch der
    chemischen Physik, Vol. 8, Sect. 1B, Becker & Erler, Leipzig
    (1943).

19. T. C. Troxell and H. A. Scheraga, Macromolecules, 4, 528
    (1971).

20. P. Moha, G. Weill, and H. Benoit, J. Chem. Phys., 61, 1240
    (1964).

21. R. D. B. Fraser, J. Chem. Phys., 21, 1511 (1953).

22. A. Wada and S. Kozawa, J. Polymer. Sci., A2, 853 (1964).

23. O. Wiener, Kolloidchem. Beih., 23, 189 (1926).

24. J. E. Mayfield, Ph.D. Thesis, University of Pittsburgh (Pa.)
    (1968).

25. J. E. Mayfield and I. J. Bendet, Biopolymers, 9, 669 (1970).

26. W. Heller, Rev. Mod. Phys., 14, 390 (1942).

27. N. A. Tolstoi, A. A. Spartakov, A. A. Trusov, and
    S. A. Shichelkunova, Biofizika, 11, 453 (1966) (in Russian);
    Biophysics (USSR) (Engl. Transl.), 11, 515 (1966).

28. N. A. Tolstoi, A. A. Spartakov, and G. I. Khil'ko, Kolloidn. Zh., 22, 705 (1960) (in Russian); Colloid J. (USSR) (Engl. Transl.), 22, 701 (1960).

29. N. A. Tolstoi and A. A. Spartakov, Kolloidn. Zh., 28, 580 (1966) (in Russian); Colloid J. (USSR) (Engl. Transl.), 28, 468 (1966).

30. N. A. Tolstoi, A. A. Spartakov, and A. A. Trusov, Kolloidn. Zh., 28, 735 (1966) (in Russian); Colloid J. (USSR) (Engl. Transl.), 28, 596 (1966).

31. N. A. Tolstoi, A. A. Spartakov, and A. A. Trusov, Kolloidn. Zh., 29, 584 (1967) (in Russian); Colloid J. (USSR) (Engl. Transl.), 29, 434 (1967).

32. N. A. Tolstoi, A. A. Spartakov, and A. A. Trusov, Issled. v Obl. Poverkhn. Sil Akad. Nauk SSSR, Inst. Fiz. Khim. Dokl. na Konf. 1966, 56 (in Russian); through Chem. Abstr. 70, 50930u. (1969).

33. F. S. Allen and K. E. Van Holde, Biopolymers, 10, 865, (1971).

34. M. Schoentjes and E. Fredericq, Biopolymers, 11, 361 (1972).

35. G. H. Beaver, E. A. Johnson, H. A. Willis, and R. G. J. Miller, Molecular Spectroscopy, Macmillan, New York (1961), p. 145.

36. H. Benoit, Compt. Rend., 228, 1716 (1949).

37. C. T. O'Konski and A. J. Haltner, J. Am. Chem. Soc., 78, 3604 (1956).

38. C. Houssier and E. Fredericq, Biochem. Biophys. Acta, 120, 113 (1966).

39. R. S. Morgan, Biophys. J., 3, 253 (1963).

40. K. Sauer and M. Calvin, J. Mol. Biol., 4, 451 (1962).

41. W. Kuhn, H. Duhrkop, and H. Martin, Z. Physik. Chem., 45B, 121 (1939).

42. F. S. Allen and K. E. Van Holde, Rev. Sci. Instr., 41, 211 (1970).

43. D. W. Juenker and E. W. Parsons, J. Opt. Sco. Am., 48, 857 (1958).

44. A. Zaromb and J. Y. Chang, J. Electrochem. Soc., 109, 1034 (1962).

45. T. C. Troxell and H. A. Scheraga, Macromolecules, 4, 519 (1971).

46. R. Mandel and G. Holzworth, Rev. Sci. Instr., 41, 755, (1970).

47. N. Gō, J. Phys. Soc. Japan, 23, 88 (1967).

48. M. Grosjean and M. Legrand, Compt. Rend., 251, 2150 (1960).

49. G. Spach, Compt. Rend., 249, 667 (1959).

50. I. Tinoco, J. Am. Chem. Soc., 79, 4336 (1957).

51. M. Tsuboi, J. Polymer Sci., 59, 139 (1962).

52. E. Charney, J. B. Milstein, and K. Yamaoka, J. Am. Chem. Soc., 92, 2657 (1970).

53. B. R. Jennings and E. D. Baily, Nature [Phys. Sci.], 233, 162 (1971).

54. M. Shirai, Abstracts of Papers, 152nd ACS Meeting, New York, Sept. 1966, IO31.

55. T. C. Troxell and H. A. Scheraga, Biochem. Biophys. Res. Commun., 1969, 913.

56. A. J. Bar and E. D. Roberts, J. Chem. Phys., 51, 406 (1969).

57. G. A. Dvorkin, Dokl. Acad. Nauk SSSR, 135, 739 (1960) (in Russian); Dokl. Biochem. Sect. (Engl. Transl.), 135, 833 (1960).

58. G. A. Dvorkin and V. I. Krinskii, Dokl. Acad. Nauk SSSR, 140, 946 (1969) (in Russian); Dokl. Biol. Sci. Sect. (Engl. Transl.), 140, 766 (1961).

59. Ye. I. Golub and G. A. Dvorkin, Biofizika, 9, 545 (1964) (in Russian); Biophysics (USSR) (Engl. Transl.), 9, 595 (1964).

60. C. Houssier and E. Fredericq, Biochem. Biophys. Acta, 88, 450 (1964).

61. C. Houssier, J. Chim. Phys., 65, 36 (1968) (in French).

62. T. Soda and K. Yoshioka, Nippon Kagaku Zasshi, 87, 1326 (1966).

63. D. G. Bradley, N. C. Stellwagen, C. T. O'Konski, and C. M. Paulson, Jr., Biopolymers, 11, 645 (1972).

64. M. Shirai, Nippon Kagaku Zasshi, 86, 1115 (1965).

65. G. A. Dvorkin and A. S. Spirin, Dokl. Acad. Nauk SSSR, 135, 987 (1960) (in Russian); Dokl. Biochem. Sect. (Engl. Transl.), 135, 270 (1960).

66. E. I. Golub and V. G. Nazarenko, Biophys. J., 7, 13 (1967).

67. C. Houssier and E. Fredericq, Biochim. Biophys. Acta, 138, 424 (1967).

68. C. T. O'Konski and A. J. Haltner, J. Am. Chem. Soc., 79, 5634 (1957).

69. S. Broersma, J. Chem. Phys., 32, 1626 (1960).

70. C. M. Paulson, Jr. and C. T. O'Konski, in press (1976).

71. C. Hornick and G. Weill, Biopolymers, $\underline{10}$, 2345 (1971).

72. J. B. Milstein and E. Charney, Biopolymers, $\underline{9}$, 991 (1970).

73. J. B. Milstein and E. Charney, Macromolecules, $\underline{2}$, 678 (1969).

74. C. T. O'Konski and S. Krause, J. Phys. Chem., $\underline{74}$, 3243 (1970).

75. J. E. Hearst, J. Chem. Phys., $\underline{38}$, 1062 (1963).

76. C. Houssier, Arch. Int. Physiol. Biochim., $\underline{75}$, 169 (1967).

77. C. Houssier and E. Fredericq, Biochim. Biophys. Acta, $\underline{120}$, 434 (1966).

78. T. Soda and K. Yoshioka, Nippon Kagaku Zasshi, $\underline{86}$, 1019 (1965).

79. W. H. Orttung, J. Am. Chem. Soc., $\underline{87}$, 924 (1965).

80. V. V. Kurbasov, Kolloidn. Zh., $\underline{26}$, 330 (1964) (in Russian); Colloid J. USSR (Engl. Transl.), $\underline{26}$, 278 (1964).

81. V. V. Kurbasov, Kolloidn. Zh., $\underline{26}$, 481 (1964) (in Russian); Colloid J. (USSR) (Engl. Transl.), $\underline{26}$, 413 (1964).

82. C. T. O'Konski and K. Bergmann, J. Chem. Phys., $\underline{37}$, 1573 (1962).

Chapter 8

ELECTRIC FIELD LIGHT SCATTERING

Barry R. Jennings

Physics Department
Brunel University
Uxbridge, Middlesex, England

I.   INTRODUCTION

The scattering of light from isolated particles suspended in
a liquid medium has long been recognized as a method for the
evaluation of their molecular mass (M) and their size or shape.
As early as 1919, Gans [1] realized that the scattering properties
of colloidal suspensions would reveal information on the structure

of the dispersed particles. Later, Cabannes [2] showed that the
optical depolarization of solutions and suspensions was related to
the optical anisotropy of the solute, and hence again to its
atomic structure.

The conventional scattering method (in the absence of applied
electric fields) has become standard as a means of measuring the
molecular mass and radius of gyration (S) of particles in suspension. The radius of gyration is evaluated if the scattering
particles have a dimension comparable to the wavelength (λ_0) of
the incident light in vacuo. The particles, which are assumed to
be both uncharged and noninteracting, are randomly positioned and
oriented throughout the body of the suspension. As the incident
light beam strikes a particle, wavelets are scattered from its
constituent parts. Owing to the overall size of the particle, the
optical path between the scattering elements approaches λ_0 and
the scattered wavelets interfere. Hence, whereas small particles,
with dimensions much less than λ_0, give a symmetric variation of
the scattered intensity (I) with the angle of observation (θ),
this is not so with larger particles (Fig. 1). For the latter,
the angular (or polar) excess scattering pattern becomes dissym-
metric about the angle θ equal to 90° and is characteristic of
the size and shape of the solute.

Theoretically, the scattered intensity is accommodated in the
equation [3]

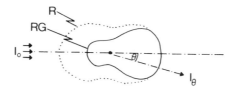

FIG. 1. Polar scattering diagram for incident, unpolarized
light (I_0). R represents the Rayleigh theory for small particles and R-G the Rayleigh-Gans theory for macromolecules whose
dimensions are of the order of the wavelength (λ_0).

$$\frac{K'c}{I} = \frac{1}{\overline{M}_w} P^{-1}(\theta) + 2Bc \tag{1}$$

where K', I, and B are an optical constant, the excess
scattered intensity of the solution over that of the solvent at
angle θ and the second osmotic virial coefficient, respectively,
of a suspension of concentration c (g ml^{-1}). A definite weight
average is involved for M. The factor $P^{-1}(\theta)$ is the reciprocal
of the large particle-scattering factor $(P(\theta))$. This is the factor which accounts for the interference effects. It is a normalized number which takes into account the scattered intensity from
all pairs of scattering elements, summed for all such pairs in the
scattering particle. In the absence of interference effects, it
has the value of unity, and Eq. (1) collapses to the regular
Rayleigh theory for small particles. Expressions for $P(\theta)$ are
quite complicated. They have been developed for many common particle shapes, too numerous to list here. Details are given in
reviews on scattering [3-6]. However, Guinier [7] noted that for
all particle shapes, the first two terms of a series expansion of
$P^{-1}(\theta)$ always have the same form when expressed in terms of the
radius of gyration about the center of mass. The Rayleigh-Gans
theory can then be generalized [3] as follows:

$$\frac{K'c}{I} = \frac{1}{\overline{M}_w} \left\{ 1 + \frac{16\pi^2 n^2}{3\lambda_0^2} <S^2>_Z \sin^2 \frac{\theta}{2} + \ldots \right\} + 2Bc \tag{2}$$

This equation forms the basis of conventional measurements to
evaluate $\overline{M}_w$ and $<S^2>_Z$ from data on I at various θ and c.
We note that, with polydisperse samples, definite weight and Z
average parameters are determined for M and S^2.

Commercial light-scattering photometers consist basically of
a light source and a collimating optical system to produce a
parallel light beam through a suitable cell. A photomultiplier
detection system is used which scans around the cell to record the
scattered intensity at various angles θ. A full description of

the various designs and experimental procedures is given in a
number of reviews [3-5]. The usual experimental procedure is to
take measurements of I at various angles θ for solutions of
different concentration and to express the data in a Zimm [8]
plot. This is essentially a plot of K'c/I against
$[\sin^2 (\theta/2) + gc]$, where g is a number. Simultaneous extrapolation is made to both zero angle and zero concentration to obtain
two limiting curves under these conditions. The zero concentration curve yields both $\bar{M}_w$ and $<S^2>_z$ from its intercept and
initial slope, while the zero angle curve has a slope from which
B is obtained. Should the particle shape be known, the radius of
gyration can be interpreted directly into relevant geometric
sizes; for example, $S^2 = 3D^2/20$ for spheres of diameter D, or
$L^2/12$ for thin rods of length L or $\Gamma^2/6$ for flexible coils
with an average end separation Γ.

Finally, before proceeding to consider the effect of electric
fields, and hence particle alignment on P(θ), it is pointed out
that Rayleigh-Gans theory is subject to the two conditions [6]

$$2h\nu^{1/3} \left| m - 1 \right| \ll 1 \quad \text{and} \quad \left| m - 1 \right| \ll 1$$

where h, ν, and m are $2\pi n/\lambda_0$, the particle volume and the
ratio of the refractive index of the particle (n_p) to that   (n)
of the surrounding medium or solvent respectively.

II. SCATTERING IN AN ELECTRIC FIELD

Particle orientation in an electric field affects the
scattering properties of the suspension in at least two ways.
First, by analogy with the other electro-optic effects mentioned
in this treatise, the inherent anisotropy in the scattering properties of the individual particles will be enhanced upon alignment. This is true even if the particles are small. Both the
depolarization ratio and the scattered intensity at any angle of
observation will be affected. A preliminary theoretical equation
has been derived by Kielich [9]. However, changes in the

scattered intensity (ΔI) will be extremely small for small particles. Second, for particles which are large enough to manifest
the internal interference effects outlined in the previous section, changes will occur in the interference pattern upon particle
alignment. Hence, the scattered intensity will also change. This
effect is relatively large. It is pictorially presented in Fig. 2
where a cylinder is depicted in two orientations relative to the
incident light beam. The optical path difference between the
wavelets scattered from the extremities of the rod at any angle
θ is, in general, very sensitive to the orientation of the rod.
Furthermore, it can be seen from this diagram that the direction
into which the particles tend to orientate when the external field
is applied can be deduced from the changes in the polar scattering
diagram. Such knowledge enables the direction as well as the magnitude of the particle dipole moments to be determined. Dipole
moments along and across the rod would be indicated by Figs. 2(i)
and (ii), respectively, for an electric field applied in the direction of $\theta = 90^{\circ}$. It is thus seen that for large particles, a
change in $P(\theta)$ is a useful and sensitive indicator of molecular
orientation and leads to scattered intensity changes which are far

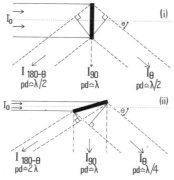

FIG. 2. Dependence of the polar scattering envelope on
molecular orientation. pd refers to the optical path difference
between rays scattered from opposite ends of the molecule whose
length is approximately λ_0. Cases (i) and (ii) represent two
orientations. (Reprinted from Ref. 10, p. 253, by courtesy of the
Society of Chemical Industry.)

larger than those due to the anisotropy effect alone. Unless
otherwise stated, the representative results cited below refer to
molecules of such size that the scattered intensity changes are
predominantly due to internal interference effects. Because of
this, the theory cannot simply be expressed in terms of the orientation factors Φ given in Chap. 3 and elsewhere [11] for birefringence, dichroism, and optical activity changes, but must be
developed independently.

Although electric field light scattering was recorded by
Bloch [12] in 1908 for ammonium chloride aerosols, by Zeeman and
Hoogenboom [13] for aerosols and vapors, and by Subrahmanya et al.
[14] for silver iodide, stearic acid, and benzopurpurine sols, it
was Wippler [15-18] who first derived suitable equations for the
scattered intensity changes with Rayleigh-Gans scatterers. His
theory was for rods or coils in relatively low-intensity fields.
Since then, disks have been considered under low-intensity fields
[19,20] and rods under high-intensity fields [21]. With all
theories it is assumed that the molecules are isolated as in an
infinitely dilute solution, that they are optically isotropic,
unchanged, and monodisperse. Further, it is presupposed that the
usual experimental correction factors [3], including the volume
and Thomson terms $\{\sin \theta/(1 + \cos^2 \theta)\}$ have been incorporated
in the data. Finally, as the intensity changes ΔI are directly
related to changes in $P(\theta)$, the equation

$$\frac{\Delta I}{I_0} = \frac{\Delta P(\theta)}{P(\theta)_0} \tag{3}$$

holds, where the subscript 0 refers to the relevant parameter in
the absence of the applied electric field.

III.   THEORY

A summary of equations is given for specific particle shapes.
The reader is referred to the original literature for details of
the rather lengthy derivations.

In all cases, it is assumed that permanent and induced
dipolar mechanisms are responsible for the molecular orientations

and that the dipole moments are coincident with the molecular geometric axes.

The basic expression for $P(\theta)$ for any molecular shape is

$$P(\theta) = \frac{1}{n^2} \sum_{i=1}^{n} \sum_{j=1}^{n} \cos[h\underline{r}_{ij} \cdot \underline{s}] = \frac{1}{n^2} \sum_{i=1}^{n} \sum_{j=1}^{n} \cos[hrs \cos \xi]$$

(4)

where h has been defined in Sec. I. The vector $\underline{s}$ is of magnitude $2 \sin(\theta/2)$ and is formed by $\underline{s} = i_\theta - i_0$ with i_θ and i_0 unit vectors in the scattered incident directions, respectively. An example is given in Fig. 3c. The vector $\underline{r}_{ij}$ joins the i and j scattering elements of which the molecule has a total of n. The angle ξ is that between the vectors $\underline{r}_{ij}$ and $\underline{s}$ and will be influenced by the molecular orientation and hence by the orientation distribution function.

A.   Continuous DC Fields

These are the easiest to generate. They correspond to the simplest theoretical interpretations as the molecules orient under their influence until they obtain static equilibrium. Under such conditions, Maxwell-Boltzmann or equilibrium statistics can be used to describe the system.

1.   Rods

With thin rods, cylinders or helices, the vector $\underline{r}_{ij}$ can be considered as being along the rod axis so that the angle ξ will always be between this axis and the $\underline{s}$ vector.

(a)   Low degree of orientation. In this case, the potential energy of the oriented molecules is much less than the thermal energy of the system. Following the method of Peterlin and Stuart [22] and Benoit [23] for birefringence, Wippler [16] considered the distribution function for a thin rod in terms of a series of Legendre polynomials. He produced the final equation

$$\frac{\Delta I}{I_0} = \frac{1}{4}(1 - 3 \cos^2 \Omega) C_R \{\beta^2 + 2\gamma\}$$

(5)

with

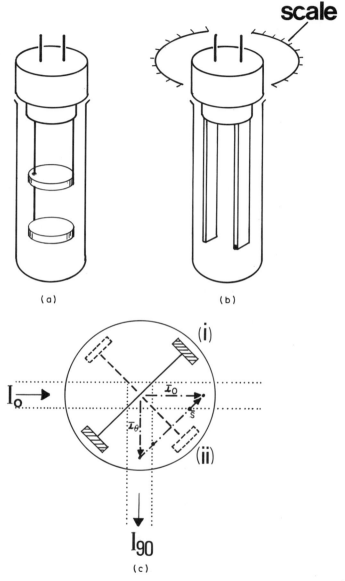

FIG. 3. Cylindrical cell electrode system. A has fixed
$\Omega = 90^{\circ}$, B has variable Ω. C shows two positions of the electrodes of cell B for rigidity measurements: (i) and (ii)
correspond to $\Omega = 0$ and 90°, respectively, when observations are
made at $\theta = 90^{\circ}$; $\bar{s}$ is the s vector for the case shown.
(Reprinted from Ref. 10, p. 255, by courtesy of the Society of
Chemical Industry.)

$$C_R = \frac{1}{3} + \frac{1}{2} \frac{1}{P_0(\theta)} \left(\frac{\sin 2x}{2x^3} - \frac{1}{x^2} \right)$$

Here, C_R is independent of the electrical properties of the molecules, $x = hL \sin(\theta/2)$ for rods of length L and Ω is the angle between the electric field and the s vector. The dipole moments (μ) and electric polarizabilities (α) are expressed in the factors

$$(\beta')^2 = \frac{\mu_a^2 - \mu_b^2}{k^2 T^2} E^2 \quad \text{and} \quad \gamma = \frac{\alpha_a - \alpha_b}{2kT} E^2$$

The subscripts $\underline{a}$, $\underline{b}$, $\underline{c}$ refer to the axes of equivalent ellipsoids. With rods, $\underline{a}$ represents the unique major axis and the $\underline{b}$ and $\underline{c}$ axes are assumed to be equal. In fact, Wippler considered only the specific case with β' proportional to μ_a (i.e., with $\mu_b = 0$). The more general expression is given here in the light of later work [24]. In the absence of orienting fields,

$$P_0(\theta) = \frac{1}{x} \int_0^{2x} \frac{\sin W}{W} dW - \left(\frac{\sin x}{x} \right)^2 \quad \text{and} \quad <S^2> = \overline{L^2}/12 \qquad (6)$$

Equation (5) has the following implications. First, ΔI depends strongly on the field direction Ω. Greatest changes are observed for $\Omega = 0$. These will be decreases about the ambient value and of twice the magnitude of the increases observed when $\Omega = 90°$. Second, ΔI can be positive or negative [24], depending on the relative magnitudes of μ_a, μ_b, α_a, and α_b in the factors $(\beta')^2$ and γ. Thus, the sign of ΔI is a useful indicator of the directions of the predominant dipole moments. Third, with direct current fields, one cannot differentiate between permanent and induced dipolar contributions. Fourth, ΔI is a quadratic function of the field strength. This provides a useful practical criterion for small orientation.

   (b) Arbitrary degrees of orientation. This situation was studied by Stoylov [25] who considered the orientational energy of the particles in terms of Maxwell-Boltzmann equilibrium functions. A general equation for molecules with both permanent and induced moments was not derived. Instead, equations were given for the two specific cases of polar, but electrically isotropic, and non-

polar, electrically anisotropic molecules. For the former,
$\beta' \neq 0$, $\gamma = 0$, and

$$\frac{\Delta I}{I_0} = \frac{1}{P_0(\theta)} \left[1 - \frac{x^2}{3} \left\{ \cos^2 \Omega - \frac{(3 \cos^2 \Omega - 1)L}{\beta} \right\} \right.$$

$$\left. + \frac{x^4}{36} \left\{ \cos^4 \Omega + \frac{2LA}{\beta'} - \frac{8B(3L - \beta')}{(\beta')^3} \right\} \right]^{-1} \qquad (7)$$

where A and B are trigonometric functions of Ω alone, i.e.,

$$A = 2 \cos^4 \Omega - 3 \sin^2 \Omega \cos^2 \Omega$$

and

$$B = \cos^4 \Omega + 3 \sin^4 \Omega/8 - 3 \sin^2 \Omega \cos^2 \Omega$$

while L is the Langevin function = $(\coth \beta' - \frac{1}{\beta'})$. For nonpolar
particles, where $\beta' = 0$, $\gamma \neq 0$,

$$\frac{\Delta I}{I_0} = \frac{1}{P_0(\theta)} \left[1 - \frac{x^2}{6} \left\{ \sin^2 \Omega + (3 \cos^2 \Omega - 1)M \right\} \right. \qquad (8)$$

$$+ \frac{x^4}{36} \left\{ \frac{3 \sin^2 \Omega}{8} + (\cos^4 \Omega - \frac{3 \sin^4 \Omega}{8})M \right.$$

$$\left. \left. - \frac{(3M + 1)B}{2\gamma} \right\} \right] - 1$$

where M is the Langevin function equal to

$$\frac{\exp \gamma}{(2\bar{\gamma} \int_0^\gamma \exp w^2 \, dW)} - \frac{1}{2\gamma}$$

These equations are valid for any angle Ω. They do not depend solely on E^2, although they reduce to Eq. (5) for small
field strengths (i.e., when β' and γ are << 1). It should be
noted that they are restricted to the condition that $x << 1$,
which implies that observations should be made at small θ if L
is comparable to λ_0. Practically, it is not feasible to overcome
this restriction by considering particles of small length L, for
then the intensity changes are too small to measure.

(c) Saturation orientation. This is when the electric field
is so large that molecular alignment is complete. Thus the
following equation can be derived from either Eq. (7) or (8) by
putting $\beta' = \gamma = \infty$, whence,

$$\frac{\Delta I}{I_0} = \frac{1}{P_0(\theta)} \left\{ \frac{\sin (x \cos \Omega)}{x \cos \Omega} \right\}^2 - 1 \tag{9}$$

As expected, the scattering depends only upon the molecular
geometry and the direction of alignment. The expression is
particularly interesting when ΔI is measured under conditions of
$\Omega = 90°$. This is the simplest experimental arrangement, whence,
for

$$\Omega = \frac{\pi}{2}, \quad \frac{\Delta I}{I_0} = \frac{1}{P_0(\theta)} - 1 \tag{10}$$

If complete orientation can be obtained or estimated, a method of
evaluating $P_0(\theta)$ is obtained [25].

In conclusion it is mentioned that Stoylov [25] has given
simplified forms of Eq. (7) and (8) for short rods. Although
these equations are more attractive in their simplicity, they represent a somewhat impracticable situation, for the smaller the
particle size the smaller are the internal interference effects
and the accompanying intensity changes.

2. Disks

Infinitely thin disks have been considered only at low
degrees of orientation. The disks are characterized by their fa-
cial diameter d, and described through the scattering parameter
$y = hd \sin (\theta/2)$, which is directly analogous to the x used in
the rod equations. Stoylov [19] produced an equation which, when
corrected for misprints, has the form of Eq. (5),
of opposite sign, i.e.,

$$\frac{\Delta I}{I_0} = (-)\frac{1}{4}(1 - 3 \cos^2 \Omega) C_D \{ (\beta')^2 + 2\gamma \}$$

but with

$$C_D = 2\left\{\frac{1}{3} - \frac{2}{P_0(\theta)} \sum_{m=0}^{\infty} \frac{(-1)^m y^{2m}}{m!(m+2)!(2m+3)}\right\}$$ (11)

which, like C_R, is independent of the electrical parameters.
For disks

$$P_0(\theta) = \frac{2}{y^2}\left[1 - \frac{\Lambda_1(2y)}{y}\right]$$ (12)

where Λ_1 is the first-order Bessel function. For small angles
of observation such that $y \ll 1$, the equations reduce to the
form

$$\frac{\Delta I}{I_0} = (-)\frac{1}{180}(1 - 3\cos^2 \Omega)\frac{y^2}{P_0(\theta)}\{(\beta')^2 + 2\gamma\}$$ (13)

which is more useful, especially when it is appreciated that
$P_0(\theta)$ and y are obtained experimentally from a Zimm plot. This
equation illustrates the following factors. First, at low E,
$\Delta I \propto E^2$ as for rods. Second, the term $(1 - 3\cos^2 \Omega)$ is again
present. Third, the intensity changes are of opposite sign to
those for rods. This is a consequence of developing the theory in
terms of the unique or symmetry axis of the particles which is the
minor geometric axis for a disk but the major axis for a rod.

3. Polymer Chains

The theoretical approach for randomly coiled molecules is
very different from that adopted for the rigid rods and disks of
the previous sections. This is due to the ability of flexible
coils to both deform and rotate in the applied field (see Chap.
4). In general, a flexible random coil is spherically symmetric
and so would show no average change in ΔI upon rotation. Adopt-
ing the model of N-jointed monomeric units or statistical segments of length a, which are assumed to follow a gaussian
distribution, equations have been developed for (a) polar, freely

jointed segments, (b) nonpolar, electrically anisotropic, freely
jointed segments, and (c) completely stiff, sterically hindered
chains consisting of polar segments. Low field strengths alone
are considered.

a. <u>Polar, flexible coils</u>. Equations were developed by
Wippler and Benoit [18,26] and Isihara et al. [27] for dipoles
confined to the bond or segment axes. Wallach and Benoit [28]
extended the treatment to off-axis moments and to chains in which
two segmental dipoles, μ_1 and μ_2 were involved, each dipole
corresponding to alternate segments along the polymer backbone.
They considered all segments to have a length "a". If one segment
has a dipole moment μ_1 directed at an angle β_1 to the segmental direction, its neighbors have dipole moments μ_2 directed off
axis by an angle β_2. Hence, the polymer has an alternate
arrangement of segments of the two polar monomer constituents.
This model is obviously relevant to the study of copolymers in
solution. It can be condensed into the simpler situation considered by previous workers [18,26,27] simply by putting $\mu_1 = \mu_2$
and $\beta_1 = \beta_2 = 0$. For the comprehensive model, equations have
been developed only for specific values of Ω.

With $\Omega = 0$,

$$\frac{\Delta I}{I_0} = \frac{N^2}{3P_0(\theta)} \left\{ \frac{ha \ sin \ ^\theta/2}{3} \right\}^2 \left(\frac{E}{kT}\right)^2 f \tag{14}$$

and for $\Omega = 90^\circ$,

$$\frac{\Delta I}{I_0} = \frac{2N}{15P_0(\theta)} \left\{ \frac{ha \ sin \ ^\theta/2}{3} \right\}^2 \left(\frac{E}{kT}\right)^2 F \tag{15}$$

where $\overline{a^2} = 6\overline{s^2}/N = \overline{\Gamma^2}/N$ for a gaussian chain with $\overline{\Gamma}$ the
average end to end separation. Also,

$$P_0(\theta) = \frac{2}{z_2} \{exp \ (-z) - (1 - z)\} \tag{16}$$

with $z = 1.5 \, h^2 \sin^2 (^\theta/2) Na^2$

$$f = \frac{\{\mu_1 \overline{\cos \phi_1} + \mu_2 \overline{\cos \phi_2}\}^2}{4} \qquad (17)$$

and

$$F = \frac{\{\mu_1^2 \overline{(3 \cos^2 \phi_1 - 1)} + \mu_2^2 \overline{(3 \cos^2 \phi_2 - 1)}\}}{4} \qquad (18)$$

Averaging over all possible segment orientations is indicated by the bar over the relevant functions of ϕ_1 and ϕ_2.

From these equations, we see that scattered intensity changes are only measurable for $\Omega = 0^o$ as ΔI is a function of N only rather than N^2 for $\Omega = 90^o$. This is important as it affords a quick method of identifying polar, flexible coils. Also, Eqs. (14) and (15) indicate that the dependence of ΔI on Ω is not governed by the factor $(1 - 3 \cos^2 \Omega)$ as in the case of rigid molecules [see Eqs. (5) and (13)]. The ratio

$$R = \frac{\Delta I_{\Omega=0}}{\Delta I_{\Omega=90}}$$

thus affords a measure of molecular rigidity [26].

Finally, the factors f and F indicate the molecular model. If the monomer units are identical and their dipoles are unidirectional along the polymer contour length, $\phi_1 = \phi_2 = 0^o$ and $\mu_1 = \mu_2$, whence $f = F = \mu^2$, and $R = - 5N/2$. If the monomers are identical, but alternate moments reverse their polarity, as in a head-to-head configuration, then $\phi_1 = \phi_2 = 0^o$ and $\mu_2 = -\mu_1$, whence $f = 0$ and no scattered intensity change will be observed. This is due to a cancellation of the dipole moments. Finally, for a homopolymer in which the segmental dipole moments are identical but directed off axis, $\mu_1 = \mu_2 = \mu$ and $\phi_1 = \phi_2 = \phi$, whence

$$f = (\mu \; \overline{\cos \phi})^2$$

In all these examples, f and hence the dipole moments may be found since the parameters $P_0(\theta)$, N and a may be found from a Zimm plot obtained in the absence of the electric fields.

(b) <u>Nonpolar, electrically anisotropic, flexible coils</u>. This case was considered by Wippler [17] who deduced the expression

$$\frac{\Delta I}{I_0} = \frac{4N}{15P_0(\theta)} (1 - 3 \cos^2 \Omega) \left\{ \frac{ha \sin \theta/2}{3} \right\}^2 \gamma \tag{19}$$

where γ refers to the electrical anisotropy of each segment. The equation shows that induced dipole moments alone will not give large intensity changes as ΔI is a function of N to the first power only. It is thus impractical to attempt measurements of ΔI, in order to evaluate γ for such molecules.

(c) <u>Polar, stiff coils</u>. The final model that Wippler considered was that of a coil whose spatial distribution of bonds or statistical elements is gaussian. Unlike the flexible chain, which continuosuly "snakes" within its molecular volume, a completely stiff or "frozen" structure was envisaged. Such a molecule cannot be deformed in an electric field. It will orientate, however, as the resultant overall dipole moment, which will be a fixed vector quantity within the molecular geometry, orientates into the electric field. The equation

$$\frac{\Delta I}{I_0} = \frac{N^2}{P_0(\theta)} (1 - 3 \cos^2 \Omega) \left\{ \frac{ha^2 \sin^2 (\theta/2)}{9} \right\}^2 (\beta')^2 \tag{20}$$

was developed where β' refers to the segmental dipole moment term ($\mu E/kT$). ΔI is large here because of its dependence on N^2 and a^4. The characteristic term $(1 - 3 \cos^2 \Omega)$ is present, as expected for a rigid molecule. In this case, as with rods and discs, the parameter R is equal to (-2).

B. Continuous Alternating Fields

Theoretical considerations have only been given to low degrees of orientation. Under this condition, all molecular models indicate a quadratic dependence on the field strength, so that ΔI is similar to the electric birefringence [22] in exhibiting a component which fluctuates at twice the angular frequency of the applied field. This "alternating" contribution is in addition to the time-independent, or "steady," contribution which is generated by the equivalent dc field strength of the applied alternating electric field.

1. Rods

Only in the case of rodlike particles has a rigorous equation been derived [29]. It may be expressed as

$$\frac{\Delta I}{I_0} = \frac{1}{4}(1 - 3 \cos^2 \Omega) C_R \{ S + A^{1/2} \sin (2\omega t + \psi) \}$$

where S is the steady component coefficient

$$S = \frac{(\beta')^2}{[1 + \omega^2/4\theta^2)]} + 2\gamma \tag{21}$$

A is the square of the alternating component coefficient, i.e.,

$$A = \frac{(\beta')^4 + 4(\beta')^2\gamma}{[1 + (\omega^2/4\theta^2)][1 + (\omega^2/9\theta^2)]} + \frac{4\gamma^2}{1 + (\omega^2/9\theta^2)} \tag{22}$$

and ψ is a phase angle. There was an error in the original publication of this angle [29]. It should have read

$$\tan \psi = \frac{2\theta}{\omega} \left\{ \frac{(\beta')^2 + 6\gamma[1 + (\omega^2/4\theta^2)] + 2(\beta')^2[1 - (\omega^2/4\theta^2]}{5(\beta')^2 + 4\gamma[1 + (\omega^2/4\theta^2)]} \right\} \tag{23}$$

In these equations β' and γ are identical to those used previously as long as the root mean square electric field strength is used; θ is the rotary diffusion constant. For a dc field, Eq. (21) collapses to Eq. (5) as expected, and A becomes zero.

The steady component must be measured independently. From Eq. (21) one sees that a frequency dispersion is predicted from which the dipolar and induced polar contributions can be isolated. Under low-frequency conditions, both $(\beta')^2$ and γ contribute

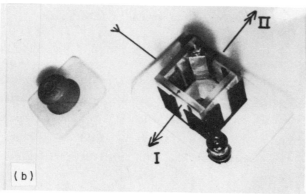

FIG. 4. Free standing cells. Cell A has a cooling jacket and horizontal electrodes, hence $\Omega = 90^\circ$. Cell B also has fixed electrodes, but $\Omega = 90^\circ$ and 0° when viewed from positions I and II, respectively. (From Ref. 35.)

to the steady component whilst at high frequencies, such that $\omega > \omega_c$ (the critical frequency), only 2γ contributes to S and ΔI. As with other electro-optical properties, Θ can be found from the steady component by considering the critical frequency, or by plotting the data as a suitable function of ω^2 and analyzing the slope [Eq. (21)]. It is to be noted that with rods (and other rigid particles), nonpolar, but electrically anisotropic particles still demonstrate an appreciable scattered intensity change. This is in contrast to the flexible molecules. Rigid, electrically anisotropic particles can be recognized by a measurable ΔI which is frequency independent. This implies that Θ cannot be obtained from the steady component in such cases. However, measurements of the alternating component are able to reveal Θ even when $\beta' = 0$. This can be appreciated from Eq. (22). Hence, although the theory is more complex and the procedure for differentiating between the contributions from permanent and induced moments is more complicated, alternating component data have a distinct advantage when dealing with nonpolar particles.

In principle, analysis of the phase angle ψ could provide an alternative means of elucidating the data. This has not yet been reported with light-scattering measurements, although measurements have been made in the author's laboratory.

2. Disks

The terms in the brace of Eq. (21) reflect the frequency and time dependence of ΔI. These terms are remarkably similar to the equivalent factors in the birefringence equations of Peterlin and Stuart [22] for continuous ac fields on rigid ellipsoids. This analogy enabled Jennings et al. [20] to predict similar equations for disks. Hence, for $\beta' \neq 0$, $\gamma = 0$,

$$
\frac{\Delta I}{I_0} = (-) \frac{1}{4}(1 - 3 \cos^2 \Omega) C_D (\beta')^2 \left\{ \frac{1}{[1 + \omega^2/4\Theta^2]} \right.
$$
$$
\left. + \frac{\cos (2\omega t - \psi^1)}{[1 + (\omega^2/4\Theta^2)]^{1/2} [1 + (\omega^2/9\Theta^2)]^{1/2}} \right\}
\tag{24}
$$

while for $\beta' = 0$, $\gamma \neq 0$,

$$
\frac{\Delta I}{I_0} = (-) \frac{1}{2}(1 - 3 \cos^2 \Omega) C_D \gamma \left\{ 1 + \frac{\cos (2\omega t - \psi^{11})}{[1 + (\omega^2/9\Theta^2)]^{1/2}} \right\}
\tag{25}
$$

with ψ^1 and ψ^{11} phase angles. Hence β', γ, and Θ can be obtained from steady and alternating component measurements as in the case of rod molecules. Equations (24) and (25) can be superposed for molecules which are both polar and electrically anisotropic, as long as the addition is made vectorially.

3. Polymer Chains

Flexible molecules have not yet been considered in alternating fields. It is clear, however, that measurements of ΔI as a function of frequency will display a relaxation dispersion, and that the relevant critical frequency will be related to a molecular relaxation time or diffusion constant. Dipole moments can only be realized by extrapolating to, or measuring at, d.c. conditions.

C. Pulsed Fields

By analogy with the development of the Kerr effect as outlined in Chaps. 1 and 3, interest has developed in the study of the transient changes in the scattered intensity when pulsed fields are applied. The method has the potential of providing all the information gained from continuous fields with reduced difficulties from the associated conductivity and electrophoretic effects.

In principle, the scattered intensity will be similar to the birefringence. ΔI will follow the three regions of a rise as the molecules orientate when the field is initiated, a steady plateau as the molecules achieve steady-state orientation in the field, and a decay region once the field has terminated. Theory has not been rigorously developed for these scattering changes; it is tempting to appropriate the equations developed for birefringence. Two cautionary remarks must be made over such action. First, as was pointed out by Stoylov and Sokerov [30], the birefringence equations are independent of particle size, while the scattering equations must still be confined to the limitations of the Rayleigh-Gans conditions [p. 278]. Second, from the foregoing theory for continuous dc fields, one can see that the scattering and birefringence equations are not always equivalent, particularly at arbitrary and high field strengths.

The following observations are made:

(i) The decay is a first-order rate process and so must follow the equation

$$\frac{\Delta I}{\Delta I^0} = \exp(-6\theta t) \tag{26}$$

where ΔI^0 is the value of ΔI at the time $t = 0$ as the pulse terminates. This equation, first assumed by Wippler [17], is valid only within the confines of the Rayleigh-Gans theory, and for x or $y < 1$.

(ii) The equation for the steady-state or plateau region of ΔI will be exactly that for an equivalent continuous field, provided that the conditions for the continuous field equations are adhered to. This is especially relevant to the equations for the E^2 region where low orientation is encountered. It is also necessary to ensure that the pulse length is such as to allow a proper plateau to be established.

(iii) In principle, various types of pulse can be used, including dc, sinusoidal bursts, reversing dc pulses, and exponentially decaying pulses to mention but a few. The most useful are dc and sinusoidal pulses [31]. These afford a method of differentiating between permanent and induced polar contributions to ΔI [cf. Eqs. (5) and (21)]. It should not be necessary to obtain a dispersion curve, as long as the frequency of the sinusoidal pulse is such that $\omega \gg \omega_c$ [31]. Knowledge of ω_c is not required to evaluate θ as the decay gives this directly [see (i) above].

(iv) The establishment transient after the pulse commences will almost certainly not follow the birefringence equations.

IV. APPARATUS

Electric field scattering experiments are especially useful if made in conjunction with conventional measurements with no applied field. The combination of the electrical data and rotary diffusion constant with the molecular weight and radius of gyration lead to a wealth of information on the scattering particles. It is thus of value to be able to measure I and ΔI at various θ and obtain Zimm plot data simultaneously.

Although instruments can be custom built, the apparatus will be described in terms of modified commercial photometers. This is because many laboratories have such instruments available and because the conversion is relatively straightforward. The basic optical arrangement for conventional photometers is detailed in a number of reviews [4,32,33]. Briefly, it consists of a lamp and collimating optics which direct a light beam through a wavelength-selecting filter, polarizer if required, a cell which contains the suspension under observation and through which the parallel light beam passes, and finally into some form of light trap to ultimately absorb the transmitted light. A receiving optical unit collects the light scattered at an angle θ and brings it to the photocathode of a photomultiplier, whose output is amplified and displayed on a suitable meter. Occasionally, a second photomultiplier is incorporated to monitor and compensate the scattered intensity for fluctuations in the incident light beam.

The generation of suitable fields is described elsewhere [10, 31,34]. One should bear in mind the field strength required, the recording of ΔI at small θ when limited by the theory to x, y, or $z \ll 1$ and the ability to vary the field direction and Ω.

A. Using Continuous Fields

With sinusoidal fields, the steady and alternating components must be detected independently. Conventional photometers often have a dc amplifier and output meter in order to eliminate noise and fluctuations from the signal. This arrangement will also lose the double frequency fluctuation. A basic instrument is thus capable of measuring the steady component with little modification. Two modifications are, however, necessary. First, commercial light sources are often run from ac supplies. It is desirable to replace any such lamp by one operated on a dc supply. Second, an electrode system must be designed for the cell. In Figs. 3 and 4 two types are shown. The first figure shows cylindrical cells, fitted with alternative types of electrodes. Figure 3A corresponds to $\Omega = 90^{\circ}$ with θ variable. Figure 3B is suitable for variable θ and Ω, but is a somewhat restrictive arrangement as the electrodes must not be allowed to enter the incident or observed light paths. Figure 3C shows cell B in plan and indicates how, with θ fixed at 90°, the electrodes can be put in positions (i) and (ii) to obtain the conditions $\Omega = 0$ and 90°, respectively. From these two measurements the rigidity of the suspended particles can be estimated through the ratio R (Sec. IIIA3). The Sofica photometer uses this type of cylindrical cell, immersed in a surrounding vat of benzene. Other photometers use freestanding cells in air. Early cells of this type, as used by the author, are shown in Fig. 4. With cell B, the electrodes are fixed, but the two conditions of $\Omega = 0$ and 90° are obtained by viewing at $\theta = 270^{\circ}$ and 90°, respectively. All cell electrodes must be anodized or painted with black-pigmented coatings on all faces except those which are diametrically across the incident light beam. This reduces spurious reflections.

The apparatus for steady component measurements is schematically presented in Fig. 5 with the switch (S) set at position a. The procedure is to record I_E and I_0 for a required set of conditions of Ω, θ and E, and obtain ΔI by subtraction [10].

By setting the switch to b in Fig. 5, one can measure the alternating component. The object of the design is to reject the dc level of the photomultiplier response together with all noisy fluctuating components except that of frequency (2ω). A convenient method of achieving this is to use a "phase-sensitive" detector. The author and his coworkers [24,35-38] have been the only group to date to measure this component. At the time of writing, detectors are available commercially which are capable of recording sinusoidal signals some 100 dB below the level of the accompanying noise in the signal. Furthermore, as the alternating component is absent when E = 0, it is not obtained as a difference reading. Hence, this component is often detected with greater ease and precision than the accompanying steady-component change. Because of this, it is suggested that the steady component could be better detected by modulating or chopping the

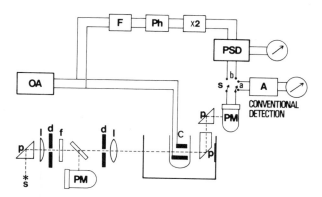

FIG. 5. Diagrammatic representation of the author's apparatus for continuous fields. Optical components are designated: s, source; p, prims; ℓ, lens; d, diaphragm; f, filter; c, cell. Electronic components are shown as: PM, photomultipliers; A, dc amplifier; OA, oscillator amplifier; F, filter; Ph, phase adjuster; X2, frequency doubler; PSD, phase-sensitive detector; S, switch. Steady and alternating components of ΔI are measured when ac fields are used and S is set at a and b, respectively.

scattered beam at some discrete frequency and using the phase-
sensitive detector to measure the modulated component intensity.

B. Using Pulsed Fields

The apparatus is presented schematically in Fig. 6. The
transient optical response is recorded by the photomultiplier,
amplified, displayed on an oscilloscope and simultaneously photo-
graphed [31,39]. It is important to note that, if the transient
is to be analyzed with reasonable precision, it must be as free
from noise as possible. The method has no inherent design factor
which reduces noise. Hence, it is essential to observe the
following. First, the light source must be run from a stable dc
supply. This will probably mean changing the lamp and supply of
commercial photometers. Second, the photomultiplier must be as
free from dark noise as possible and its dynode chain loading must
be considered with care.

It is convenient to trigger synchronously the oscilloscope
with the firing of the pulse generator, leaving the camera shutter

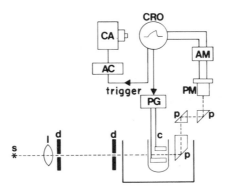

FIG. 6. Diagrammatic representation of the author's apparatus for pulsed electric fields. Optical components are designated
as in Fig. 5. In addition, the following apply: PG, pulse
generator; PM, photomultiplier (low noise); AM, amplifier (not
dc); CRO, oscilloscope; AC, actuator; CA, camera.

open. Using polaroid film, the transient is obtained extremely quickly, although the data analysis is a longer procedure than in the case of continuous fields. Also, the accuracy is not very high.

V. ILLUSTRATIVE RESULTS

Examples of electric field light scattering experiments are given below, not with the prime intention of characterizing the specific material, as was often the object of the original experimenters, but rather to illustrate procedures, advantages, and typical results of the method. Discussion of particular studies will not convey all of the information deduced by the original workers. Readers who are interested in the specific colloidal or polymer systems should consult the original articles referenced below.

A. Continuous Fields

1. Rods, Cylinders, and Helices

This class has probably been the most studied by the method. Generally thin helices are treated as rods. A comprehensive list of molecular types has been investigated, which includes viruses, clay minerals, helical polymers, and nucleic acids. Such a wide range of materials has enabled a sound appraisal to be made of the theories, particularly at low degrees of orientation.

An idea of the magnitude of the changes which may be expected at low orientation is given in Fig. 7 where data are presented for tobacco mosaic virus (TMV). Steady-component intensity changes of the order of 8% were reported while remaining in the region where ΔI depended on E^2. For this material, one sees no noticeable concentration dependence of ΔI, an observation which is in agreement with other reported data [37,40,41]. With

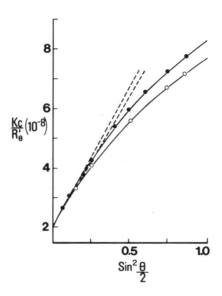

FIG. 7. Change in the reciprocal angular envelope for TMV
solutions (extrapolated to c = 0). Broken lines indicate initial
slopes, closed and open circles represent E = 0 and E = 190
V cm^{-1}. (From Ref. 35.)

poly-benzyl-L-glutamate (PBLG) in the helix promoting solvent
ethylene dichloride, fields of 1.8 kV cm^{-1} give a 1% intensity
change for polymer of 2.5×10^5 MW. This is well within the linear region where ΔI is proportional to E^2.

There are two procedures for evaluating $\{(\beta^1)^2 + 2\gamma\}/E^2$.
The first is to plot a graph of $\Delta I/I$ as a function of E^2 and
to obtain the slope of the linear region. This is then used in
Eq. (5). In order to evaluate C_R, it is usual to simultaneously
obtain a Zimm plot from which $P_0(\theta)$ and x are deduced.

An alternative procedure, which does not require the Zimm
plot to be drawn, is to expand both $P_0(\theta)$ and x in Eq. (5) as
functions of $\sin^2(\theta/2)$. In this case, one can show that for
rods, and small angles of observation, the steady component can be
analyzed in the expression

$$\frac{m_E - m_0}{m_0} = \frac{[(\beta')^2 + 2\gamma]}{15} (1 - 3 \cos^2 \Omega) \tag{27}$$

Where m represents the initial slope of the zero concentration Zimm plot curves. Experimental data are given in Fig. 7 by way of example.

Differentiation between the permanent and the induced dipolar contributions is made through frequency dependence measurements. In Fig. 8, data are presented from Wallach and Benoit's [42] experiments on three PBLG samples. From these graphs, one can deduce that the material has a permanent dipole moment whose contribution has relaxed out by a frequency of 10 kHz. Furthermore, as

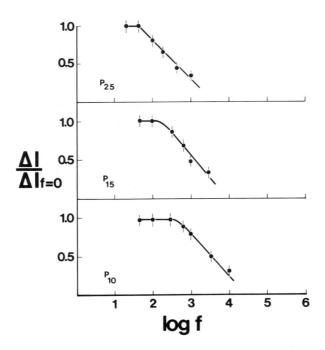

FIG. 8. Steady component frequency dispersion. Data for samples P_{25}, P_{15}, and P_{10} of polybenzylglutamate with $M = 3.2$, 2.2, and 1.7×10^6, in ethylene dichloride; $\Omega = 0°$, $\theta = 90°$, $E = 500$ V cm^{-1}. (Reprinted from Ref. 42, p. 46, by courtesy of Wiley-Interscience Inc.)

ΔI decreases with increasing frequency, β' is positive, but β' can be expressed as $(\beta')^2 = (\mu_a^2 - \mu_b^2)(E^2/k^2T^2)$. This shows that the dipole moment is predominantly along the helix backbone (axis a). Also, the larger the molecule, the smaller the critical frequency at which the relaxation occurs. Thus the figure indicates that sample P_{25} and P_{10} have the greatest and smallest helix lengths (or molecular weights), respectively. This was the case. Finally, the contribution from the induced moment was less than 10% of the total, indicating that this moment was relatively insignificant. From this type of frequency-dependence curve, the proportions of $(\beta')^2$ and 2γ are determined.

An interesting effect arises when the particles have predominant dipole moments across, rather than along, the rod or helix major axis. Then, $\mu_b^2 > \mu_a^2$ and $(\beta')^2$ becomes negative and of opposite sign to 2γ. In Fig. 9, data are presented for a suspension of the clay mineral attapulgite. In suspension, the particles are cylindrical rods or needles [43]. The induced contribution is positive and independent of frequency. It thus

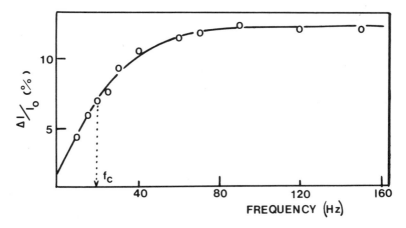

FIG. 9. Frequency dispersion for particles with dipoles along a minor geometric axis. Data for attapulgite, $\theta = \Omega = 90°$, $c = 9.2 \times 10^{-6}$ g ml^{-1}, and $E = 65$ V cm^{-1}; f_c is the critical frequency, whence $\theta = 60$ sec^{-1}. (Reprinted from Ref. 24, p. 1756, by courtesy of the Institute of Physics.)

forms the high-frequency asymptote for ΔI. At lower frequencies, the permanent moment contributes to ΔI, but in a negative sense and of greater importance the lower the frequency. Thus, one sees a dispersion effect which apparently <u>increases</u> with increasing frequency. This immediately indicates the transverse direction of the predominant dipole moment. The relative magnitudes of the permanent and induced contributions are obtained from the high frequency (where ΔI is a function of 2γ) and the zero frequency [where ΔI is a function of $2\gamma + (\beta')^2$] limits of the curve. It should be noted that, unlike Fig. 8, the induced contribution is greater than that from the permanent moment.

Finally, from such dispersion curves, one can locate that frequency ω_c (or $f_c = \omega_c/2\pi$), for which the permanent dipolar contribution is half-completed or relaxed. By considering Eq. (21),

$$f_c = \theta/\pi$$

from which the rotary diffusion constant is obtained. In general, the corresponding rod length from the Broersma equation [44] is in reasonable agreement with the Zimm plot value.

To the summer of 1971 some dozen or so rodlike materials were studied in suspension by analysis of the steady component of ΔI. Measurements of the alternating component have been less common. With attapulgite sols, excellent agreement was obtained for both the steady and alternating components in that $\theta = 60 \text{ sec}^{-1}$ fitted both sets of data (Fig. 9 and Ref. 24). Similar concord was evident from experiments on hectorite sols where the alternating and steady components indicated values of $\theta = 170 \text{ sec}^{-1}$ and $165 \pm 35 \text{ sec}^{-1}$, respectively [38]. These two sets of experiments may be taken as verification of Eq. 21.

Although experimentally easier to realize, data for arbitrary field strengths are difficult to analyze at the present time. It is especially difficult to develop criteria to differentiate between polar and anisotropic effects as the theory exists only for

each of these electrical properties when in the absence of the
other. In practice, it is most common to find that both permanent
and induced moments are present together. In spite of these
difficulties, Stoylov [41] has attempted to make an appraisal for
the theory from measurements on aqueous suspensions of TMV and the
clay mineral palygorskite. Graphs were obtained of ΔI as a
function of E^2 in which the linear region was definitely exceeded. Attempts were made to fit the experimental data to various theoretical curves for which (a) it was assumed that the
particles had no dipole moment, and (b) particle sizes were obtained from electron microscopic data. Only when the particle
polydispersity was considered [45] was the comparison of experimental and theoretical data good. However, to the writer's knowledge, such theoretical curves have not been used to determine
colloidal or molecular parameters.

Of more interest is the attempt by Stoylov et al. [46] to
measure $P_0(\theta)$, and hence the particle length, by increasing the
field strength to completely saturate the molecular orientation.
Under such conditions Eq. (10) should hold. Data are presented
in Fig. 10 for bentonite clay disks and E. coli bacterium filaments. The bentonite curve is the nearest to saturation yet reported for any material. Attempts on TMV and palygorskite sols
[47] did not produce saturation within the practically attainable
field strengths. This underlines the major difficulty of the method. It is to be hoped that (a) sufficiently high fields may
soon be developed to enable true saturation conditions to be
reached, and (b) the material will not suffer damage under such
conditions. The saturation method has value in that, whereas
changes of 10 or 20% may be observed for $\Delta I/I$ at low degrees of
orientation, they may rise to 100% or more with orientation saturation.

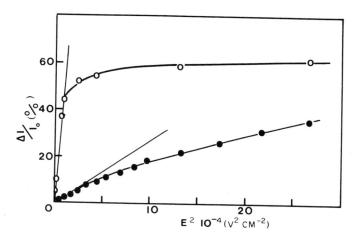

FIG. 10. Approach to saturation conditions. For bentonite disks (open circles) saturation is almost obtained. The curve with closed circles is for <u>E</u>. <u>coli</u> bacterium filaments. (Reprinted from Ref. 46, p. 1166, by courtesy of the authors.)

2. Disks

A cursory thought might be that this limiting shape of an ellipsoid deserves little special consideration. This is not so. Apart from the relative ease of deriving theories for disks, owing to the reduction of one dimension to negligible proportions, it has been found that a number of clay minerals exist as very thin platelets in suspension. They may often be approximated to disks.

Experimentally, the procedure for disks is similar to that for rods. At low degrees of orientation the equations have all the characteristics of those for rods, but differ from them in their numerical coefficients. Thus, one can analyze data by plotting ΔI as a function of E^2, of $\sin^2(\theta/2)$ or of the frequency f. By expanding $P_0(\theta)$ in terms of $\sin^2(\theta/2)$ and reducing the resulting series of Eq. (11) to $y \ll 1$ as for small angles of observation, the equivalent expression to Eq. (27) is [20]

$$\frac{m_E - m_0}{m_0} = \frac{(1 - 3 \cos^2 \Omega)}{30} \left\{ \beta^2 + 2\gamma \right\} \tag{28}$$

The relative contributions from β^2 and 2γ must be determined from a dispersion curve, which is again similar to that for rod molecules [cf. Eqs. (24) and (21)]. Typical curves are reproduced in Fig. 11 where data are shown on fresh and aggregated sols of montmorillonite. Auxiliary scattering studies proved the fresh dispersion to be represented by thin, approximately rectangular platelets of 600 nm × 125 nm. The aged sample consisted of more spherical particles, which were identified as gel structures

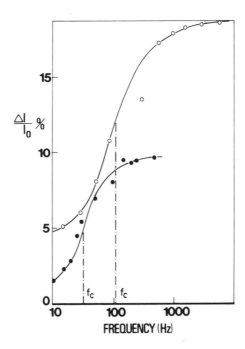

FIG. 11. Frequency dependence for montmorillonite sols. Open and full circles represent individual disklike particles and the more spherical-like agglomerated particles, respectively. The agglomerates formed on standing. The different critical frequencies indicate not only the presence of aggregation with standing, but also enable rotary diffusion constants to be determined. (Reprinted from Ref. 48, p. 450, by courtesy of Academic Press.)

of 9 or 10 individual montmorillonite platelets. This study is
briefly elaborated later (Sec. VC). Because the transverse disc
axes (i.e., directions b = c) are the major dimensions of a
disk, a permanent dipole moment along the unique axis, that is
through the disk face, corresponds to a negative β. Hence, with
disks, dispersion curves which increase with frequency are the
more common. The magnitudes of the relative intensity changes are
also of the order of 10 to 20% for low degrees of orientation.

3. Polymer Coils

To date, little experimental work has been undertaken on this
class of material. This is because the relative changes are small
at low orientation for which the only theory exists. Measurable
changes in $P(\theta)$ require that molecular dimensions be of the
order of λ_0. For visible light, this corresponds to polymer
coils which exceed 300 nm in size and are thus of relatively high
molecular weight. However, Wippler [18] has studied solutions of
poly-DL-phenylalanine in benzene, which he concluded had a permanent moment directed along the chain contour, while Wallach and
Benoit [28] studied solutions of nitrocellulose of various molecular weight in ethyl acetate.

An interesting and useful facet of the nitrocellulose study
was that measurements were undertaken at various field strengths,
angles of observation and frequencies. From both the field
strength and the angular dependence, Eq. (14) was used to obtain
independent values for the segmental dipole moments. Although
going beyond the scope of the theory of Sec. IIIA3, measurements
at various frequencies demonstrated the presence of a discrete
relaxation effect which was remarkably similar to the Debye dipolar orientation hypothesis implicit in the equations presented
in this review. Wallach and Benoit, analyzed the critical frequency, obtained a rotary diffusion constant, and then used the
Zimm [49] equation to show that the value of θ was within an
order of magnitude of that calculated from the molecular

dimensions. Wallach and Benoit obtained a rigidity parameter of
-0.5. As this was not equal to -2, they assumed that the system
was freely flexible, and so interpreted the amplitude of their
intensity changes according to Eq. (14). They reported apparently
good values of about 3 D for the segmental dipole moments. Recently, Jennings and Schweitzer [50] have verified Wallach and
Benoit's measurements, but find the original dipole moment calculations to be in error. The assumption that the only alternative
to a rigid molecule is a completely flexible one is also wrong,
for a flexible coil would not give a rigidity parameter of -0.5
but of $-\infty$. It appears that if the data are reinterpreted in terms
of a very stiff, highly extended molecule, one can obtain consis-
tency with the theory and a correctly calculated monomer dipole
moment which is close to the 3 D predicted from the chemical
structure.

A final illustrative study is that outlined by Jennings [51]
in which the helical poly-L-proline form II was studied in order
to determine the flexibility of the helix. Measurements were made
for $\Omega = 0^{\circ}$ and 90° with the result that measurable changes
were only detectable for $\Omega = 0^{\circ}$. This suggested that Eq. (14)
was applicable, for any class of particle other than polar, flexible chains would have shown negligible or measurable values of
ΔI for both $\Omega = 0^{\circ}$ and 90°. The helical backbone of the
polyproline suggested a unidirectional dipole moment while the
intensity changes, observable for $\Omega = 0$ alone, showed the helix
to be so flexible that it assumed an approximately random coil
conformation in solution.

B. Pulsed Fields

Pulsed fields have only relatively recently been employed in
electric field light scattering. However, the method has already
been used on a number of colloidal suspensions. To 1971, results
have been reported only on essentially rigid particles although
current work is in progress in the author's laboratory on flexible
systems.

The initial experiments were reported by Stoylov and Sokerov
[30,52] for TMV solutions. These authors were concerned with the

determination of Θ from the field-free decay transient. The latter was analyzed in terms of Eq. (26). In these very early studies, Eq. (21) could not be used to analyze the transient amplitude as the field strength required to manifest the transients was well in excess of the quadratic region of ΔI on the field strength. Furthermore, the decay analysis through Eq. (26) is strictly only valid for monodisperse solutes. The procedure of plotting a semilogarithmic plot of the normalized value of ΔI against the time, is identical to that used for other transient optical data.

In order to demonstrate the approximate similarity of the decay process, concurrent birefringence and transient scattering experiments were made on hectorite suspensions by Brown et al. [34]. By remaining in the region of low orientation, these authors showed that, with reasonably monodisperse material, both the electrical parameters and diffusion constants were consistent from both methods. The transient scattering decay can thus be evaluated to yield Θ. The birefringence and scattering averages differ, however, when polydisperse systems are used [53].

Analysis of the steady amplitude of the transient response is less straightforward. Theories for the establishment of birefringence transients (Chap. 3) show that all the available information should be contained in the response for a dc rectangular pulse. Unfortunately, these theories cannot be transposed directly into the scattering analogue. However, a method has been introduced which is based on the use of sinusoidal pulsed fields. Although the exact nature of the equations governing the steady maximum of the transient scattering change is not known, the relative amplitudes of ΔI for data at different frequencies and small field strengths will be governed by the factor S of Eq. (21). Schweitzer and Jennings [31] have thus proposed a method based on the use of two pulses, one at very high and the other at low or zero frequency, in order to realize the limiting conditions of dispersion graphs such as Figs. 8 and 9. Figure 12 shows the response of a suspension of Laponite clay under two sinusoidal pulses. At the lower frequency, the response demonstrates the

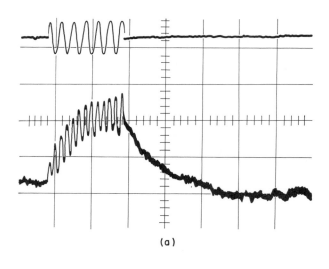

(a)

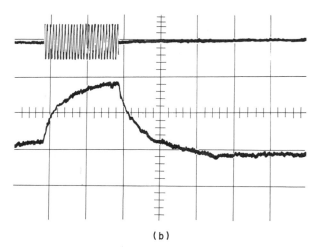

(b)

FIG. 12.   Transient responses to pulsed sinusoidal fields,
for laponite synthetic clay, $c = 7.5 \times 10^{-4}$ g ml^{-1}, $\theta = 30^\circ$,
$\Omega = 90^\circ$, $E = 80$ V cm^{-1},   and field frequencies of 60 and 250 Hz
for (A) and (B), respectively. The presence of the double frequency component and the accompanying phase change is noticeable
at lower frequencies. The response in (B) has been attenuated by
a factor of 2 relative to (A). (Reprinted from Ref. 31, p. 303,
by courtesy of the Institute of Physics.)

double-frequency, this component has fallen to negligible proportions [see Eq. (21)]. The greater magnitude of ΔI at the higher
frequency (note the amplitude scale), indicates that the Laponite
rods have a dipole moment perpendicular to their symmetry axis.
Dipole moments can be obtained by calibrating the oscilloscope
amplitude and using Eq. (21) for the relative magnitudes of $(\beta')^2$
and 2γ.

More work is desirable on the effect of particle
heterogeneity and polydispersity on electric field scattering.
For instance, it is not yet clear as to what type of polydispersity average should be associated with the dipole moments and the
diffusion constants obtained from dispersion measurements or
pulsed transients. Some recent work [53,54] has been concerned
with the analysis of transient decay using Eq. (26). If the
initial slope of the plot of the natural logarithm of $(\Delta I/\Delta I^0)$
vs. t is analyzed, this leads to a value close to $\overline{\theta}_{(Z+1)}$ for
flexible, polar random coils. Such a discrete average is not indicated with rod molecules, however. Indeed, the result obtained
depends upon both θ and λ_0. However, as the initial slope
corresponds to the most reliable experimental data in the decay,
the method shows promise in the study of chain polymers. A further advantage of the pulsed method is that, by analogy with the
method proposed for birefringence studies [55] the applied field
amplitude and duration may be varied so as to effectively select
or reject the full contribution from all sizes of particles in a
polydisperse sample. Obviously, a field of sufficient amplitude
to orient all the particles but of insufficient duration to allow
the larger particles to orient completely, effectively weights
ΔI toward the contribution of the smaller particles. Some initial experiments [52] have demonstrated the validity of this
approach.

Finally, there is one advantage of the transient scattering
method over related birefringence measurements. When the permanent
and induced dipolar mechanisms work in quadrature, electro-optic
reversals occur [23,20,56]. Quadratic, pulsed birefringence cannot demonstrate reversal as the ambient condition is one of zero

birefringence. Hence, when a reversal effect is encountered the
birefringence falls to zero and rises again (Fig. 13). The
scattering ambient value is not zero. Thus, reversal effects,
which immediately reveal the quadrature nature of the dipole moments, are directly manifest through the scattering changes as
demonstrated in Fig. 13.

C. Merits and Uses of Electric Field Scattering

Electro-optical procedures lead to the evaluation of dipole
moments and rotary diffusion constants. The major advantage of
scattering experiments is that they provide in addition, the
molecular mass and radius of gyration. The method has been used
to study the geometrical and electrical structure of a number of
materials once the equations have been verified on model systems,
of which TMV solutions and clay suspensions have been prominent.
This useful combination of dipole moments and structural geometry
has been particularly demonstrated for polybenzylglutamate where

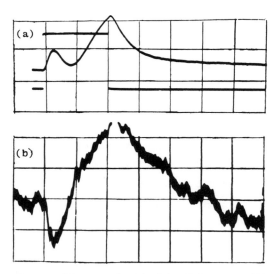

FIG. 13. Comparative quadratic birefringence (A) and
scattering (B) transients for laponite clay sols. The scattering
transient demonstrates a reversal in the optical behavior.
Concentration of 1.5×10^{-3} g ml^{-1}, E = 200 V cm^{-1}, and for (b),
$\theta = 90^{0}$, $\Omega = 0^{0}$. Time scale of 5 msec per division. (Reprinted
from Ref. 20, p. 616, by courtesy of Academic Press.)

the α-helical conformation was simultaneously verified in terms of helix length and dipole moment [42,57]. Furthermore, the proposed helical conformation of polyhexylisocyanate was confirmed [58] by demonstrating the molecular rigidity and showing that from a determination of both Θ and $(\overline{s^2})$, the only compatible structure was rodlike. A more spectacular example of the combined measurements of Θ, $(\overline{s^2})$, rigidity and dipolar direction has recently been reported [48] for montmorillonite gels. A fresh suspension of the mineral, presumably ungelled, was shown to be compatible with a suspension of approximately rectangular platelets of 600 nm × 125 nm × 1 nm, each of which had a permanent dipole moment which was directed out of the plate face. An aged sample, which the particle mass showed to consist on average of at least 9 of the original plates, behaved as a suspension of more nearly spherical particles, again with the dipolar axis perpendicular to the major geometric axis. By considering some 11 models for the arrangement of 9 or 10 plates, Schweitzer and Jennings [48] were able to show that only one model gave consistent dipolar direction and compatible values for Θ, and $(\overline{s^2})$. The resulting structure consisted of the original plates, associated with their neighbors along their long edges in a face-to-edge manner, so as to form a cardhouse type stacking of the particles. This was the first experimental verification of the face-to-edge association of clay particles, and illustrates the versatility of the scattering method.

Another advantage is that colored solutions can be studied [59]. Normal scattering theory must be modified if either the particles or the medium absorb the incident radiation. Many colloidal dispersions are strongly absorbing over wide or narrow spectral ranges. Nevertheless, as long as the particles scatter sufficient of the incident radiation, ΔI may be measured. Rotary diffusion constants are evaluated from relative and not absolute intensities, so that particle dimensions may still be determined. Typical transients are shown in Fig. 14.

Stoylov et al. [60] have shown that ΔI is sensitive to changes in 2γ and hence of the electric polarizability of the

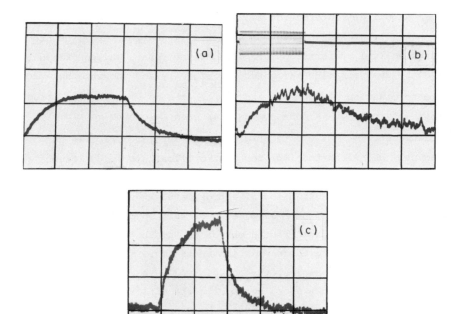

FIG. 14.   Transient scattered intensity changes from absorbing solutions.   (A) vanadium pentoxide in methanol, λ_0 = 436 nm, E = 635 V/cm; (B) benzopurpurine in methanol, white light, E = 1.13 kV/cm; (C) TMV in water, λ_0 = 310 nm,   E = 320 V/cm. Time scales of 20 msec, 15.5 msec, and 10 msec per division for (A), (B), and (C), respectively.   Pulsed sinusoidal fields of 5 kHz frequency were used.   The field duration is indicated in frame (B) for that particular case.   (Reprinted from Ref. 59, p. 166, by courtesy of I. P. C. Business Press Ltd.)

particles.   Remarkable changes in ΔI for palygorskite suspensions were observed as small quantities of the surface active substance cetyl pyridinium chloride were added to them.   This work demonstrates the value of scattering experiments for studying the surface and interaction properties of molecules and particles.

Finally, a word is due on the presence and importance of other orienting mechanisms than the dipolar orientation system used for the theory.   Recent studies by Picot et al. [61] involving birefringence and electric-field scattering measurements on suspensions of polymer single crystals and by Schweitzer and Jennings [31] on aqueous Laponite suspensions, indicate that

different values may be obtained for Θ from the critical frequency and transient decay curves (Fig. 15). This is qualitatively compatible with the suggestion that particle polarization is at
least partially due to the presence of mobile charge carriers
either within a solvent ion atmosphere or along the surface of the
particles themselves (Chap. 3, 17 and 19). It would appear that
electric field scattering experiments have shown a need for the
theories which account for molecular orientation in electric
fields by Maxwell-Wagner and other associated polarizing processes in which mobile charges are involved.

In conclusion, the scattering method gives a host of structural information for particles in suspension, and is readily
realized experimentally by adapting existing commercial scattering
photometers.

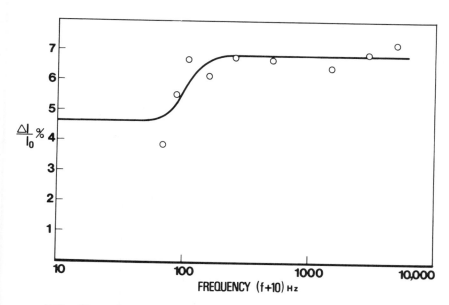

FIG. 15. Dispersion data for laponite sols, obtained from
the steady amplitude of transient responses for sinusoidal pulsed
fields. The critical frequency does not give the same rotary
diffusion constant $(\Theta = 290 \text{sec}^{-1})$ as the decay of the transients
$(\Theta = 3 \text{ to } 7 \text{ sec}^{-1})$, thereby indicating the probable presence of
mobile charge carriers in the system. (Reprinted from Ref. 31,
p. 305, by courtesy of the Institute of Physics.)

D. Limitations of the Method

There are two important limitations when compared to other
electro-optic methods. First, and by far the most serious, is the
requirement that the scattering particles be at least comparable,
and preferably larger than, the wavelength of the incident light
beam. This is very limiting with synthetic polymers. Moreover, a
molecule of a given contour length will in general give larger
intensity changes if it assumes a highly asymmetric conformation
that if it coils into a random chain. This is good for the study
of helical polymers. However, the majority of polymers are more
or less randomly coiled. At low degrees of orientation relative
intensity changes of the order of a few percent are common for
material of the order of 300,000 MW. Second, apart from being
restrictive in particle size, the method is not so sensitive as
the electric birefringence or dichroism methods. If one is simply
after the evaluation of (β'), γ, and Θ alone (without $\overline{M}$ and
$\overline{S^2}$) one would measure birefringence and avoid the problems
associated with particle size limitations.

REFERENCES

1. R. Gans, Ann. Phys., $\underline{62}$, 331 (1919).

2. J. Cabannes, La Diffusion Moleculaire de la Lumière, Presses
    Universitaires, Paris (1929).

3. K. A. Stacey, Light Scattering in Physical Chemistry,
    Butterworths, London (1956).

4. C. Tanford, Physical Chemistry of Macromolecules, Wiley, New
    York (1961).

5. M. Kerker, The Scattering of Light, Academic, New York (1969).

6. H. C. van de Hulst, Light Scattering by Small Particles,
    Wiley, New York (1957).

7. A. Guinier, Ann. Phys., $\underline{12}$, 161 (1939).

8. B. H. Zimm, J. Chem. Phys., 16, 1099 (1948).

9. S. Kielich, Acta Phys. Polon., A37, 447 (1970).

10. B. R. Jennings, Brit. Polymer J. 1, 252 (1969).

11. K. Yoshioka and H. Watanabe, Physical Principles and Techniques of Protein Chemistry, Part A, (S. Leach, ed.), Academic Press, New York (1969).

12. E. Bloch, Compt. Rend., 146, 970 (1908).

13. P. Zeeman and C. H. Hoogenboom, Physik Z., 19, 913 (1912).

14. R. S. Subrahmanya, K. S. G. Doss, and B. S. Rao, Proc. Indian Acad. Sci., 19, 405 (1944).

15. C. Wippler, J. Phys. Radium, 14, 65 (1953).

16. C. Wippler, J. Chim. Phys., 51, 122 (1954).

17. C. Wippler, J. Chim. Phys., 53, 316 (1956).

18. C. Wippler, J. Polymer Sci., 23, 199 (1957).

19. S. P. Stoylov, Izv. Inst. Fizkhim., 6, 79 (1967).

20. B. R. Jennings, B. L. Brown, and H. Plummer, J. Colloid Interface Sci., 32, 606 (1970).

21. A. Scheludko and S. P. Stoylov, Kolloid Z., 199, 36 (1964).

22. A. Peterlin and H. A. Stuart, Handbuch und Jahrbuch der chemischen Physik, 8, section 1B (1943).

23. H. Benoit, Ann. Phys. (Paris), 6, 561 (1951).

24. H. Plummer and B. R. Jennings, Brit. J. Appl. Phys. (J. Phys. D.), Ser 2, 1, 1753 (1968).

25. S. Stoylov, Coll. Czech. Chem. Commun., 31, 2866 (1966).

26. C. Wippler and H. Benoit, Makromol. Chem., 13, 7 (1954).

27. A. Isihara, R. Koyama, N. Yamada, and A. Nishioka, J. Polymer Sci., 17, 341 (1955).

28. M. L. Wallach and H. Benoit, J. Polymer Sci., A2, 4, 491 (1966).

29. H. Plummer and B. R. Jennings, J. Chem. Phys., 50, 1033 (1969).

30. S. P. Stoylov and S. Sokerov, J. Colloid Interface Sci., 24, 235 (1967).

31. J. Schweitzer and B. R. Jennings, J. Phys. D., 5, 297 (1972).

32. F. W. Billmeyer Jr., Textbook of Polymer Science, Wiley, New York (1962).

33. B. R. Jennings, Lab. Equip. Dig., 10, 77 (1972).

34. B. L. Brown, B. R. Jennings, and H. Plummer, Appl. Opt., 8, 2019 (1969).

35. B. R. Jennings, Ph.D. thesis, Southampton University, U.K. 1964.

36. B. R. Jennings and H. G. Jerrard, J. Chem. Phys., 42, 511 (1965).

37. B. R. Jennings and H. G. Jerrard, J. Chem. Phys., 44, 1291 (1966).

38. B. R. Jennings and H. Plummer, J. Colloid Interface Sci., 27, 377 (1968).

39. S. P. Stoylov, Adv. Colloid Interface Sci., 3, 45 (1971).

40. C. Wippler, J. Chim. Phys. 53, 328 (1956).

41. S. Stoylov, Coll. Czech. Chem. Commun., 31, 3052 (1966).

42. M. L. Wallach and H. Benoit, J. Polymer Sci., 57, 41 (1962).

43. R. E. Grimm, Clay Mineralogy, McGraw-Hill, New York (1953).

44. S. Broersma, J. Chem. Phys., 32, 1626 (1960).

45. S. P. Stoylov, J. Polymer Sci., C, 2435 (1967).

46. S. P. Stoylov, S. Sokerov, I. Petkanchin, and N. Ibroshev, Dokl. Akad. Nauk SSSR, 180, 1165 (1968).

47. S. P. Stoylov, J. Colloid Interface Sci., 22, 203 (1966).

48. J. Schweitzer and B. R. Jennings, J. Colloid Interface Sci., 37, 443 (1971).

49. B. H. Zimm, J. Chem. Phys., 24, 269 (1956).

50. B. R. Jennings and J. Schweitzer, European Polymer J., 10, 459 (1974).

51. B. R. Jennings, Br. Polymer J., 1, 70 (1969).

52. S. P. Stoylov and S. Sokerov, J. Colloid Interface Sci., 27, 542 (1968).

53. J. Schweitzer and B. R. Jennings, Biopolymers, 11, 1077 (1972).

54. J. Schweitzer and B. R. Jennings, Biopolymers, 12, 2439 (1973).

55. C. T. O'Konski, K. Yoshioka, and W. H. Orttung, J. Phys. Chem. 63, 1558 (1959).

56. M. J. Shah and C. M. Hart, IBM Res. Dev., 7, 44 (1963).

57.   B. R. Jennings and H. G. Jerrard, J. Phys. Chem., $\underline{69}$, 2817
      (1965).

58.   H. Plummer and B. R. Jennings, Eur. Polymer J., $\underline{6}$, 171
      (1970).

59.   B. R. Jennings and J. Schweitzer, Polymer, $\underline{13}$, 164 (1972).

60.   S. P. Stoylov, I. Petkanchin, and S. Sokerov, Proc. Cong.
      Surface Activity, 5th Barcelona, $\underline{2}$, 163 (1968).

61.   C. Picot, C. Hornick, G. Weill, and H. Benoit, J. Polymer
      Sci., Part C $\underline{30}$, 349 (1970).

Chapter 9

ELECTROPHORETIC LIGHT SCATTERING

W. H. Flygare[*] and Steven L. Hartford[**]

Department of Chemistry, School of Chemical Sciences
University of Illinois
Urbana, Illinois

and

B. R. Ware[†]

Department of Chemistry
Harvard University
Cambridge, Massachusetts

---

[*]Guggenheim Fellow, 1972.
[**]Current affiliation: Ames Research Center, Moffett Field,
California.
[†]Alfred P. Sloan Fellow.

I. INTRODUCTION

The frequency analysis of the laser light scattered from macromolecules in solution has proved useful for the rapid and accurate measurement of the diffusion coefficients of these macromolecules. The basic principle is that the random translational (and rotational) motion of the macromolecules in solution gives rise to the creation and decay of concentration (and orientation) fluctuations, which creates a frequency distribution of the scattered light. The width of the spectrum of the scattered light can be measured, and from it the translational (and in some cases the rotational) diffusion coefficient can be directly calculated.

The addition of an electric field to a solution of charged macromolecules will cause them to move with a constant drift velocity, which is equal to the applied field times the electrophoretic mobility of the ion. This constant velocity causes the entire spectrum of the light scattered from these macromolecules to be Doppler-shifted, and the magnitude of the Doppler shift can be used to determine the magnitude of the velocity and hence the electrophoretic mobility.

The purpose of this chapter is to describe the results obtained by a frequency analysis of the light scattered from a solution of ions under the influence of an electric field. This electro-optic effect is called electrophoretic light scattering because the technique combines electrophoresis methods with laser beat frequency spectroscopy [1].

II. THEORY

A. Light Scattering Theory

Light scattered from two different scattering centers (labeled 1 and 2) is shown in Fig. 1. As we are interested in the scattering which leads to constructive interference at a detector, we write the field of scattering center 2 relative to the position at 1 by

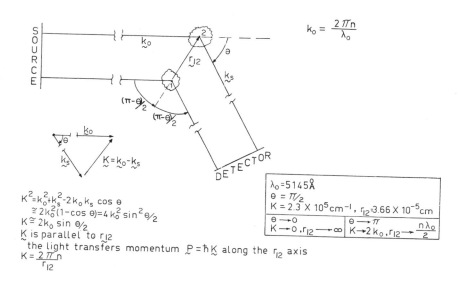

FIG. 1. Schematic diagram of the geometry of the light
scattering experiment.

$$E_2(t) = A_2 e^{i\underline{K}\cdot\underline{r}_{12}} e^{-i\omega_0 t} \qquad (1)$$

where A_2 is the time-independent geometric amplitude factor,
which is also proportional to the molecular polarizability. $\underline{K}$ is
the scattering vector described in Fig. 1 and $\underline{r}_{12}$ is the position vector between scattering centers 1 and 2. The magnitude of
K gives the Bragg law for constructive interference as shown in
Fig. 1:

$$K = 2k_0 \sin \theta/2 \qquad (2)$$

It is evident from Fig. 1 that the distance between scattering centers which is probed in the fluid depends on the scattering
angle θ. When θ is small (forward scattering) large spatial
differences in the fluid are probed. When θ is large (back
scattering) distances are probed which approach half the wavelength
of the incident radiation, $\lambda/2$. It is intuitively clear that
concentration fluctuations in a solution of macromolecules will
decay (or be created) at a slower rate for larger distances in the
fluid. Thus, low-angle scattering will lead to narrower-width

frequency distributions than high-angle scattering. We now proceed to derive equations for the frequency distribution and to determine how it depends on the scattering vector $\underline{K}$ and the translational diffusion coefficient of the macromolecule, D.

There are several methods of evaluating the spectrum of the scattered light from the distribution of scattering fields in Eq. (1) [2-4]. We shall proceed by a simplified statistical approach and consider only translational motion for a spherical macroion. According to the Wiener-Khintchine theorem [5], the autocorrelation function, $C(\tau)$, is related to the power spectrum, $I(\omega)$, of the scattered light by the Fourier transform.

$$I(\omega) = \frac{1}{2\pi} \int_{-\infty}^{\infty} C(\tau) e^{i\omega\tau} \, d\tau \tag{3}$$

where the autocorrelation function, $C^*(\tau) = C(-\tau)$ is a measure of the characteristic lifetime of concentration fluctuations in a solution of macromolecules. The autocorrelation function is related to the scattered field by

$$C(\tau) = \langle E_s^*(t) E_s(t + \tau) \rangle \tag{4}$$

where the brackets indicate a time average. The total field scattered into the detector is given by a sum over all N scattering centers, with wave vector $\underline{K}$ (relative to the common origin) and position $\underline{r}_j$ (relative to the common origin).

$$E(t) = \sum_{j=1}^{N} A_j e^{i\underline{K}\cdot\underline{r}_j} e^{-i\omega_0 t} \tag{5}$$

Substituting Eq. (5) into Eq. (4) gives the following autocorrelation function for independent spherical scatterers:

$$C(\tau) = NA^2 e^{-i\omega_0\tau} \langle e^{-i\underline{K}\cdot\underline{r}(t)} e^{i\underline{K}\cdot\underline{r}(t+\tau)} \rangle \tag{6}$$

According to the ergodic hypothesis, we can replace the time average by an ensemble average given by [4]

$$C(\tau) = NA^2 e^{-i\omega_0\tau} \langle \int W^c(\underline{r}_0, t | \underline{r}_0 + R, t+\tau) e^{-i\underline{K}\cdot\underline{r}_0} e^{i\underline{K}\cdot(\underline{r}_0 + R)} d^3 R \rangle \tag{7}$$

$W^c(r_0, t \mid r_0+R, t+\tau)$ is the conditional probability that a scattering center located at r_0 at time t will be at $r_0 + R$ at time $t + \tau$. The brackets indicate the ensemble average. If we assume the system is homogeneous, isotropic, and stationary, W^c is independent of r_0 and t giving

$$C(\tau) = NA^2 e^{-i\omega_0\tau} \int G_s(R,\tau) e^{i\underline{K}\cdot\underline{R}} d^3R \tag{8}$$

G_s is the ensemble averaged conditional probability that a particle located at the origin at time $t = 0$ will be in a unit volume element at R at time τ.

We now assume that microscopic fluctuations in concentration in a solution of macromolecules are created and decay according to the macroscopic diffusion equation which describes the ensemble-averaged behavior of the microscopic system. Thus the time evolution of G_s is given by

$$\frac{\partial G_s}{\partial t} = D\nabla^2 G_s \tag{9}$$

where D is the translational diffusion coefficient. This equation can be solved by transforming to K space, solving for the standard $G_s(t = 0)$ delta function initial condition, and then transforming back to give

$$G_s(R,t) = \frac{1}{4\pi Dt}^{3/2} e^{-R^2/4Dt} \tag{10}$$

Substituting this equation into Eq. (8) gives the correlation function for a solution of macromolecules.

$$C(K,\tau) = NA^2 e^{-i\omega_0\tau} e^{-K^2 D\tau} \tag{11}$$

The autocorrelation function can be measured directly, and for many purposes this is the usual technique of choice. The equivalent information is contained in the frequency spectrum, and for electrophoretic light scattering the frequency spectrum is often much easier to interpret. We therefore confine our discussion to frequency analysis. The power spectrum corresponding to the autocorrelation function in Eq. (11) is found by Fourier transformation to be:

$$I(\omega) = NA^2 \left[\frac{DK^2/\pi}{(\omega-\omega_0)^2 + (DK^2)^2} \right] \qquad (12)$$

Thus, the scattered light has a Lorentzian frequency distribution with a half width $\Delta\omega_{1/2} = DK^2$.

According to Eq. (12) the spectrum of the scattered light will be centered at the incident light frequency ω_0. Using an optical laser and typical macromolecules with $D \stackrel{\sim}{=} 10^{-7}$ cm^2/sec it is easy to show that the half-widths will be from 10^2 to 10^4 Hz for $\theta = \pi/2$. Optical resolution of this high degree is presently not possible with grating or interferometric spectrometers. Thus, the technique of optical mixing (beat frequency spectroscopy) [6,7] is employed to reduce the analysis to the lower frequencies. A thorough analysis of optical mixing methods has been given by Cummins and Swinney [8].

In heterodyne beat-frequency spectroscopy, the scattered light is mixed with the incident light in a nonlinear detector (photodiode or photomultiplier tube) and the difference (beat) frequencies can then be analyzed by conventional audiofrequency apparatus. The result is to reduce the Lorentz line shape in Eq. (12) to a distribution of the same form about zero frequency. Cummins et al. [9] first applied the heterodyne mixing techniques to the measurement of the diffusion coefficient of a macromolecule (polystyrene latex spheres) in solution.

An alternate technique is to mix the scattered light in the photocell with itself, self-beat-frequency spectroscopy, which also reduces the frequency distribution to an analysis of audiofrequencies. The self-beat method leads to a frequency distribution which is equivalent to the expression in Eq. (12) being convoluted with itself. The result is a Lorentzian line shape with twice the half-width, $\Delta\omega_{1/2} = 2K^2D$. The self-beat method [10,11] has led to the measurement of diffusion coefficients in a number of large macromolecules, many of which are of biological interest. There are several reviews which have listed some of the results and applications of beat frequency spectroscopy to the measurement of

the diffusion coefficient of macromolecules in solution [12,13]
which includes the measurement of the rotational diffusion constant as well as rates of exchange between internal degrees of
freedom in a molecule. More recent work has involved the analysis
of larger molecules such as the coliphages T4, T5, T7, and λ
[14], vescicular stomatitis virus and its defective T particles
[15], and motile organisms [17,18].

The use of the technique of beat-frequency spectroscopy to
evaluate diffusion coefficients is now well established and the
recent applications to the use of D to evaluate molecular
weights [14-16] and molecular dimensions [15,19,20] indicate that
this technique will be quite useful in analyzing the hydrodynamic
properties of molecules in single component mixtures. However,
experimental studies of the light scattered from mixtures of
different macromolecules in reactive fluctuations or in static
polydisperse systems have not been so successful. The reason for
the difficulty is that the light scattered from a mixture of molecules A and B in solution will yield a sum of Lorentzians with
half-widths related to their diffusion coefficients D_A and D_B
and integrated intensities related to their relative concentrations. The data must be least-squares fit to a sum of two
Lorentzians, a process which is statistically unreliable unless
D_A differs greatly from D_B.

In the next section we show that by applying an electric
field, the spectral distributions of a mixture of different macroions can be shifted to different centers in frequency if the
electrophoretic mobilities of the molecules are different, thus
providing a reliable means of analyzing multicomponent solutions
by laser light scattering.

B. The Effect of an Electric Field

With only minor changes, the equations derived in the
previous section apply as well to solutions of macromolecules

which have an electric charge. However, if an electric field is applied to the solution, the molecules will migrate with a fixed drift velocity, and the spectrum will be Doppler-shifted by an amount proportional to the velocity of the charged macroion in solution. In this section we proceed by the method of Sec. IIA to calculate the predicted spectrum of light scattered from a solution of charged macromolecules which are migrating in an electric field. This calculation was first performed by Ware and Flygare [1] at the time when they reported the first experimental observation of this effect.

The application of an electric field to a solution of charged macromolecules causes the macromolecules to migrate toward the electrode of opposite polarity. We shall arbitrarily choose the direction of this motion to be the x axis. The drift velocity, v_d, is equal to the electrophoretic mobility, u, of the macromolecule times the electric field strength

$$v_d = Eu \tag{13}$$

The time evolution of G_s [see Eq. (9)] under the conditions of both random translational motion <u>and</u> the drift velocity along the x axis is

$$\frac{\partial G_s}{\partial t} = D\nabla^2 G_s + uE \frac{\partial G_s}{\partial x} \tag{14}$$

The solution of this equation is obtained by standard Fourier transform techniques as before and the result is:

$$G_s(R,t) = \left(\frac{1}{4\pi D\tau}\right)^{3/2} \exp \frac{-[(x + uEt)^2 + y^2 + z^2]}{4Dt} \tag{15}$$

where $R^2 = x^2 + y^2 + z^2$ as in Eq. (10). Substituting Eq. (15) into Eq. (8), and integrating, gives the correlation function with the electric field along the x axis.

$$C(K,\tau) = NA^2 e^{-i\omega_0\tau} e^{-iK_x uE} e^{-K^2 D\tau} \tag{16}$$

The Fourier transform of this correlation function [see Eq. (3)] gives the power spectrum of the scattered light from the solution of macromolecules under the influence of the electric field along the x axis.

$$I(\omega) = NA^2 \left[\frac{DK^2}{(\omega - \omega_0 + K_x uE)^2 + (K^2 D)^2} \right] \tag{17}$$

This equation shows that the Lorentz line shape of the Rayleigh scattered light has been shifted in angular frequency by $\Delta\omega = K_x uE$ under the influence of the electric field. In the absence of any broadening effects other than diffusion, the half-width remains unchanged and is still related to the diffusion coefficient, $\Delta\omega_{1/2} = K^2 D$. The resolution of the electric field is defined as the ratio of the Doppler shift to the diffusion-controlled width. According to Eqs. (12) and (17)

$$R = \frac{\Delta\omega_{shift}}{\Delta\omega_{1/2\ width}} = \frac{K_x uE}{DK^2} \tag{18}$$

Because K_x is the projection of $\underline{K}$ on the axis of $\underline{v}_d$, the expression $K_x uE$ can also be written $\underline{K} \cdot \underline{v}_d$, so that the resolution can be written as

$$R = \frac{\underline{K} \cdot \underline{v}_d}{DK^2} \tag{19}$$

The vector dot product will be maximized if the scattering geometry is arranged to make $\underline{K}$ parallel to $\underline{v}_d$, i.e., along the x axis. The denominator DK^2 is minimized by performing the experiment at very low scattering angles. Therefore, the experimental conditions which maximize the resolution of electrophoretic light scattering are to perform the experiment at low scattering angles with the electric field (x axis) oriented perpendicular to the incident light. In this configuration $K_x = K \cos \theta/2$ and Eq. (19) can be rewritten to give

$$R = \frac{KuE \cos \theta/2}{DK^2} = \frac{\lambda_0 uE \cos \theta/2}{4 \ Dn \sin \theta/2} \tag{20}$$

At low angles this reduces to

$$R = \frac{uE\lambda_0}{2\pi nD\theta} \tag{21}$$

Equation (21) shows that the resolution of electrophoretic light scattering increases with increasing electric field strength and with decreasing scattering angle. Equation (21) also shows that the resolution is proportional to the electrophoretic mobility divided by the diffusion coefficient, which in the simplest electrophoretic theories is given by

$$u/D = \frac{Ze}{kT} \cdot f(K) \tag{22}$$

where Z is the number of unit charges on the macroion, e is the value of a unit charge in esu, k is Boltzmann's constant, T is the absolute temperature, and $f(K)$ is some function which accounts for the screening of the charge on the macromolecule by the counterions [21]. Electrophoretic light scattering is therefore equally applicable to molecules of all sizes, but the technique is more easily applied to very large macromolecules which can carry high charges. For these macromolecules it may be very interesting to combine the mobility and diffusion constants obtained from electrophoretic light scattering to calculate the charge on the macroions.

Because electrophoretic light scattering involves measuring a frequency which is shifted from the incident frequency, the detection must be accomplished by mixing the shifted spectrum with the incident light, which is the heterodyne method described above. The resulting signal can be analyzed by measuring either its autocorrelation function or its frequency spectrum. A summary of the equations and anticipated heterodyne experimental correlation functions and spectra both in the absence of the field and for an electric field oriented perpendicular to the incident light is presented in Fig. 2.

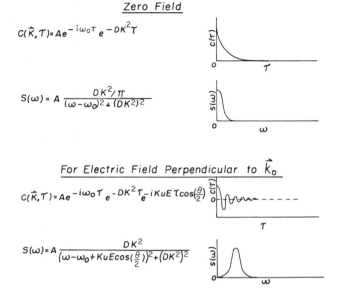

Zero Field

$$C(\vec{K}, T) = A e^{-i\omega_0 T} e^{-DK^2 T}$$

$$S(\omega) = A \frac{DK^2/\pi}{(\omega - \omega_0)^2 + (DK^2)^2}$$

For Electric Field Perpendicular to $\vec{K}_0$

$$C(\vec{K}, T) = A e^{-i\omega_0 T} e^{-DK^2 T} e^{-iKuET\cos(\frac{\theta}{2})}$$

$$S(\omega) = A \frac{DK^2}{(\omega - \omega_0 + KuE\cos(\frac{\theta}{2}))^2 + (DK^2)^2}$$

FIG. 2. Summary of the equations for the autocorrelation functions and power spectra of the light scattered from a solution of macromolecules. The diagrams are the anticipated forms of the autocorrelation functions and power spectra for a heterodyne experiment.

The theory clearly indicates that the measurement of the spectrum of light scattered from a solution of charged macromolecules which are transporting in an electric field can provide a simultaneous determination of the electrophoretic mobilities and diffusion coefficients of the macromolecules. In the next sections we will discuss the experimental observation of this phenomenon both for single- and multicomponent macromolecular systems.

C. Multicomponent Systems and Reacting Systems in the Presence of an Electric Field

For n noninteracting species in a mixture, each will contribute to the spectrum independently giving the following expression for the power spectrum:

$$I(\omega) = \sum_{a=1}^{n} N_a A_a^2 \frac{D_a K^2/\pi}{(\omega - \omega_0 \pm K_x u_a E)^2 + (D_a K^2)^2} \qquad (23)$$

From such a spectrum the diffusion coefficient of each species can be determined from the linewidth of each peak. Also, the relative abundance of each species can be determined from the integrated intensities of the individual peaks. Some multicomponent systems have been studied by electrophoretic light scattering and will be discussed in a later section.

The spectrum becomes much more complicated if the n species are interacting or reacting. In this case, n coupled equations analogous to Eq. (14) must be solved:

$$\frac{\partial G_s(\underline{R},\tau)_1}{\partial t} = D_1 \nabla^2 G_s(\underline{R},\tau)_1 + u_1 E \frac{\partial G_s(\underline{R},\tau)_1}{\partial t} + k_1 G_s(\underline{R},\tau)_1$$
$$+ k_2 G_s(\underline{R},\tau)_2 + \ldots + k_n G_s(\underline{R},\tau)_n \qquad (24)$$

$$\frac{\partial G_s(\underline{R},\tau)_2}{\partial t} = D_2 \nabla^2 G_s(\underline{R},\tau)_2 + u_2 E \frac{\partial G_s(\underline{R},\tau)_2}{\partial t} + \ldots$$

$$\vdots$$

etc.

The kinetics of interacting species in transport has been studied by a variety of techniques. This work is reviewed in a recent book by Cann [22].

There have been some calculations describing the electrophoretic light scattering spectra of simple systems of interacting species. Phillies [23,24] has considered a system of two species which can interconvert by charge transfer or isomerization:

$$A \underset{k_b}{\overset{k_a}{\rightleftarrows}} B \qquad (25)$$

The solutions to the two coupled equations analogous to Eq. (24)

depend on the electrophoretic mobilities, reaction rates, and diffusion coefficients in a complicated manner. In a review of this technique, Ware [25] has presented a computer simulation of the spectrum for one system given in Eq. (25). Figure 3 shows the simulated electrophoretic light scattering spectra for different values of the rate constants.

Another two-species system, representing a dimerization reaction, has been discussed by Berne and Gininger [26]:

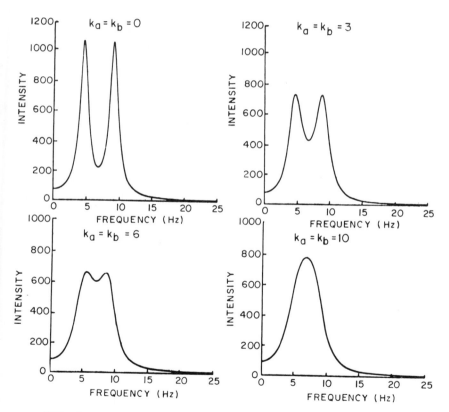

FIG. 3.   Predicted electrophoretic light scattering spectra for the system $A \rightleftarrows B$. Experimental conditions assumed were $D_A = D_B = 6 \times 10^{-7}$ cm^2 sec^{-1}, $u = 1 \times 10^{-4}$ cm^2 sec^{-1}V^{-1}, $u_A = 2 \times 10^{-4}$ cm^2 sec^{-1}V^{-1}, $E = 100$ V cm^{-1}, $\theta = 1^\circ$, $\lambda_0 = 5.145 \times 10^{-5}$ cm. (From Haas and Ware, unpublished.)

$$2A_1 \underset{k_b}{\overset{k_f}{\rightleftarrows}} A_2 \tag{26}$$

The solution to the two coupled equations for the dimerization reaction is similar to the previous reaction discussed. Figure 4 shows computer simulations of the electrophoretic light-scattering spectra for a dimerization reaction.

Although the expressions for the spectra of the reacting systems are very complicated, in both cases the physical significance is rather simple. Electrophoretic light-scattering studies on reacting systems are analogous to exchange rate studies in NMR [27]. If the reaction rates are very slow, the two peaks will be unperturbed and distinct, provided their electrophoretic mobilities are sufficiently different. For faster reaction rates, on the order of the time scale $K_x uE$ and DK^2, the two peaks broaden and move toward each other. For very fast reactions, the two peaks coalesce into one. These trends are clearly seen in Figs. 3 and 4.

To date, no one has reported any experimental results of electrophoretic light scattering on interacting systems. Such experiments could yield information about the electrophoretic mobilities, diffusion coefficients, and reaction rates, simultaneously.

In a related work, Friedhoff and Berne [28] have made an irreversible thermodynamic analysis of electrophoretic light scattering. By developing a fluctuation theory of electrolyte solutions by the coupling of the equations of irreversible thermodynamics with the Poisson equation, they show that in a binary solution (two ionic species), the measured electrophoretic shift is not proportional to the mobility of either ion but depends instead on the rate of change of transfer number with salt solution. All of the experiments described in this Chapter will involve scattering of a macroion in the presence of an electrolyte; thus, there are at least three ionic species present and we expect that the measured electrophoretic mobilities of the macroion are unperturbed by this effect.

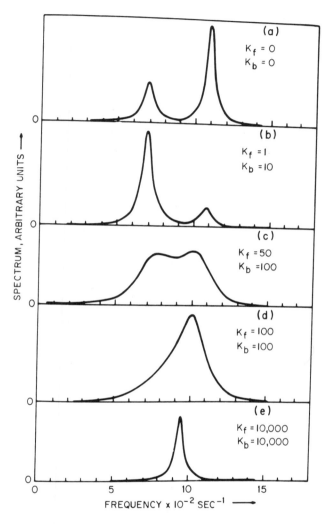

FIG. 4.   Predicted electrophoretic light-scattering spectra for the case of dimerization $2A_1 \overset{\rightarrow}{\underset{\leftarrow}{}} A_2$. The parameters used in the calculation are   T = 283°K, η = 0.01 P,   E = 500 V/cm,   q (effective charge on the monomer) = 5,   a (molecular radius of the monomer) = 20 Å, $K_{eq} = 10^4$, θ = 3°, λ = 5000 Å.   (From Ref. 26.)

III. EXPERIMENTAL METHODS AND RESULTS

A. General Method

The apparatus used to observe electrophoretic light scattering is a combination of a laser light scattering spectrometer with an electrophoresis assembly. A diagram of a typical apparatus is shown in Fig. 5. The apparatus includes a laser, a focusing lens, a light scattering chamber with electrodes and a temperature control, a method of generating and applying the electric potential to the electrodes, collection and detection optics, a photomultiplier tube, an amplifying system, and a spectrum analyzer or autocorrelator for calculating the spectrum or autocorrelation function of the photocurrent.

The incident laser light is normally polarized in the zy (vertical) plane as shown in the figure. The beam may be focused to increase the incident intensity, which improves the signal-to-noise ratio. However, focusing the beam with a spherical lens may reduce the beam diameter to a distance comparable to or even smaller than the typical distance moved by a macro-ion during a single analysis time. The resolution of the experiment is then decreased by the transit-time broadening, and some compromise between signal-to-noise ratio and spectral resolution must be made. Alternatively, the beam can be focused in the vertical direction only by means of a cylindrical lens, which will leave the beam dimension along the path of migration unaffected. The scattered light is collected at a fixed angle defined by the collection optics which, after a correction for refraction, determines the scattering angle θ.

The optical detection system must include provision for the collection of unshifted laser light to act as the heterodyne local oscillator. At sufficiently low scattering angle, the heterodyne light can be collected without difficulty by including in the view of the detection optics the spot formed by the entry or exit of the laser beam from the chamber windows. This light is scattered by stationary particles in the glass and is therefore a suitable local

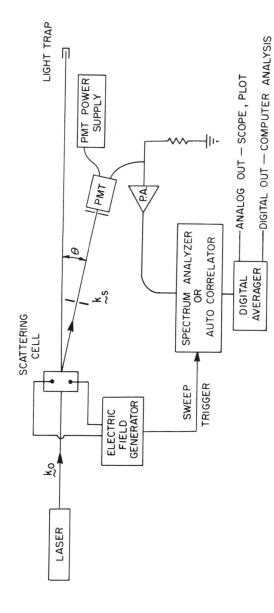

FIG. 5. Schematic representation of the electrophoretic light-scattering apparatus. See text for complete description.

oscillator. The method has the disadvantages of forcing the detection to include the solution in contact with the windows and of not allowing the adjustment of the intensity of the local oscillator. The latter disadvantage can be important when the laser light includes substantial power supply modulation frequencies or vibrational modulations. Workers in the laboratory of Ware utilize a reflex real-image optical system which permits both the convenient alignment of the detection optics and the precise control of the local oscillator position and intensity [29]. In their system a biconvex lens focuses a magnified image onto the photocathode, with the distances selected so that the scattering region in the chamber is in focus at the photocathode. At the focal point of the lens is an adjustable iris to control the optical depth of field. In front of the photocathode is a reflex mirror which rotates into position to deflect the image onto a screen for viewing by the experimenter. On the screen is a cross-hair which corresponds to the center of the photocathode. The alignment of the optics is simply performed by placing the desired image on the cross-hair and adjusting the slit to view the desired width of the image. Heterodyne light is introduced by placing a fine wire or rod between the focusing lens and the scattering chamber. The wire is illuminated by unshifted laser light which is deflected from the incident beam. The wire is positioned so that it is out of focus to an extent just sufficient to make its apparent image at the screen the same size as the scattering region. The intensity of the local oscillator is easily adjusted by changing the amount of light deflected onto it. Usually a ratio of about five or ten to one (heterodyne light to scattered light) produces the optimal signal. Using this system it is easy to locate a precise region in the scattering chamber and to avoid viewing the spurious signals which are produced near the walls.

Several designs for electrophoretic light scattering chambers have been reported. The relevant design considerations are the minimization of the effects of joule heating and the resulting convection, avoidance of electrode products or polarization artifacts,

and the control or elimination of electroosmosis in the chamber. In
addition it is often useful to minimize the required solution volume
and to maximize the achievable electric field strength. The first
chamber used by Ware and Flygare [1] was a modified Tiselius
electrophoresis chamber with a provision to pass laser light through
the chamber perpendicular to the direction of migration of the ions.
The Ag/AgCl electrodes were immersed in the solution to be analyzed,
which necessarily contained Cl^- ions. The scattering region was in
thermal contact on one side with a reservoir of circulating coolant.

    Hartford and Flygare [30] have used a modified version of the
original Ware-Flygare chamber which has improved thermal conduction
and electrode isolation. Their chamber is shown in schematic form
in Fig. 6. The scattering region of this chamber is a spectrophoto-
metric cuvette with both ends open and each end connected to a larger chamber as shown in the figure. There is a free path between
the two end chambers through the cuvette. The incident laser beam
passes through the scattering region perpendicular to the plane of
the figure. The electrodes are platinized platinum. A dialysis
membrane prevents the macromolecules in the scattering region from
coming into direct contact with the electrodes and prevents any
bubbles or electrode products from reaching the scattering region.
The region between the electrode and the dialysis membrane is filled
with the buffer which was used in the dialysis of the macromolecules.
The temperature of the chamber is regulated by passing coolant above

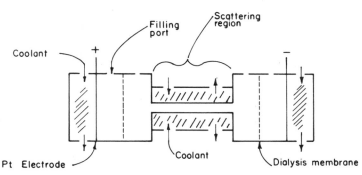

FIG. 6. Cross section of the electrophoretic light-scattering
cell used by Hartford and Flygare [30].

and below the scattering region and on the outside of both electrodes as shown. The electric field strengths which can be obtained before the onset of intolerable Joule heating range from 40 V/cm at .1 M ionic strength to 200 V/cm at .001 M ionic strength. A chamber similar to the one shown in Fig. 6 has been described by Mohan et al [37].

A distinctly different chamber design has been developed by Uzgiris and co-workers [33,34]. They use a small pair of parallel-plate platinized platinum electrodes which they immerse in the solution to be studied. The electrode spacing is very narrow, usually less than 1 mm, so the application of a voltage which is low enough to avoid significant bubbling at the electrodes will result in an electric field which is adequate for many analyses. No additional temperature control is attempted, but the solution being heated is of such small volume that the thermal contact with the solution around the electrodes is usually sufficient to maintain acceptably constant temperature. This arrangement does have the disadvantage of being open at the top, so that solution which has been heated slightly will rise more readily and result in the onset of convection at lower field strengths than if the top of the solution were bounded by a stationary wall. Under most experimental conditions, the narrow spacing of the electrodes also requires that the potential of the electrodes be reversed to avoid substantial solution inhomogeneity and possible electrode polarization. A major advantage of the Uzgiris design is the absence of electroosmosis. Several of these features will be discussed in greater detail below.

Recently a new type of chamber has been described by Haas and Ware (Analytical Biochemistry, in press). This chamber consists of a small conduction gap between two hemicylindrical electrodes. The gap is formed by dielectric inserts which are inserted between the electrodes. This arrangement has the advantages of low solution volume (.4 ml), short thermal time constant, and a low and uniform current density at the electrodes. In addition this chamber can be centrifuged at forces up to 15,000 g for removal of particulate contamination, making it especially suitable for analysis of dilute solutions of proteins. Electrophoretic light scattering experiments on hemoglobin, including determinations of the isoelectric point, have been performed, and the results have been used to assess the accuracy of electrophoretic light scattering experiments and to evaluate the relative importance of distorting effects such as Joule heating, transit time broadening, and electroosmosis. Modified versions of this chamber have been used for experiments on larger particles [29].

The method of application of the electric field to the scattering region is another important consideration. The electric field at the scattering region must be constant and its magnitude must be determined accurately. The error in electrophoretic mobility will be a combination of errors in determination of velocity and electric field. Moreover, the choice of the means of application of the electric field can affect the magnitude of the Joule heating and the degree to which the heating introduces error into the experiment. One advantage of electrophoretic light scattering over conventional electrophoresis is the ability to use pulsed electric fields. Ware and Flygare [1] originally suggested that heating effects could be reduced by using a pulsed electric field and by making the cross section of the scattering region small so that the large surface-to-volume ratio allows more effective cooling. The pulses are usually of alternate polarity to prevent the formation of any concentration gradients or boundaries. Using short pulses, useful data have been obtained for fields as high as 400 V/cm as compared to the typical values for electrophoresis of 5-10 V/cm. Unfortunately, the pulses cannot be made arbitrarily short, because the measurement time, and hence the pulse length, must be at least as great as the reciporcal of the desired bandwidth. Moreover, because the resolution of electrophoretic light scattering is greatest at low angles corresponding to narrow linewidths (10-100 Hz), the desired bandwidths are very small (1-10 Hz), so that pulse times must be long (0.1-1.0 sec). One result of this is that the loss in

resolution with increasing scattering angle is partially made up
for by the fact that at higher angles one can use shorter pulses
and commensurately higher fields. When the pulse repetition rate
exceeds the linewidths of the electrophoretic spectra, side bands
start to appear which greatly complicate the interpretation of the
spectra. Bennett and Uzgiris [31] reported electrophoretic light
scattering experiments using the narrow gap parallel-plate electrode
arrangement described above with square-wave electric field of various modulation frequencies. The calculated spectrum is a series of
peaks of differing intensities which occur at multiples of the ac
field. The shifted peak is reflected in the envelope of the spectrum, with the maximum of the envelope being the Doppler-shift frequency that would be observed if a dc field of the same voltage was
applied. Relative intensities of the spectral peaks are a complicated function of the diffusion linewidth and Doppler-shift frequency. Figure 7 shows an experimental and calculated

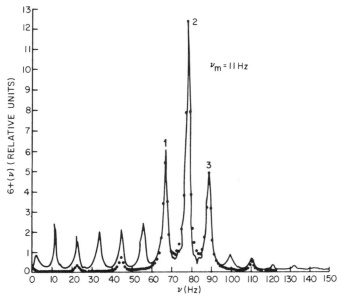

FIG. 7. Electrophoretic light scattering spectrum of 0.82
μm polystyrene latex sphere with a 35 V/cm electric field applied
as a square wave with an 11 Hz modulation frequency: Solid curve,
experimental; points, calculated. (From Ref. 31.)

electrophoretic light scattering spectrum for 0.82 µm polystyrene
latex spheres with an 11-Hz square-wave electric field. Notice that
the calculated spectrum shows peaks at only even multiples of the
modulation frequency while the experimental spectrum has peaks at
all multiples of the modulation frequency. The experimental spectrum
can be calculated completely if all of the phases of the square wave
electric field are averaged in the calculation (i.e., t = 0 may be
chosen at any point during an oscillation of the square wave and
these different possible phases must be averaged). Also, notice in
Fig. 7 that the envelope of the experimental data is not symmetric as
the theory predicts. The excess intensity on the low-frequency side
of the envelope is probably the result of low-mobility components in
the solution.

     Whether the electric field is applied in pulses, square waves,
or as a constant, it is usually best to use a constant-current power
supply. Although the early work was performed using constant-
voltage power supplies, this method requires the assumption that the
solution between the electrodes is a constant, uniform ohmic conduc-
tor. This assumption will generally be untrue, as solution
inhomogeneities and electrode polarization invariably ensue to some
degree during the duration of a pulse. When a constant-current
power supply is used, the electrode polarizations and solution
inhomogeneities are automatically compensated for to keep the current, and hence the electric field, constant in the scattering region. The electric field in the scattering region, E (V/cm), may
be calculated from the current flowing in the region, i (amperes),
the specific conductivity of the solution κ (ohm^{-1}cm^{-1}), and the
cross-sectional area of the region A (cm^2) [32]:

$$E = \frac{i}{A\kappa} \tag{27}$$

Clearly the determination of the electric field from the current
introduces errors in determination of the solution conductivity and
the chamber cross-section, but these errors are generally far less
than the error introduced by the assumption that constant voltage
would lead to constant field. Rimai et al have discussed the use

of constant-current square wave modulation in the parallel plate
configuration and have shown that in their apparatus the use of
constant-voltage square waves would lead to serious errors [40].
The use of constant-voltage power supplies introduces substantially
less error in chamber configurations which have large electrode area
and isolation of electrodes from the scattering region.

The inevitable heating problem caused by the passage of the
current through the scattering region must be considered in the
choice of solvent and in the choice of a power supply, as well as in
the design of the chamber. The current passing through the chamber
develops heat at a rate given by:

$$H = \frac{i^2}{A^2 \kappa} \ W/cm^3 \tag{28}$$

Equation (28) can also be written:

$$H = E^2 \kappa \tag{29}$$

which shows that the maximum field strength for a particular heating
rate is proportional to $1/\sqrt{\kappa}$. For this reason it is most advantageous to perform experiments at low ionic strengths. Classical
Tiselius electrophoresis methods are generally less reliable at low
ionic strengths, because the boundaries and concentration gradients
inherent in the method are less stable at low salt. Electrophoretic
light scattering has no such limitation, though there are practical
limits to which the salt concentration may be reduced in order to
keep the ionic strength and pH known and constant. A more serious
limitation in many experiments is the fact that many samples,
particularly those of biological origin, are not stable at low ionic
strength and must be analyzed in the presence of substantial ion
concentration. The adverse effects of heating are twofold: the
electrophoretic mobility is changed by a change in temperature,
largely through the change in viscosity, and the thermal gradients
result which can lead to convection. Convection often occurs in any
chamber at a particular heating rate, so that if the heating rate

can be kept below the critical value, the effects of convection are insignificant. The change in electrophoretic mobility occurs for any heating rate, and it has the dual ill effects of changing the accuracy of the mobility determination and of broadening the spectral peaks by moving the Doppler-shift frequency as a measurement is made. Both ill effects are obviated considerably by the use of a constant-current power supply; as the solution warms, the conductivity changes in about the same way as the electrophoretic mobility, so that the constant current produces a slightly lower field in roughly the same proportion as the increase in the electrophoretic mobility, maintaining a constant Doppler shift at the proper value for the initial temperature and conductivity. Using a constant-current power supply, reliable electrophoretic light scattering experiments have been performed at salt concentrations as high as physiological (.17 M) (Ware and Smith, unpublished results).

Electroosmosis may be a serious problem for electrophoretic light scattering experiments which are performed in chambers in which the scattering region is bounded by walls which bear an electrical charge. Ions of opposite sign will tend to concentrate near the wall, and when the electric field is applied, they will migrate toward the electrode of opposite polarity, carrying with them a small quantity of solvent. This mass transport creates a pressure across the chamber which induces a back-flow of solution. The resulting parabolic electroosmosis flow profile is superimposed on the electrophoretic velocity, which is constant across the chamber cross-section. Electroosmosis can be eliminated either by eliminating the walls in contact with the scattering region, as is done in the parallel-plate electrode configuration, or by neutralizing the wall charge. The latter is normally accomplished by coating the walls with an uncharged substance such as methylcellulose. When analyzing biological samples it is sometimes possible to reduce the wall charge by the use of a compatible substance such as hemoglobin or serum, though this procedure depends greatly on the adherence properties of the wall material. If electroosmosis is present, the

measurements must either be made at the point in the flow profile at
which the electroosmotic velocity is zero (the so-called "stationary
layer") or must be corrected for the magnitude of electroosmosis at
the point at which they are made. Calibration for electroosmosis is
easily accomplished by measurement of the flow profile in the chamber,
preferably using a sample of known electrophoretic mobility, such as
erythrocytes. However, the calibration must be performed often, because the charge on the walls is usually affected by the substances
being analyzed.

The photocurrent which contains the frequency shifts of the
scattered light is analyzed by either a spectrum analyzer or autocorrelator. When several peaks are present in the spectrum, the
autocorrelation function is much less perspicuous, and frequency analysis provides the experimenter with the best presentation of the
data. The frequencies of interest in this experiment are on the order of Hz to hundreds of Hz, which is the frequency range of common
vibrations. Therefore, mechanical and acoustical shielding and damp-
ing, particularly of the heterodyne source, are indispensable. The
data should be processed with real-time efficiency, which means
effectively that all of the data should be collected and processed
simultaneously. In the frequency domain this is the equivalent of
analyzing a signal with a parallel bank of filters so that all frequencies are measured all of the time. This is contrasted with
conventional swept-oscillator spectrum analyzers which measure only
one frequency band at a time. The difference in signal-to-noise
ratio for the two methods is a factor of $\sqrt{N}$, where N is the number of resolution elements in the spectrum. An even more important
consideration is that the electric field pulse can be N times
shorter for real-time analysis. A number of spectrum analyzers of
both the "time-compression" and Fourier transform types are commercially available. These devices operate with real time-efficiency in
the frequency domain of interest for electrophoretic light scattering.
The definition of real-time analysis may be altered somewhat for
electrophoretic light scattering experiments, in which a pulsed

electric field is used, because the system has a substantial "down time" after each pulse, usually of duration five to fifteen times the length of the pulse, during which data may be processed. This makes it feasible to program a laboratory digital computer to calculate either the complete spectrum or the complete autocorrelation function for each measurement. Digital processing and plotting of the data are usually performed after the experiment in any case, so the use of a programmable computer for data computation and analysis could reduce the total time required to obtain data in final form.

It should be evident from the foregoing discussion that electrophoretic light scattering is a rather complicated experimental technique. There are a number of critical components and adjustments, and there are several potential sources of spurious data. Nevertheless, in the short time since its inception, the technique has been successfully employed in several laboratories for the analysis of a wide variety of systems. An important advantage of the technique is that once the components are assembled and the adjustments are made, many samples can be analyzed within a short period of time, so that absolute accuracies can be checked and relative mobilities can be measured with confidence.

B. Macroion Systems

The application of electrophoretic light scattering has been successfully demonstrated on a number of macroion systems. We will discuss these individually below.

1. Proteins

The first successful electrophoretic light-scattering experiment was reported by Ware and Flygare in 1971 [1]. The macromolecule studied was bovine serum albumin (BSA) and a modified Tiselius electrophoresis cell was used along with the pulsed electric field technique as discussed previously.

The experiment was performed in the following way. The sample to be studied was dialyzed to a known pH and ionic strength and was then filtered into the scattering chamber through millipore filters in order to remove large particles. Once the chamber was inserted into the apparatus, the electrodes and coolant tubes were connected. After the alignment procedure was completed, the operation of the instrument was fully automatic. A switching circuit initiated the electric potential at the electrodes and simultaneously sent a trigger pulse to the autocorrelator or spectrum analyzer so that data were accumulated only when the electric potential was applied. A slight delay was allowed so the system could reach a steady state before data were accumulated. This process takes an amount of time roughly equal to the reciprocal of the bandwidth of the measurement (usually 1-10 Hz). The potential was then removed and the system equilibrated for an amount of time normally equal to 5 to 10 measurement times. A single measurement was repeated for a preselected number of collections. A typical experiment took from 5 min to 1 h as compared to 12-48 hr for a normal electrophoresis experiment. In the early studies by Ware and Flygare, the electrophoretic mobility was calculated from the frequency of the oscillation of the cosine part of the autocorrelation function and the diffusion coefficient was calculated from the decay constant of the exponential part of the correlation function as observed on an oscilloscope display of the data [1]. Although the method of data analysis was somewhat crude, a reproducibility of 10-15% was achieved for both electrophoretic mobility and diffusion coefficient, and the results were in good agreement with accepted values. The experiments were performed at a variety of pH and ionic strength conditions to verify the proper dependence of electrophoretic mobility on solution conditions. It was also verified that no effect was observed at the isoelectric point. Experimens were performed at various angles to verify that the shift was proportional to $\sin$ as the theory predicts $[\underline{K} \cdot \underline{v}_d = K v_d \cos \theta/2 = (4\pi n/\lambda_0) \sin \theta/2 \cos \theta/2 = (2\pi n/\lambda_0) \sin \theta]$. These angular dependence data are shown in Fig. 8.

High-resolution experiments on BSA were reported by Ware and Flygare in 1972 [35]. In these experiments the autocorrelation

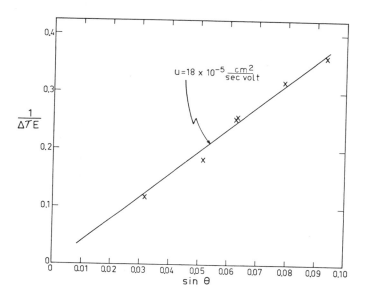

FIG. 8. A plot of the Doppler shift per unit field as a
function of the sine of the scattering angle. The dependence is
linear as the theory predicts, and the slope of the line leads to an
average value for the mobility of 14×10^{-5} cm^2 sec^{-1}V^{-1}, which is corrected to 20 C to give 18×10^{-5} as shown. These data were all taken
on a solution of bovine serum albumin at pH 9.0 and ionic strength
0.01. (From Ref. 1.)

functions were recorded on digital paper tape and Fourier transformed.
Data for an experiment on BSA are presented in Fig. 8. The three
experiments shown were performed at zero field, 100 and 175 V/cm,
respectively. The autocorrelation functions are the actual data, and
the corresponding spectra below were obtained by digital Fourier transforms. In the two electric field experiments it is evident that there
are two shifted peaks rather than one. This is a reproducible feature
of high-resolution spectra on BSA and is attributable to the fact that
commercial freeze-dried preparations of BSA contain 20-30% BSA dimers
[36]. Thus these data also demonstrate the utility of this technique
for the analysis of multicomponent systems; the resolution shown in
these data exceeds the resolution of electrophoresis for this system.

    The data in Fig. 8 taken at 175 V/cm are included to demonstrate
the effects of heating. Note that the decay constant of the

correlation function is considerably smaller and the spectral lines are significantly broader. Also the dimer-to-monomer ratio is less reliable, presumably due to convection and line-broadening. This problem is avoided by taking the data only from spectra at sufficiently low fields not to show these effects. The electrophoretic mobilities taken from the spectra such as are shown in Fig. 9 were reproducible to within 5% for both monomer and dimer, and the monomer results were in good agreement with electrophoresis data. In the same report Ware and Flygare reported multicomponent spectra from a solution of BSA, BSA dimers, and fibrinogen.

One of the first contributions of electrophoresis was the characterization of human blood plasma, and this system has been a subject of several electrophoretic light scattering investigations. Ware studied the electrophoretic light scattering of human blood plasma in 1971 with promising results. One of the interesting features of the electrophoretic light scattering spectrum is that the globulins are considerably more intense with respect to albumin than they are in a typical electrophoresis experiment, the reason being that the globulins have higher molecular weights than albumin and hence scatter light with a commensurately higher intensity. A typical spectrum obtained in these early studies is shown in Fig. 10. The highest-mobility component is attributable to albumin. It has the proper mobility for albumin, and the mobility was very reproducible in experiments performed at a number of electric fields and scattering angles. The lower-mobility components are presumably attributable to the various globulins, which are known to be electrophoretically distinct, but the assignment of these peaks by performing experiments on purified samples has not been done. Spectra at higher fields showed a resolution of more electrophoretically distinct components, but the assignment of these peaks by performing experiments on purified samples has not been done; and the absolute position and relative intensity of these peaks were not sufficiently reproducible to make any assignments.

In a recent publication Mohan, Steiner, and Kaufmann [37] described experiments on human serum and on bovine serum albumin with

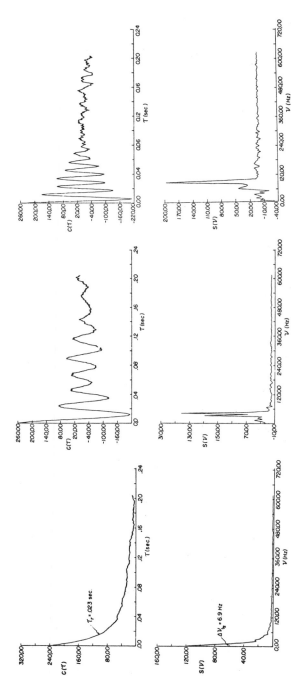

FIG. 9. Representative data for an experiment on a bovine serum albumin solution at zero field, 100 V/cm, and 175 V/cm. (From Ref. 35.)

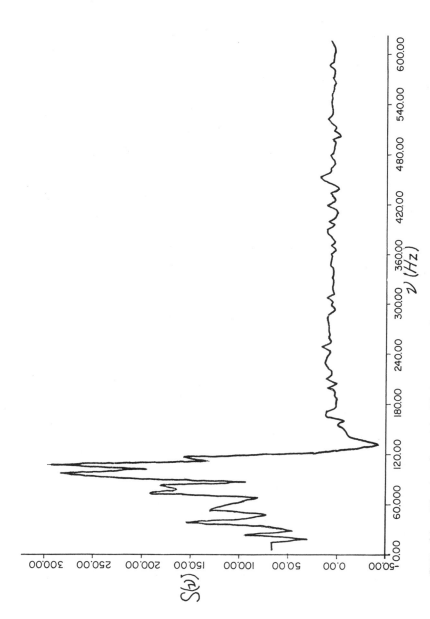

FIG. 10. Electrophoretic light scattering spectrum of human plasma at pH 8.4 and ionic strength 0.0060. The scattering angle was 7° 30' and the electric field strength was 150 V/cm.

results similar to the results obtained by Ware and Flygare in 1971. In the analysis of the multicomponent spectra from human serum, the signal-to-noise ratio was enhanced by digital smoothing, and the resulting spectra showed features very similar to Fig. 10. The mobility of the albumin was correct, and the exact features of the globulin spectra were assigned to identified components of human serum by comparison with electrophoresis experiments of serum on disc gels. Both the relative mobilities and the integrated intensities, corrected for the known molecular weights of the globulins, were used to assign the various peaks in the electrophoretic light scattering spectrum.

The analysis of protein solutions is one of the most difficult electrophoretic light scattering applications for two reasons. First the intensity of scattered light is relatively low for these low-molecular-weight macromolecules, making the potential interference from dust and other particulate contamination even more severe. Secondly, the spectral peaks from these solutions are broadened by diffusion, so that relatively high electric fields are required to achieve spectral resolution of several components. Larger macromolecules and particles not only have much higher light scattering cross-sections, but also have much lower diffusion coefficients, so that diffusion broadening is much less severe and resolution of electrophoretically distinct components can be achieved at much lower electric fields. Indeed for particles in the micron size range, such as blood cells, the broadening due to diffusion is insignificant, and the resolution is virtually independent of the electric field strength. For these systems the electrophoretic light scattering spectrum is an accurate representation of the true electrophoretic distribution of the components.

2. Nucleic Acids

Hartford and Flygare [30] have measured the electrophoretic mobility of calf thymus DNA as a function of ionic strength using electrophoretic light scattering. The data were taken with a real-time spectrum analyzer of the time compression type. The frequency

shift in the heterodyne spectrum caused by the electrophoretic migration
of the molecules was measured directly from the signal averaged spectrum. Figure 11 represents a typical spectrum of calf thymus DNA averaged 100 times, which required ∿5 min. The intensity at low frequency
is probably due to a combination of amplitude modulation of the laser
beam caused by mechanical vibration and impurities in the sample.

The electrophoretic mobility for calf-thymus DNA was examined over
ionic strengths from 0.004 to 0.10 M NaCl and the results were in agreement with earlier classical work; the mobility is proportional to the
square root of the ionic strength for low ionic strength.

Hartford and Flygare [30] have also measured the difference in
electrophoretic mobility between the native and denatured forms of calf
thymus DNA. After making measurements on the native DNA the sample was
irreversibly denatured by heating it to 95°C for 1 h and then immersing
it in an ice bath [38]. Measurements were then immediately taken from

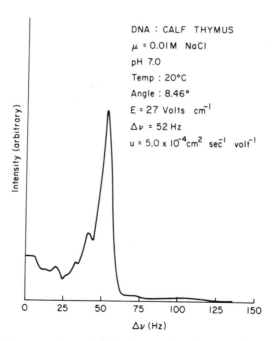

FIG. 11. Electrophoretic light-scattering spectrum of calf
thymus DNA under conditions shown.

the denatured sample. Figure 12 shows a comparison of the
electrophoretic frequency shift of the native and denatured DNA as a
function of the electric field. The observed difference in mobility
is approximately fifteen percent and the shift is linear in electric
field as expected. The measurements are in good agreement with previous values obtained using the Tiselius electrophoresis apparatus
[39].

In these experiments on calf thymus DNA, no attempt was made to
measure the diffusion coefficient from the linewidth of the spectrum,
because the commercial preparations of the DNA have a molecular weight
range of 50,000-2,000,000. The linewidth is a weighted average of the

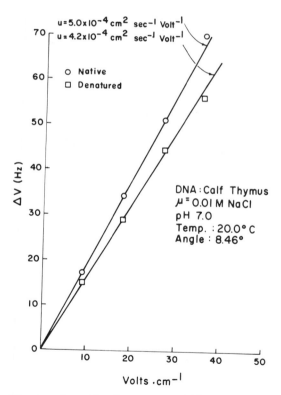

FIG. 12. Electrophoretic frequency shifts as a function of
electric field for the native and denatured forms of calf thymus DNA
under the conditions shown.

contributions from all of the DNA molecular weight components. From
the single well-defined peak observed in Fig. 11, it is clear that the
electrophoretic mobility is independent of the DNA chain length.

3. Viruses

Hartford and Flygare [30] have reported the electrophoretic light
scattering determination of the mobility of tobacco mosaic virus (TMV)
as a function of ionic strength. The results were consistent with the
Debye-Hückel screening law. Rimai et al [40] have reported the
characterization of a number of RNA tumor viruses using electrophoretic
light scattering. The viruses studied were avian myeloblastosis (AMV)
murine leukemia (MuLV), murine mammary tumor (MuMTV), and feline
leukemia (FeLV) viruses. These viruses have similar electrophoretic
mobilities near pH 7, but become quite distinct at acid pH, with greatly different mobilities at pH 3, indicating that the surface chemistry
of the viruses is a means of distinguishing among them. Using the
Debye-Hückel model, these investigators calculated the surface charge
density and the number of charges on the surface of each virus at each
pH studied. The viruses were negatively charged at all pH's studied,
with the exception of the FeLV, for which the sign of the charge at pH
3 was ambiguous.

4. Synthetic Polymers

Polystyrene latex spheres (PSLS) are a common standard in light
scattering measurements, and they have been used for this purpose in
electrophoretic light scattering as well. Uzgiris [34] reported the
first electrophoretic light scattering experiments on PSLS, observing
that the mobility for a given lot of spheres is uniform, but that it
may vary substantially from lot to lot. The ionic strength dependence
of the mobility of the spheres was determined, and general agreement
with the Debye-Hückel law was obtained. In the low ionic strength
limit between .05 and .001, the electrophoretic mobility was proportional to the inverse of the square root of the ionic strength as expected, but below .001, the electrophoretic mobility was found to be

independent of the ionic strength, an effect which is to be expected but which had not previously been reported.

Uzgiris and Costaschuk [41] have used electrophoretic light scattering to study the reaction of PSLS, which are normally negatively charged, with the cationic polymer polyethyleneimine (PEI). As increasing amounts of PEI are added to the PSLS, the surface charge of the PSLS, and hence the electrophoretic mobility, is reduced and then changes sign, reaching a plateau value of positive charge at a concentration of PEI which is about 1.4 times the concentration required for a monolayer coverage. In the region near charge neutralization the spectra were substantially broader, indicating charge heterogeneity. Figure 13 shows a plot of the electrophoretic mobility of the PSLS as a function of the added concentration of PEI.

In a related study Uzgiris [42] has developed electrophoretic light scattering as a sensitive assay for the antigen-antibody reaction. The assay is performed by coating the polystyrene spheres with antigen and then monitoring the electrophoretic mobility of the coated spheres as antibody is added. In his experiments Uzgiris used bovine serum albumin (BSA) as the antigen and the anti-BSA antibodies were

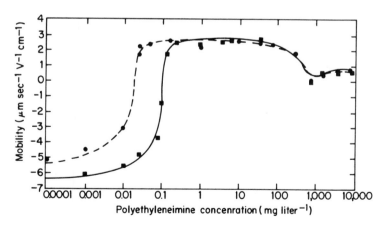

FIG. 13. Electrophoretic mobilities, deduced from light scattering in an electric field of 33 V/cm and a $6°50'$ scattering angle, as a function of polyethyleneimine concentration: Solid line for latex sphere concentration 21 mg/liter and dashed line for 4 mg/liter. (From Ref. 41.)

produced in rabbits. Concentrations of antibody as low as 5-10 ng/ml could be detected by this method in 30 min. The assay assumes that the antibody has a surface charge which is substantially different from the antigen, which is often the case since immunoglobulins tend to have quite low electrophoretic mobilities. However, the assay is not speci-fic in the sense that any agent which altered the surface charge of the spheres would be detected as antibody.

Yoshimura, Kikkawa, and Suzuki [43] have reported several studies of polystyrene spheres by electrophoretic light scattering. They have studied the size dependence of the electrophoretic mobility in the range .1 μm to 1 μm, with the result that the different sizes have different mobilities. This is the range of sizes for which the Debye-Huckel theory predicts a size effect on mobility. When the product of the Debye-Huckel constant times the particle radius becomes much greater than one, the electrophoretic mobility is independent of par-ticle size and is proportional only to the surface charge density at the hydrodynamic surface of shear, the so-called zeta potential. For typical salt concentrations, the electrophoretic mobility is expected to be independent of size for particles in the micron range and larger. Below that range the size and geometry of the particles can affect the mobility. In a recent report these workers report that two sizes of PSLS which are electrophoretically distinct in salt solutions move at a single average velocity when dissolved in pure water. This is presumably a confirmation of the irreversible thermodynamics analysis of electrophoretic light scattering of two-component systems which has been given by Friedhoff and Berne [28] and was discussed earlier. Another effect which must be taken into account in the electrophoretic analysis of large particles is the hydrodynamic interaction between the particles, which often places a practical limit on the concentra-tion at which two large particles of different mobilities can be distinguished.

5. Living Cells

The electrophoretic properties of living cells have been studied for many years using the classical technique of microelectrophoresis,

in which the experimenter observes the motion of individual cells in
an electric field and times the transit over a measured distance. The
electrophoretic mobility of a blood cell is related directly to its
surface charge density, or zeta potential, and is one of the few physical parameters which can be used to characterize the unperturbed condition of the cell. Cell surfaces under physiological conditions bear
negative charges, and the magnitude of the charge can be affected by
reaction with any charged species or by reaction with particles which
would mask or unmask some of the charged groups on the cell surface or
by any internal change in the cell which would cause it to alter the
surface density of charged groups on the outer membrane. These several possibilities include a large portion of the reactions of interest
at cell surfaces, and a number of interesting studies have been performed. Unfortunately, there have been numerous instances of con-
flicting results in this area. Microelectrophoresis has the
disadvantage that it requires human observation of cells one at a
time, so that it is difficult to get a statistical sample within the
short period of time over which the electrokinetic properties of cells
may be assumed constant. On the other hand, electrophoretic light
scattering is an extremely rapid technique which permits the complete
electrophoretic characterization of a sample of hundreds of cells in
a matter of seconds, with an apparatus that is fully automatic and
therefore much more likely to provide reproducible, unbiased results.
It is not surprising that the electrophoretic characterization of living cells has become a major area of application of electrophoretic
light scattering.

Uzgiris [33] has reported electrophoretic light scattering
experiments on living cells. Human erythrocytes, which are an
excellent electrophoretic standard, were studied in suspensions of
physiological saline concentration, and the accepted value for the
electrophoretic mobility was obtained, though the experimental uncertainty in those early experiments was rather high ($\sim$10%). The
bacterium Staphylococcus epidermidis was also studied, and the Doppler
shift frequency was shown to be linearly proportional to the applied
electrode voltage. The linewidth of the Doppler-shifted peak due to
S. epidermidis was too broad to be accountable to diffusion, as

evidenced by experiments on polystyrene spheres which produced a
narrower linewidth even though they were much smaller than the bacteria
and would hence be expected to have a higher diffusion coefficient and
substantially broader linewidth. The discrepancy in linewidth was
attributed by Uzgiris to electrophoretic heterogeneity of the bacteria.
Subsequent experiments have established that electrophoretic light
scattering is an excellent means for characterizing the shape and width
of the electrophoretic distribution of a suspension of large particles.
Figure 14 shows the comparison of the power spectrum for the
electrophoretic light scattering experiments on S. epidermidis, which
had a mean radius of .7 μm, and the PSLS, which had a mean radius of
.18 μm.

Uzgiris and Kaplan [44] have extended the original work on blood
cells to include further study on human erythrocytes and leukocytes.
The experimental uncertainty in these measurements is estimated to be
around 5%, and the mobilities measured agree with those of
microelectrophoresis measurements where comparable data exist. The
titration behavior of erythrocytes and lymphocytes was determined by

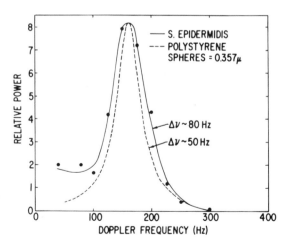

FIG. 14. Comparison of the power spectrum of He-Ne laser light
scattered from S. epidermidis and from polystyrene latex spheres in
distilled water. The excess with the solid curve is attributed to the
nonuniform mobilities of the S. epidermidis. The interpretation of
the low frequency tail is somewhat obscured by the presence of a size-
able residual homodyne spectrum component for the S. epidermidis
experiments (From Ref. 33.)

measuring the electrophoretic mobility of these cells in isotonic
media of low ionic strength in the pH range of 3 to 10. At physiological pH measurements have been made as a function of ionic strength,
with the interesting result that although erythrocytes and lymphocytes
have the same electrophoretic mobility in the high ionic strength
media normally used for electrokinetic characterization, they have
substantially different mobilities in isotonic media of low ionic
strength (media were made isotonic by addition of sucrose). Using
standard cell separation procedures, these workers have analyzed
several subpopulations of leukocytes in pH 7.3, ionic strength .005
solvent. Mononuclear leukocytes separated by Ficoll-Hypaque
centrifugation were found to have a fast and slow component which have
not yet been identified but which may be T and B lymphocytes. Nylon-
purified lymphocytes, monocytes, and granulocytes were all found to be
electrophoretically distinct from each other and from erythrocytes.

Kaplan and Uzgiris [45] have studied the alteration of the
electrophoretic mobilities of human lymphocytes induced by incubation
with the mitogenic plant lectins phytohemagglutinin (PHA) and
Concanavalin A (Con A). Incubation with either agent for 90 min at
$37^{\circ}C$ resulted in stable and reproducible decreases in electrophoretic
mobility and increases in the isoelectric point. These changes were
interpreted as the result of binding to the cell surface and not as a
result of transformation. In the case of Con A, washing the cells
with a specific inhibitor for Con A binding, methyl-α-D-glucoside
(MAG), resulted in complete reversal of the electrokinetic changes.

Josefowicz and Hallett [46] have used electrophoretic light
scattering to study the binding of pokeweed mitogen to the surfaces
of rat thymus lymphocytes. Fifteen minutes after introducing pokeweed
mitogen to the suspension of thymocytes, their mobility decreased by
more than a factor of two. In addition there was evidence that the
distribution of velocities was greater after binding the mitogen,
which may be explained by a different affinity of various subsets of
the thymocyte population for pokeweed. However, the autocorrelation
functions were not collected for a sufficient pulse duration to obtain
adquate spectral resolution to characterize the change in the velocity
distribution.

Smith, Ware, and Weiner [29] have recently reported a study of the electrophoretic mobilities of mononuclear white cells from the peripheral blood of normal human subjects and patients with acute lymphocytic leukemia. Both fresh and cryopreserved samples of each type were analyzed, with the result that cryopreservation had no distinguishable effect on the electrophoretic distributions of either normal or leukemic cells. Mode (most probable) mobilities of cells from nine leukemic patients were 7% to 28% below the average mode mobility of normal cells. The standard deviation among normal samples was 2%. Normal cells populations had an asymmetric, often bimodal, electrophoretic distribution, while leukemic cells consistently had a single symmetric distribution. The electrophoretic distributions from normal and leukemic cells are shown in Fig. 15. This is the first instance in which electrophoretic light scattering has been used to detect a pathological condition of living cells. Further work to develop the potential of electrophoretic light scattering for applications in diagnosis, characterization, and monitoring of disease and treatment is in progress.

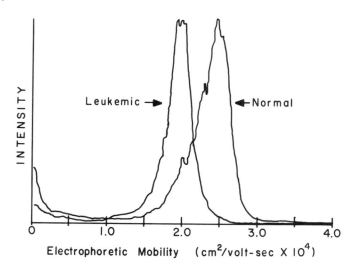

FIG. 15. Overlay of the electrophoretic light scattering data for normal and leukemic cells. (From Ref. 29.)

IV.   CONCLUSION

Electrophoretic separation and analysis in various forms has been the most useful of all approaches to the characterization of macromolecular systems. The use of gels and other supporting media have greatly expanded the analytical capabilities of electrophoresis, and it is a rare laboratory that studies charged macromolecules without employing some form of electrophoresis. Electrophoretic light scattering is the most recent in a series of important advances in electrophoresis methodology, and in this final section we shall attempt to assess briefly the potential significance of this technique.

Electrophoretic light scattering is strictly an analytical technique. Though it cannot be used as a means of separation of species with differing electrophoretic mobilities, it may be useful as a rapid assay of the electrophoretic separability of various species. The laser Doppler approach of electrophoretic light scattering is applicable to particles of all sizes, but the difficulty of preparing solutions of small macromolecules which are free from contamination by large particles will continue to be a substantial limitation. Moreover, the diffusion broadening which is inevitably present in the analysis of small macromolecules will always be a limiting factor for achieving electrophoretic resolution among different species. We cannot reasonably expect electrophoretic light scattering to be an attractive alternative to electrophoresis on gels and other supporting media for such systems. However, electrophoretic light scattering is the technique of choice for such systems when the magnitude of the electrophoretic mobility in solution is an important parameter.

Electrophoretic analysis for larger particles (.1 μm and larger) is much more difficult and laborious by classical means and is much easier for electrophoretic light scattering. Biological particles in this size range include cells, viruses, organelles, vesicles, and nucleic acids. For many of these particles the characterization of the surface charge and the study of surface reactions is an area of substantial interest and promise, and we anticipate that this will be the area in which electrophoretic light scattering will make the greatest contribution.

In the short time since its inception [1], electrophoretic light scattering has been refined and developed to the point that rapid measurements on a routine basis are now possible. Several research groups are now using this technique to pursue a number of different applications, and it is reasonable to expect further refinements as a number of investigators make independent contributions. The time may not be far away when an electrophoretic light scattering apparatus will be available as a routine research or clinical instrument which can be used along with other techniques to further our understanding of the myriad of charged particles and their surface reactions.

ACKNOWLEDGMENT

This work was supported by ARPA Grant DAHC-15-67-G10 to the Materials Research Laboratory, University of Illinois.

REFERENCES

1. B. R. Ware and W. H. Flygare, Chem. Phys. Letters, $\underline{12}$, 81 (1971).

2. L. I. Komarov and I. Z. Fisher, Soviet Physics JETP, $\underline{16}$, 1358 (1963).

3. R. Pecora, J. Chem. Phys., $\underline{40}$, 1604 (1964).

4. H. Z. Cummins, F. D. Carlson, T. J. Herbert, and G. Woods, Biophys. J., $\underline{8}$, 518 (1969).

5. C. Kittel, Elementary Statistical Physics, Wiley, New York (1958).

6. A. T. Forrester, W. E. Parkins, and E. Gerjouy, Phys. Rev., $\underline{72}$, 728 (1947).

7. A. T. Forrester, R. A. Gudmundsen, and P. O. Johnson, Phys. Rev., $\underline{99}$, 1691 (1955).

8. H. Z. Cummins and H. L. Swinney, Prog. Opt., $\underline{8}$, 133 (1969).

9. H. Z. Cummins, N. Knable, and Y. Yeh, Phys. Rev. Letters, $\underline{12}$, 150 (1964).

10. S. B. Dubin, J. H. Lunacek, and G. B. Benedek, Proc. Nat. Acad. Sci., 57, 1167 (1967).

11. F. T. Arecchi, M. Giglio, and U. Tartari, Phys. Rev., 163, 186 (1967).

12. B. Chu, Ann. Rev. Phys. Chem., 21, 145 (1970).

13. G. B. Benedek, Polarization, Matter and Radiation, Presses Universitaire de France, Paris (1969).

14. S. B. Dubin, G. B. Benedek, F. C. Bancroft, and D. Freifelder, J. Mol. Biol., 54, 547 (1971).

15. B. R. Ware, T. Raj, W. H. Flygare, J. A. Lesnaw, and M. E. Reichmann, J. Virol., 11, 141 (1973).

16. N. C. Ford, Jr., Adv. Chem. Ser., 125, 25 (1971).

17. P. Berge, B. Volochine, R. Billard, and A. Hamelin, Compt. Rend. Acad. Sci., Ser. D, 265, 889 (1967).

18. R. Nossal, Biophys. J., 11, 341 (1971).

19. M. J. French, J. C. Angus, and A. G. Walton, Science, 163, 345 (1969).

20. S. H. C. Lin, R. K. Dewan, V. A. Bloomfield, and C. V. Morr, Biochemistry, 10, 4788 (1971).

21. R. A. Alberty, J. Chem. Ed., 25, 426, 619 (1948).

22. J. R. Cann, Interacting Macromolecules, Academic, New York (1970).

23. G. D. J. Phillies, M. S. Thesis, Dept. of Physics, Massachusetts Institute of Technology, Cambridge (1971), unpublished.

24. G. D. J. Phillies, J. Chem. Phys., 53, 2613 (1973).

25. B. R. Ware, Adv. Colloid Interface Sci., 4, 1 (1974).

26. B. J. Berne and R. Gininger, Biopolymers, 12, 1161 (1973).

27. Abragam, The Principles of Nuclear Magnetism, Oxford University Press, London (1961).

28. L. Friedhoff and B. J. Berne, Biopolym., 15, 21 (1976).

29. B. A. Smith, B. R. Ware, and R. S. Weiner, Proc. Nat. Acad. Sciences, 73, 2388 (1976).

30. S. L. Hartford and W. H. Flygare, Macromolecules, 8, 80 (1975).

31. A. J. Bennett and E. Uzgiris, Phys. Rev., A8, 2662 (1973).

32. C. C. Brinton, Jr. and M. A. Lauffer, Electrophoresis: Theory, Methods and Applications, Vol. 1, (M. Bier, ed.) Academic, New York (1959), p. 427.

33. E. E. Uzgiris, Opt. Commun., 6, 55 (1972).

34. E. E. Uzgiris, Rev. Sci. Instrum., $\underline{45}$, 74 (1974).

35. B. R. Ware and W. H. Flygare, J. Colloid and Interface Science, $\underline{39}$, 670 (1972).

36. K. O. Pedersen, Arch. Biochem. Biophys., Suppl., $\underline{1}$, 157 (1962).

37. R. Mohan, R. Steiner, and R. Kaufmann, Anal. Biochem., $\underline{70}$, 506 (1976).

38. L. Constantino, A. M. Liquori, and V. Vitogliano, Biopolymers, $\underline{2}$, 1 (1964).

39. B. M. Olivera, P. Baine, and N. Davidson, Biopolymers, $\underline{2}$, 255 (1964).

40. L. Rimai, I. Salmeen, D. Hart, L. Liebes, M. A. Rich, and J. J. McCormick, Biochem., $\underline{14}$, 4621 (1975).

41. E. E. Uzgiris and F. M. Costaschuk, Nat. Phys. Sci., $\underline{242}$, 77 (1973).

42. E. E. Uzgiris, J. Immunol. Methods, $\underline{10}$, 85 (1976).

43. T. Yoshimura, A. Kikkawa, and N. Suzuki, J. Appl. Phys., $\underline{11}$, 1797 (1972); Jap. J. Appl. Phys., $\underline{14}$, 1853 (1975), Optics Commun., $\underline{15}$, 277 (1975).

44. E. E. Uzgiris and J. H. Kaplan, Anal. Biochem., $\underline{60}$, 455 (1976) and E. Uzgiris and J. H. Kaplan, J. Colloid and Interface Science, $\underline{55}$, 148 (1976).

45. J. H. Kaplan and E. E. Uzgiris, J. Immunol. Methods, $\underline{7}$, 337 (1975).

46. J. Josefowicz and F. R. Hallett, FEBS Lett., $\underline{60}$, 62 (1975).

Chapter 10

CIRCULAR DICHROISM AND OPTICAL ROTATION IN AN ELECTRIC FIELD

Ignacio Tinoco, Jr.

Department of Chemistry
University of California
Berkeley, California

A container of optically active molecules (those not identical to their mirror images) will rotate the plane of polarization of incident light. In an absorption band they will also cause incident plane-polarized light to become elliptically polarized. An equivalent statement is that they will show preferential absorption for right and left circularly polarized light. These effects can obviously perturb the measurement of electrical birefringence and electrical dichroism in optically active materials. The perturbation of the electrical birefringence and dichroism is usually negligible, but the phenomenon of optical activity is of direct interest.

The theory of optical activity has been described at length [1,2]. Understanding of optical activity has been a very useful tool in determining molecular structure. We will only consider

here the change in circular dichroism (CD) and optical rotatory
dispersion (ORD) of a solution caused by orientation of molecules
by an electric field. The effect will thus depend on even powers
of the electric field, as in electric birefringence and electric
dichroism. There is no CD or ORD effect, analogous to the Faraday
effect, which is linear in the field [3]. In principle, rotatory
power can be induced in a solution by an electric field gradient.
An electric field gradient, acting on a quadrupole allowed transition, can produce circular dichroism. We will not discuss this
effect, but the equations have been mentioned [4] in a discussion
of internal field-induced optical rotation in molecules.

To understand the effect, we need only consider that the CD
of a molecule is different in different directions. It is easiest
to visualize a helical molecule and intuitively conclude that the
CD for light incident parallel to the helix axis will be quite
different from that for light incident perpendicular to the helix
axis. It is also important to realize that it is the direction of
incidence of the light that is important, not the direction of
polarization of the light. In fact, for measuring CD the incident
light is circularly polarized and for measuring ORD the plane of
polarization changes throughout the solution.

In this chapter we will present the optical rotation tensor
which characterizes the direction-dependent rotation or CD of a
molecule. This tensor has a particularly simple form for helices.
Furthermore the small amount of experimental work in this area
seems to have been done for helical polypeptides. Orientation of
helices by an electric field, or by other methods, greatly simplifies the interpretation of the CD and has helped in the understanding and development of the theory of circular dichroism.

I.   THEORY

Circular dichroism (CD) is the difference in absorption for
right and left circularly polarized light. The usual units of

measurement are either molar extinction:

$$\varepsilon_L - \varepsilon_R = \frac{A_L - A_R}{\ell c} \qquad (1)$$

where

A_L, A_R = absorbance for left (right) circularly polarized light

ℓ = pathlength in centimeters

c = concentration in mol/liter

$\varepsilon = \dfrac{\varepsilon_L + \varepsilon_R}{2}$ = molar extinction coefficient for unpolarized light

or molar ellipticity:

$$[\theta] = \frac{100\ \theta}{\ell c} \qquad (2)$$

where θ = ellipticity in degrees of the elliptically polarized light induced by the sample on incident linearly polarized light. The ratio of the minor axis to major axis of the ellipse is the tangent of θ. The molar ellipticity is related to the CD.

$$[\theta] = 3298(\varepsilon_L - \varepsilon_R) \qquad (3)$$

ORD is the rotation of linearly polarized light. The molar rotation is defined as

$$[\phi] = \frac{100\ \alpha}{\ell c} \qquad (4)$$

where α = rotation of the incident plane polarized light in degrees. It can be related to the circular birefringence $(n_L - n_R)$

$$[\phi] = \frac{1800}{\lambda c}\ (n_L - n_R) \qquad (5)$$

where

n_L, n_R = refractive index for left (right) circularly polarized light

λ = wavelength of light in centimeters

c = concentration in mol/liters

Both the CD and ORD can be related to molecular structure through a parameter called the rotational strength, R. Each transition $(0 \rightarrow A)$ in a molecule will have a rotational strength R_{OA} defined as:

$$R_{OA} = Im \; \underline{\mu}_{OA} \; \underline{m}_{AO}$$
$\underline{\mu}_{OA}$ = electric dipole transition moment
$\underline{m}_{AO}$ = magnetic dipole transition moment
Im = imaginary part

The contribution of each transition to the CD and ORD is then of the form:

$$\varepsilon_L - \varepsilon_R = R_{OA} f(\nu, \nu_{OA})$$
$$[\phi] = R_{OA} KK[f(\nu, \nu_{OA})] \tag{6}$$

The function $f(\nu, \nu_{OA})$ is the band shape and is usually approximated as a gaussian or lorentizian shape centered at the transition frequence, ν_{OA}. $KK[f(\nu, \nu_{OA})]$ is the Kronig-Kramers transform of the band. For optical rotation far from absorption bands, the bands can be replaced by delta functions which lead to Drude terms for the ORD:

$$[\phi] = \frac{48N_o \nu^2}{\hbar c} \sum_A \frac{R_{OA}}{\nu_{OA}^2 - \nu^2} \tag{7}$$

where

N_o = Avogadro's number
$\hbar$ = Planck's constant/2π
c = speed of light in centimeters per second
ν = frequency in hertz

To understand the CD and ORD of oriented molecules we need to know the direction dependence of the rotational strength, that is, the rotational strength tensor instead of the scalar [5]. Because the rotational strength is a pseudoscalar (it changes sign on

inversion of the molecule), the rotational strength tensor is a
third-rank tensor. For simplicity we will only consider here
molecules with cylindrical symmetry, such as helices. Then the
equations are [2]:

Light incident parallel to the symmetry axis:

$$(R_{33})_{OA} = \frac{3e}{2mc} \, \text{Im} [\mu_{2OA} (r_3 p_1)_{AO} - \mu_{1OA} (r_3 p_2)_{AO}] \tag{8}$$

Light incident perpendicular to symmetry axis:

$$(R_{11})_{OA} = (R_{22})_{OA} = \frac{3e}{2mc} \, \text{Im} [\mu_{3OA} (r_1 p_2)_{AO} - \mu_{2OA} (r_1 p_3)_{AO}]$$

$$= \frac{3e}{2mc} \, \text{Im} [\mu_{1OA} (r_2 p_3)_{AO} - \mu_{3OA} (r_2 p_1)_{AO}] \tag{9}$$

where

  1,2,3 label components along three perpendicular axes in the
     molecule with 3 as the symmetry axis
  $\underline{\mu}$ = electric dipole moment operator
  $\underline{p}$ = linear momentum operator
  $\underline{r}$ = position operator
  e,m = charge and mass of an electron
  c = speed of light

The average rotational strength is

$$R_{OA} = \frac{1}{3} [(R_{11})_{OA} + (R_{22})_{OA} + (R_{33})_{OA}]$$

As the components of the rotational strength will usually have
different signs, the average rotation may be small and difficult
to interpret, whereas each component may be very characteristic.

II.   ELECTRICAL ORIENTATION

   One way of measuring the components of the rotational
strength tensor is to orient the molecules with an electric field.
To avoid the effects of linear birefringence and linear dichroism,

it is necessary to have the measuring light parallel to the
electric field; the linear effects are then zero. One can calculate the combined effect of linear and circular birefringence and
dichroism on the incident light [6] and show that the linear effects would dominate the phenomena [7]. The first published
papers [8] reporting the change of optical rotation in an electric
field are incorrect, probably because of linear birefringence
artifacts.

We can calculate the change of CD or ORD due to electrical
orientation of the molecules in analogy with the other optical
effects discussed earlier in the book. Once the electrical orientation function is known, any optical property can be calculated.
We will give equations for the rotational strength and remind the
reader that the CD and ORD can be obtained by summing the contributions from all the rotational strengths for the molecule.

For light incident parallel to the orienting electric field
the appropriately averaged rotational strength in the presence of
the electric field is:

$$R_E = \int_0^{2\pi} \int_0^{2\pi} \int_0^{\pi} (\underline{e} \cdot \underline{R} \cdot e) f(\theta,\chi,\phi) \sin\theta \, d\theta \, d\chi \, d\phi \qquad (10)$$

where

> $\underline{R} = R_{11}(\underline{ii} + \underline{jj}) + R_{33}\underline{kk}$, where $\underline{i}, \underline{j}, \underline{k}$ are molecule-
> fixed unit vector with $\underline{k}$ the symmetry axis
>
> $\underline{e}$ = A unit vector along the direction of propagation of the
> light and the direction of the static electric field
>
> $f(\theta,\chi,\phi)$ = normalized electrical orientation function.
>
> $\theta, \chi, \phi,$ are the Eulerian angles relating the molecule
> fixed axes to space fixed ones

For cylindrically symmetric molecules Eq. (10) can be simplified
to [9,10]:

$$R_E = R_o - \frac{1}{3}(1 - 3 \langle\cos^2\theta\rangle)(R_{33} - R_{11}) \qquad (11)$$

where θ is the angle between the symmetry axis of the molecule and the direction of incidence of the light (which is parallel to the static electric field). The brackets $<\cos^2 \theta>$ represent the appropriate average of $\cos^2 \theta$ in the electric field. In the absence of the electric field $<\cos^2 \theta> = \frac{1}{3}$ and $R_E = R_0$. In the limit of infinite field or zero absolute temperature, $<\cos^2 \theta> = 1$ and $R_E = R_{33}$. For intermediate fields the dependence of $<\cos^2 \theta>$ on E has been discussed earlier in this book. In the low field limit $<\cos^2 \theta>$ is proportional to E^2 and Eq. (11) can be written as [9]

$$R_E = R_0 + \frac{2E^2}{45} \left[\frac{\mu_3^2}{(kT)^2} + \left(\frac{\alpha_{33}}{kT} - \frac{\alpha_{11}}{kT} \right) \right] (R_{33} - R_{11})$$

$$R_0 = \frac{2R_{11} + R_{33}}{3} \tag{12}$$

where

E = magnitude of the static electric field in esu volts per centimeter

μ_3, α_{33} = components of the permanent dipole moment and polarizability of the molecule along the symmetry axis

$\alpha_{11} = \alpha_{22}$ = corresponding components of the polarizability perpendicular to symmetry axis

k = Boltzmann constant

T = absolute temperature

It is clear that in order to obtain individual components of the optical rotation tensor (and therefore of the CD and ORD along different directions in the molecule), the orientation factor must be determined. Either the limit of infinite electric field must be measured, or it is possible to obtain the orientation factor from studies of the electrical birefringence of the same solution.

In the following sections we will discuss some of the published work on polypeptides and polynucleotides. It is worthwhile to summarize here the theoretical conclusions. The CD and ORD of

any molecule depend on the direction of propagation of light
through the molecule. In an unoriented solution the average CD or
ORD is measured. As the CD and ORD can be positive or negative,
the individual components will usually be larger and more
structurally significant than the average.

III.   ORIENTED POLYPEPTIDES AND NUCLEIC ACIDS

The optical activity of helical polymers has been the subject
of many theories and calculations [11] since Moffitt's pioneering
paper [12]. The theorists now seem to agree with the following
conclusions; their vocabulary may be different, but the results
are the same. Each allowed transition in the monomer will be
split into as many transitions as there are monomer units in the
polymer. Selection rules which depend on polymer geometry and
orientation relative to the incident light determine which transitions contribute to the absorption and which contribute to the CD.
Forbidden monomer transitions not significantly split in the polymer may also be seen in the CD and ORD spectra. Explicit calculations of all these terms have been made for polyalanine, the
simplest optically active polypeptide [13,14]. From measurements
of the average CD or ORD it is impossible to sort out all the
bands unambiguously. Measurements on oriented polypeptides, however, can give much more definite information.

The first electric field studies were done on poly-γ-benzyl-
L-glutamate dissolved in ethylene dichloride [15]. This polypeptide forms right-handed helices in solution. The cell was Teflon
with conducting glass electrodes. The electrodes were stannous
oxide-coated glass commercially obtained from Corning. The surface conductivity was about 10 ohms/square and the transmittance
was about 50% in the visible. The rotation was found to be linear
in the square of the electric field as expected from Eq. (12) and
the magnitude was easily measurable. For a 98 g/liter solution of

64,000 MW polybenzylglutamate in a 1-cm cell at 380 nm, 2000 V/cm caused a $0.15°$ change in rotation. In the wavelength region from 340 to 550 nm the rotation for light incident parallel to the helix axis was large and positive, whereas the rotation perpendicular to the helix axis was negative. This behavior was expected for right-handed helices [13], but a quantitative comparison was difficult to make. The presence of the benzene chromophores on the side chains of the polypeptides complicates calculations. If the benzene ring is uniformly oriented with respect to the helix axis, it can make a large contribution to ORD, but if it is fairly random its effect may be negligible. Measurements made on polybenzylglutamate in dioxane and chloroform showed that the ORD was indeed sensitive to solvent [16]. For poly-β-benzyl-L-aspartate (also believed to be right-handed), both the rotation parallel and perpendicular to the helix were negative [16]. Changes in orientation of the phenyl group could account for these results.

One way to avoid side-chain contributions is to measure the CD in a wavelength region out of their absorption bands. The early work was limited by the conducting glass electrodes to wavelengths above 300 nm. Hoffman and Ullman [10] made electrodes by depositing a thin coat of stannous oxide on optical silica. The surface conductivity was about 10^6 ohms/square, but they could be used down to 210 nm. The CD and ORD of poly-γ-benzyl-L-glutamate in ethylene dichloride and tetrahydrofuran were measured in the region around 220 nm. This is where the forbidden $n \to \Pi^*$ transition in the carbonyl of the amide is the dominant contribution. They found a ratio of $[\theta]_{33}/[\theta]_{11} = 1/6.5$ for the $n \to \Pi^*$ transition at 221 nm in good agreement with the predicted [13] ratio of rotational strengths of $R_{33}/R_{11} = 1/4.5$. The calculated ratio depends mainly on the angle the carbonyl of the amide makes with the helix axis. If the calculation could be tested or calibrated sufficiently, this would make a good probe of helix geometry in polypeptides.

The most thorough study of oriented CD has been made by
Mandel and Holzwarth [17]. They chose poly-γ-methyl-L-glutamate
in hexafluoroisopropanol and they used stainless steel grid electrodes for orientation. This combination allowed them to make
measurements to 185 nm and avoided the complication of phenyl side
chains. They measured the absorption and CD, then oriented the
molecules and measured the linear dichroism and circular dichroism. A set of bands was obtained which best fit all four sets of
measurements; the contribution of these bands to the CD is shown
in Fig. 1. The most significant result was the necessity of including the band labeled θ_H to get agreement with the data.
This band had been predicted by all the theoretical calculations
[11,13,14], but experimental evidence for it had not been obtained. Its existence provides strong support for the theories
and calculations of the CD of helical molecules.

The original skepticism about the presence of the θ_H band
existed because it could not be seen in absorption, and in CD it
was buried under the larger $\theta(\lambda_\perp)$ and $\theta(\lambda_{\parallel})$ bands. However,
it is the main contribution to the CD for light incident along the

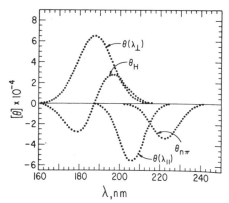

FIG. 1. The contribution of various bands to the average CD
of a polypeptide. The bands labeled $\theta(\lambda_\perp)$, $\theta(\lambda_{\parallel})$, and θ_H all
come from one $\pi \to \pi^*$ transition in each amide unit which is
split in the polymer. The $n \to \pi^*$ band is a locally forbidden
transition in the carbonyl group of the amide. (Reprinted with
permission from Ref. 17.)

helix axis near 190 nm. If, for simplicity, we only consider one $\Pi \rightarrow \Pi^*$ transition in each amide monomer unit, the following optical properties are predicted for an infinite helix. Two bands will be seen in absorption: one will be polarized parallel to the helix axis at $\lambda_{\parallel}$ and the other perpendicular to the helix axis at $\lambda_{\perp}$. However, the perpendicular one is actually two very slightly split bands. For light incident parallel to the helix axis, the electric vector is perpendicular to this axis and only the $\lambda_{\perp}$ band contributes to the CD. Therefore θ_{33} is just $3\theta_H$, which is centered at $\lambda_{\perp}$ as shown in Fig. 1. For light incident perpendicular to the helix axis the electric vector will have components both perpendicular and parallel to the helix axis. Therefore both $\lambda_{\perp}$ and $\lambda_{\parallel}$ bands will contribute to the CD and $\theta_{11} = (3/2) [\theta(\lambda_{\perp}) + \theta(\lambda_{\parallel})]$.

The magnitudes of the bands found experimentally [17] are in good agreement with those calculated [13,14] except that θ_H is only 1/4 the size predicted for an infinite helix. Mandel and Holzwarth [17] suggest the discrepancy could be due to flexibility in their polypeptide. The calculated optical properties of course depend on chain length [18]. The chain length dependence is different for the various bands and an effective chain length of 10 would be consistent with the experiments [13].

Nucleic acids are double stranded helices of polynucleotides which should show the same pattern of bands seen in polypeptides. In fact B-form DNA has the chromophoric bases essentially perpendicular to the helix axis and should show only a CD band of the kind labeled θ_H. This is indeed what is seen for DNA near 260 nm. If only one band were important in each base in the 260-nm region, one would expect that the CD for light incident parallel to the helix axis would just be three times the unoriented CD. Although CD measurements with electrical orientation of nucleic acids have not been reported, flow orientation [19] does give results in agreement with predictions [20].

IV. TRANSIENT BEHAVIOR

It should be obvious that the time-dependent properties of
the CD or ORD after application of an electric field pulse should
be the same as the other electro-optical effects. That is, rotary
diffusion coefficients are obtainable as before. For cylindrically symmetric rigid molecules no new information is obtained. The
decay of the effect will be [9]

$$\Delta R = \Delta R_0 e^{-6Dt} \tag{13}$$

where D is the rotary diffusion coefficient for rotation of the
symmetry axis. However, for less symmetric molecules or flexible
molecules different results may be expected. The measurable
relaxation times depend on the orientation of characteristic optic
axes relative to molecular axes. The optic axes will, in general,
be different for CD and absorption. Qualitatively, one can see
that certain motions of a flexible or asymmetric molecule might
change the CD greatly without changing the absorption. To study
these effects one would have to measure transients in both linear
and circular dichroism on the same sample. Experiments of this
kind have been made on polybenzylglutamate in ethylene dichloride
by Jennings and Bailey [21]. They have also extended the earlier
work done on this system in a static electric field. The decay
of the electric birefringence gave a single relaxation with a
rotary diffusion coefficient of about 100 sec^{-1}. The decay of the
electric rotation showed this relaxation, but in addition gave
clear evidence of a faster relaxation with a coefficient of about
400 sec^{-1}. The explanation of this is not known; it might be
caused by polydispersity or flexibility.

REFERENCES

1. D. J. Caldwell and H. Eyring, The Theory of Optical Activity,
    Wiley-Interscience, New York (1971).

2. I. Tinoco, Jr., Adv. Chem. Phys., 4, 113 (1962).

3. H. F. Hameka, Chem. Phys. Letters, 7, 157 (1970);
    A. D. Buckingham, C. Graham, and R. E. Raab, Chem. Phys.
    Letters, 8, 622 (1971).

4. J. A. Schellman and P. Oriel, J. Chem. Phys., 37, 2114 (1962).

5. N. Go, J. Chem. Phys., 43, 1275 (1965); J. Phys. Soc. Japan,
    21, 1579 (1966), 23, 88 (1967); A. D. Buckingham and
    M. B. Dunn, J. Chem. Soc. (A), 1971, 1988.

6. W. A. Shurcliff, Polarized Light, Harvard University Press,
    Cambridge, Mass. (1962).

7. I. Tinoco, Jr. and C. R. Cantor, Meth. Biochem. Anal., 18,
    93 (1970).

8. J. Kunz and A. McLean, Nature, 136, 795 (1935); J. Kunz and
    S. H. Babcock, Phil. Mag., 23, 616 (1936); J. Kunz and
    R. G. LeBaw, Nature, 140, 194 (1937).

9. I. Tinoco, Jr. and W. G. Hammerle, J. Phys. Chem., 60, 1619
    (1956).

10. S. J. Hoffman and R. Ullman, J. Polymer Sci., C31, 205 (1970).

11. Two recent papers which can be used to find the earlier
    references are C. W. Deutsche, J. Chem. Phys., 52, 3703
    (1970), and W. Rhodes, J. Chem. Phys., 53, 3650 (1970).

12. W. Moffit, J. Chem. Phys., 25, 467 (1956).

13. R. W. Woody and I. Tinoco, Jr., J. Chem. Phys., 46, 4927
    (1967); R. W. Woody, J. Chem. Phys., 49, 4797 (1968).

14. F. M. Loxsom, L. Tterlikkis, and W. Rhodes, Biopolymers, 10,
    2405 (1971).

15. I. Tinoco, Jr., J. Am. Chem. Soc., 81, 1540 (1959).

16. K. Yamaoka, Ph.D. Thesis, University of California, Berkeley
    (1964).

17. R. Mandel and G. Holzwarth, J. Chem. Phys., 57, 3469 (1972).

18. I. Tinoco, Jr., R. W. Woody, and D. F. Bradley, J. Chem.
    Phys., 38, 1317 (1963).

19. S-Y. Wooley and G. Holzwarth, J. Am. Chem. Soc., 93, 4066
    (1971).

20. W. C. Johnson and I. Tinoco, Jr., Biopolymers, 7, 727 (1969).

21. B. R. Jennings and E. D. Bailey, Nature, 228, 5278 (1970);
    J. Polymer Sci. (Polymer Symposia) C42, 1121 (1973).

Chapter 11

ELECTRO-OPTICS IN THE INFRARED REGION

Elliot Charney

Laboratory of Chemical Physics
National Institute of Arthritis, Metabolism and Digestive Diseases
National Institutes of Health
Bethesda, Maryland

I.   INTRODUCTION

Both electro-optic linear birefringence and linear dichroism in the infrared region of the spectrum differ in a singular aspect from that in other regions of the spectrum, a difference which would be trivial were it not that the potentially available information may be extremely useful. Its utility lies in at least two areas: the area of structural research, for which transition moment directions of normal vibrational modes may provide useful information, and the area of theoretical research on potential functions and on the perturbations caused by external fields, for which frequency and intensity changes caused by the applied fields in vibration-rotation lines are uniquely valuable. The triviality

arises from the fact that, except in details describing the origin
of infrared intensity and the phenomena of linear infrared birefringence and dichroism, the theory is essentially identical with
the familiar treatments of these effects for regions of the electronic spectrum.*

The phenomena associated with electronic transitions are
treated in considerable detail in other chapters of this volume,
most notably in those by A. D. Buckingham, C. T. O'Konski,
S. Krause, and W. Liptay. In this brief chapter, we intend to
outline a few of the details specific to infrared vibration-
rotation spectra and describe briefly a few of the highlights of
the very sparse literature on experimental results.

II. TRANSITION MOMENT DIRECTIONS

Electromagnetic radiation is absorbed along molecule-fixed
directions, so that in oriented systems the direction of maximum
absorption of polarized radiation may be used to determine the
transition-moment directions. The absorption along two different
molecule-fixed directions may be related in a variety of ways to
describe the dichroism [1]. Since linear birefringence is related to linear dichroism by a Kronig-Kramers transformation, the
same information can, in principle, be obtained from either experiment. In practice, one or the other of the techniques is prefer-
able, depending on the conditions. For infrared bands the
transition moment directions are fixed by the normal-mode directions for fundamental vibrations and by a variety of symmetry
considerations for overtone and combination bands. The first
suggestion that these directions could be related to the so-called

*There are, of course, very low-energy electronic transitions fre-
quently associated with liganded metal atoms which appear in the
infrared region. For these cases, the treatment would, of course,
be that associated with the electronic transition, and the fact
that it appears in the infrared region is of no special consequence.

"anomalous electro-optic birefringence" in the vicinity of
resonant absorption frequencies appears to have been made by
Kastler [2]. Charney and Halford [3] demonstrated this with Kerr
birefringence measurements in regions of the higher frequency
fundamental vibrations of nitrobenzene and developed a semiclas-
sical theory to relate the transition moments to the birefringence.
The result for the birefringence, Δn, in the vicinity of a
vibrational transition designated by the index k was related to
the associated transition moment and to polarizabilities associated with transitions at higher energies by

$$\Delta n(\omega) \sim \frac{N\ell}{\lambda \varepsilon_0} \{<\alpha_p(\omega)> - <\alpha_s(\omega)> + [<(\mu_p)_k^2> - <(\mu_s)_k^2>]\alpha'(\omega)_k]\}$$

where N is the number of molecules in unit volume, ℓ is the
pathlength of the radiation of wavelength λ in the medium, and
ε_0 is the permitivity of free space; $\alpha_p(\omega)$ and $\alpha_s(\omega)$ are the
polarizability components from higher energy bands parallel (p)
and perpendicular (s) to the direction of the applied orienting
field and the brackets < > are used to designate averages over
the orientation distribution function; μ_p and μ_s are the
corresponding transition moments defined by the integral
$<\nu|\mu|\nu'>$ where ν and ν' are wavefunctions for the vibrational
ground and excited states associated with the k'th vibrational
transition; $\alpha'(\omega)_k$ is the frequency-dependent, orientation-
independent part of a resonant term in the refractive index and
thus gives rise to the familiar anomalous dispersion, i.e., to a
peak and trough on either side of the absorption maximum. It is
obvious that the sign of the contribution to this dispersion is
controlled by the relative magnitudes of the components of the
transition moments parallel and perpendicular to the applied
field. Charney and Halford [3] showed, by averaging over all
orientations of a dipolar molecule in the applied field, that this
contribution is positive when the transition moment and permanent
molecular dipole moment are parallel and negative when they are

perpendicular to each other. If the direction of the permanent
moment is known a straightforward measurement of electro-optic
birefringence established the relative orientation of the transition moment. Under favorable conditions, e.g., in molecules of
fairly high symmetry, the possible transition-moment directions
are limited to major symmetry axes permitting unambiguous determination of the transition-moment direction from the infrared Kerr
effect measurement. In molecules of lower symmetry, limits may be
set to the relative orientations. Experimentally, the infrared
overtone and higher frequency fundamentals (to ~ 1850 cm^{-1}) of
nitrobenzene were examined and several bands found which gave rise
to the predicted effects [3].

A more extensive examination of the frequency dependence of
the electric birefringence in regions of absorption bands, which
included a quantum theoretical treatment of infrared vibration-
rotation effects, was made by Buckingham [4]. The same result was
predicted with regard to the dependence of the sign of the birefringence on the relative orientation of the transition moments
and permanent dipole moment, but, in addition, a potentially
significant quantitative test of perturbation theory was developed, in particular of the applicability of perturbation theory to
the interaction of static fields with covalent molecules.

III.   ELECTRIC FIELD-INDUCED PERTURBATIONS

Buckingham's [5] results for the effect of the applied field
on the individual lines of the P and R branches of the
vibration-rotation spectrum of a polar diatomic molecule are,
respectively,

$$\Delta n = \frac{2\pi^2 N_{vJ} E^2}{h^2 \lambda (\nu_i - \nu)^2} \cdot \frac{<\nu|\mu|\nu'>^2}{5(2J-1)(2J+1)^2} \left\{ \frac{\mu_e^2}{hcB_e} - 2J^2 (\alpha_{\parallel}^{(o)} - \alpha_{\perp}^{(o)})_e \right\}$$

$$\Delta n = - \frac{2\pi^2 N_{\nu J} E^2}{h^2 \lambda (\nu_i - \nu)^2} \frac{<\nu|\mu|\nu>}{5(2J+1)^2(2J+3)} \left\{ \frac{\mu_e^2}{hcB_e} - 2(J+1)^2 (\alpha_\parallel^{(o)} - \alpha_\perp^{(o)})_e \right\}$$

This treatment predicts an almost inverse dependence of the birefringence on the third power of the rotational quantum number J except at high temperatures since $N_{\nu J}$ increases as $2J + 1$. Intuitively, the decrease in Kerr birefringence for molecules excited to higher rotational quantum states is not unexpected, but the quantitative prediction of the rapid decrease with increasing angular momentum has yet to be verified. The experiment is a very difficult one to perform. The really superior polarizers necessary to measure the small birefringence are not readily available for this spectral region.

Condon [6] had earlier predicted another electro-optic effect for the infrared region. Symmetry forbideness of infrared bands arises from the null value of transition dipole moment, due to the absence of a finite value of the dipole moment derivative associated with a symmetric vibration of a nonpolar group of atoms, the simplest such being, of course, the vibration of a homonuclear diatomic molecule such as hydrogen. Condon predicted that in strong enough static electric fields, a moment P_ε would be induced in the direction of the light vector which would give rise to a finite value of the transition moment and thus to induced absorption in the infrared region. This effect went unobserved until 1953, when Crawford and Dagg [7] made the first observation. This was followed by a more detailed observation [8] at fields of the order of 100,000 V/cm and hydrogen densities of about 84 Amagat. Under these conditions a band appears at the location of the forbidden fundamental vibration at 4161 cm^{-1}.

Still other perturbations by strong electric fields have been found. Terhune and Peters [9] and Maker [10] have observed Stark effect shifts of a few hundredths of a reciprocal centimeter in the vibration rotation spectra of simple molecules with fields of

the order of tens of thousands of volts per centimeter. More
recently [11], this effect has again been observed in infrared
absorption of the hydroxyl group in alcohols. The most recent
work on electric field effects in this spectral region has been
an investigation of field perturbations of the symmetry in which
an ion finds itself in aqueous solutions of ionic salts. In the
work reported by Lockwood and Irish [12], the doublet of the
$v_3(E^1)$ mode of the nitrate ion in the Raman spectrum was examined
to determine whether weak fields at frequencies which produce
dielectric responses (200-500 MHz) would show up as a change or
smearing of the split doublet. No such effect was observed under
the conditions of the experiment.

IV.   MEASUREMENTS ON LARGE MOLECULES

Polymers, and biopolymers in particular, represent a special
case in which the large degree of orientation achievable with
fields of reasonable strength makes it possible to study the
structure of the molecule through the behavior of its dichroic
absorption. Several other chapters in this book deal directly
with related investigations in electronic spectral regions.
Measurements in the infrared may offer an advantage for two reasons. One is the wealth of absorption bands due to fundamental
vibrations which may be largely attributed to isolated groups of
atoms. The interpretation is thus sometimes more straightforward
for infrared than for UV absorption bands. Second, it is frequently possible from symmetry considerations to specify with
considerable confidence the transition moment direction with respect to the atomic configuration for such fundamental vibrations.
In cases where this is possible, measurement of the dichroism will
permit an analysis of the orientation of such groups in the overall molecular structure in a straightforward way from the dichroism or anomalous birefringence measurements. Unfortunately the

experimental problem of making such measurements in solutions is extraordinarily difficult because of strong absorption by the solvent. As a result very few experimental results have been reported thus far.

One such study was made by Spach [13]. This gave only qualitative results, but the results were in accord with assumed α-helical structure of the poly-γ-benzyl-L-glutamate under the conditions of the experiment. The polymer should orient with its helical axis parallel to the field direction. Consequently for any transitions polarized parallel or nearly parallel to this direction, there will be greater absorption when the polarization direction of the light coincides with the direction of the applied field. Correspondingly, transitions perpendicular to the helical axis will exhibit less absorption. The N-H hydrogen bonded band at 3300 cm^{-1} and the CO amide I band at 1655 cm^{-1}, both of which should be polarized nearly parallel to the α-helix, did indeed become more intense for radiation polarized parallel to the field. The complex amide II band, which should have its transition more nearly perpendicular to the helix axis, became less intense. In a more recent investigation [14] of the electric dichroism of the same molecule, excellent agreement was found between the orientation of the transition moment calculated from the dichroism results and that calculated from theoretical considerations for the α-helical structure. The angle which the transition moment of the N-H hydrogen bonded band makes with the helix axis was found to be approximately 28-29°; that of the Amide I band, approximately 34-36°. Spach's earlier results are in qualitative agreement with these.

To our knowledge, no further measurements of electric-field induced dichroism in the infrared have been reported. The potential for information from this source has thus scarcely been touched.

V. SUMMARY

The special characteristics of birefringence and dichroism produced by strong electric fields when the illuminating radiation is infrared are examined and reviewed. These characteristics are (1) the relatively easily determinable symmetry of molecular vibrations which makes transition moment directions available for structural investigations and (2) the ability to resolve the fine rotational structure of infrared vibration rotation bands of small molecules which should make it possible to examine critically the coincidence between theory and the actual behavior of a molecular system in strong electric fields. The former of the two characteristics should be of considerable advantage for structural determination on repeating polymers in solution.

REFERENCES

1. E. Charney, Proc. Nucleic Acid Res., $\underline{2}$, 176 (1971).

2. A. Kastler, Compt. Rend., $\underline{230}$, 1596 (1950).

3. E. Charney and R. S. Halford, J. Chem. Phys., $\underline{29}$, 221 (1958).

4. A. D. Buckingham, Proc. Roy. Soc. (London), $\underline{A267}$, 271 (1962).

5. The fact that the sign of the expression for the R branch was incorrectly given in the original paper was noted in a subsequent paper by A. D. Buckingham and D. Dows, Discussions Faraday Soc., $\underline{35}$, 48 (1963). In these equations, ν and ν' are the ground and excited state vibrational wavefunctions, μ_e is the equilibrium dipole moment, B_e is the rotational constant in the ground vibrational state, and $\alpha_{\parallel}^{(o)}$ and $\alpha_{\perp}^{(o)}$ are, respectively, the static polarizability parallel and perpendicular to the axis of the axially symmetric molecule.

6. E. U. Condon, Phys. Rev., $\underline{41}$, 759 (1932).

7. M. F. Crawford and I. R. Dagg, Phys. Rev., $\underline{91}$, 1569 (1953).

8. M. F. Crawford and R. E. MacDonald, Can. J. Phys., $\underline{36}$, 1022 (1958).

9. T. W. Terhune and C. W. Peters, J. Mol. Spectry., $\underline{3}$, 138 (1959).

10. P. D. Maker, Ph.D. Dissertation, University of Michigan, Ann
    Arbor (1961).

11. P. Handler, and D. E. Aspnes, J. Chem. Phys., $\underline{47}$, 473 (1967).

12. D. J. Lockwood and D. E. Irish, Chem. Phys. Letters, $\underline{24}$, 123
    (1974).

13. G. Spach, Compt. Rend., $\underline{249}$, 667 (1959).

14. A. Maschka, G. Bauer, and Z. Dora, Monatsh. Chemi., $\underline{101}$, 1516
    (1971).

Chapter 12

NONLINEAR ELECTRO-OPTICS

Stanislaw Kielich

Nonlinear Optics Department
Institute of Physics
A. Mickiewicz University at Poznań
Poznań, Poland

I.   INTRODUCTION

The classical electro-optical effects have been extensively discussed in the chapters of this book by O'Konski, Buckingham, O'Konski and Krause, Stellwagen, Yoshioka, Tinoco and Jennings. The microscopic mechanisms underlying these effects give rise to yet other, nonlinear electro-optical effects.   The two principal processes are essentially these:

1. Voigt's distortional process [1]:   the molecules, macromolecules, or particles undergo a nonlinear distortion of their electron shells proportional to the square of the applied electric field intensity.

2. Langevin's [2] and Born's [3] statistical-orientational process, due to anisotropy of the polarizability and permanent electric dipole moment interactions with the applied electric field.

In the classical electro-optical effects, the light wave plays the part of a measuring probe, whereas the dc or ac electric field is the agent inducing anisotropy or nonlinearity in the naturally isotropic body.   With lasers [4], we have at our disposal continuous and pulsed sources of monochromatic light, coherent in time and space, conveying energy fluxes of immense density in highly parallel beams.   The high intensity of these beams permits their use as an agent for modifying the properties of matter, besides their use as measuring agent.   Phenomenologically speaking, a quantity Q, defining a physical property of the naturally isotropic body, becomes a function of the intensity I of the intense incident light wave:

$$Q = Q(I) \tag{1}$$

The incident beam of intensity I perturbates the initial state of the system and induces a nonlinear variation in the quantity Q, and in accordance with the classical perturbation method we write:

$$Q(I) = Q_0 + Q_1 I + Q_2 I^2 + \ldots = \sum_{n=0}^{\infty} Q_n I^n \tag{2}$$

where Q_0 is the physical quantity in the absence of the perturbating light beam and the expansion coefficients Q_1, Q_2, $\ldots$
describe the structure and thermodynamical state of the body when
modified by the action of the beam.

Very intense light, when traversing an isotropic body, causes
a change in its refractive index [5] and electric permittivity
[6], i.e., laser light causes an isotropic medium to become optically nonlinear and, at the same time, anisotropic. This has been
experimentally established by Mayer and Gires [7] following
theoretical suggestions of Buckingham [5] as optical birefringence
induced in liquids by a ruby laser beam. This novel effect of
optically induced optical birefringence, or optical Kerr effect,
has been studied thoroughly in several liquids by Paillette [8] as
well as Martin and Lalanne and coworkers [9]; it provides, for
example, valuable data on the nonlinear optical polarizability of
atoms and molecules [10], molecular correlations [11], and very
fast processes [12]. Moreover, intense laser light causes self-
induced effects, such as intensity-dependent rotation of the
polarization ellipse [13], self-focusing and self-trapping of
laser beams [13-15], optical harmonics generation [16], stimulated
light scattering [17,18], Rayleigh and Raman harmonic scattering
[19,19a,20], light intensity-dependent optical rotation [21-23],
and various other nonlinear optical effects rendering apparent the
statistical properties of the electromagnetic field and matter
[24-26].

A new class of nonlinear electro-optical effects is made
accessible by applying simultaneously an intense laser beam and a
dc or ac electric field. In the first place, laser light reveals
the changes in electric permittivity predicted by Piekara and
Kielich [6] and observable as optically induced electric anisotropy [27]. Centrosymmetric bodies acted on by a dc or ac electric
field generate the second harmonic of laser light, as observed in

cubic crystals [28-30], liquids [28-31], and gases [31-33].

By utilizing the optical electric field of laser light and, in addition, electric field in electro-optical investigations, much is to be gained with respect to technique. First, laser technique permits the electrodeless application of a strong electric field to the substance under study, enabling us to achieve complete alignment of macromolecules and particles [34]. This is particularly important in the study of suspensions of conducting particles, and in cases when we want to eliminate perturbating effects like the linear electrophoretic effect, dielectric breakdown, etc., from optical measurements. Obviously, in applying laser techniques to colloid solutions, one has to choose a wavelength of the laser beam for which the substance is transparent. Caution is needed when using very intense laser radiation, which not only causes new nonlinear processes but, upward of some critical power threshold, can destroy the substance [35].

Although nonlinear optical effects have not as yet been studied experimentally to any considerable extent in macromolecular and colloidal substances, we nevertheless intend in this chapter to deal with the essential achievements of this rapidly developing branch of science. In the next few years, it should be possible to develop investigations of the nonlinear electro-optics of molecules and macromolecules, particularly of crystal polymers and liquid crystals.

II.   PHENOMENOLOGICAL DESCRIPTION
OF LINEAR AND NONLINEAR PROCESSES

A.   Strong and Weak Electric Fields

Whether the polarization of the medium shall be linear or nonlinear hinges decisively on the ratio of intensities of the externally applied electric field E and internal field F specific to the material. For instance, the average electric field

intensity with which the nucleus of a hydrogen atom acts on the
electron is of the order of $F_e \approx 10^9$ V/cm. In dense matter,
these fields are weaker than in isolated atoms, amounting in crystals to 10^8 V/cm and, in semiconductors, to 10^7 V/cm. In liquids, the Onsager reaction field acting on the electric dipole of
a molecule is of order 10^7 V/cm. Hence, internal fields are
generally large compared with the electric fields of order
$E \approx 10^5$ V/cm which are able to cause linear electric polarization
of matter in accordance with Lorentz's theory [36] in which electrons perform harmonic vibrations. Considerable nonlinear
polarization can be obtained with static applied fields of
$E \approx 10^6$ V/cm and more, but this is not generally feasible with
regard to electric breakdown.

The light emitted by conventional sources and low power gas
lasers conveys intensities in the range of $I \approx 1$ to 10^2 W/cm^2
and thus electric fields of $E \approx 1$ to 10^2 V/cm. Fields like
these are weak in comparison with the strong internal fields and,
if employed as analyzing agent, do not effect the properties of
the body investigated. Crystal lasers of high power nowadays provide light beams of intensity $I \approx 10^8$ to 10^{12} W/cm^2 and electric fields of $E \approx 10^5$ to 10^8 V/cm comparable to the internal
fields and consequently cause nonlinear electric polarization in
matter. In an electric field of such intensity, by the classical
theory of Voigt [1], the harmonicity and isotropicity of vibrations of the bound electrons undergo an impairment, and we have to
deal with anisotropic oscillators performing anharmonic, nonlinear
vibrations.

B. Electric Polarization,
and the Electric Permittivity Tensor

For generality, we represent an electric field depending
harmonically on the time t and spatial coordinates $\underline{r}$ by writing it in the form of a Fourier series, as follows:

$$\underline{E}(\underline{r},t) = \sum_a \underline{E}(\omega_a,\underline{k}_a) \exp \{i(\underline{k}_a \cdot \underline{r} - \omega_a t)\} \tag{3}$$

where ω_a is the angular oscillation frequency and $\underline{k}_a$ the wave vector of the ath mode. Summation in (3) extends over all frequencies and wavevectors, both positive and negative, with $\omega_{-a} = -\omega_a$, $\underline{k}_{-a} = -\underline{k}_a$. For the field amplitude one has $E^*(_a,\underline{k}_a$ $E^*(\omega_a,\underline{k}_a) = \underline{E}(-\omega_a,-\underline{k}_a) = E_A/2$.

Provided the electric field intensity (3) is not excessive, the electric polarization induced in the medium at the space-time point $(\underline{r},t)$ can be expanded in a power series:

$$\underline{P}(\underline{r},t) = \underline{P}^{(1)}(\underline{r},t) + \underline{P}^{(2)}(\underline{r},t) + \underline{P}^{(3)}(\underline{r},t) + \ldots = \sum_{n=1}^{\infty} \underline{P}^{(n)}(\underline{r},t)$$

$$\tag{4}$$

where the polarizations of successive orders depend linearly, quadratically, cubically, etc., on the electric field intensity (3).

A naturally isotropic body has the scalar electric permittivity ε, whereas that of a naturally anisotropic body is a tensor of the second rank ε_{ij}, where the indexes i and j label the permittivity components along the Cartesian axes x, y, z. In matrix notation we have

$$\varepsilon_{ij} = \begin{vmatrix} \varepsilon_{xx} & \varepsilon_{xy} & \varepsilon_{xz} \\ \varepsilon_{yx} & \varepsilon_{yy} & \varepsilon_{yz} \\ \varepsilon_{zx} & \varepsilon_{zy} & \varepsilon_{zz} \end{vmatrix} \tag{5}$$

with, in general, 9 tensor elements, namely, 3 diagonal and 6 nondiagonal ones. If the tensor is symmetric, $\varepsilon_{ij} = \varepsilon_{ji}$, as is the case in the absence of absorption and optical activity, 6 of the 9 nonzero elements are mutually independent. A further reduction in number of the permittivity tensor elements (5) depends on the symmetry class of the body [37]. In the simplest case of an isotropic body, the 6 nondiagonal elements of the tensor matrix (5) will vanish, and the 3 remaining diagonal ones will be equal to one another, so that we can write:

$$\varepsilon_{ij} = \varepsilon \delta_{ij} \tag{6}$$

δ_{ij} being the unit Kronecker tensor with elements equaling 1 for $i = j$ and 0 for $i \neq j$.

In the general case, we have the following tensorial relation:

$$(\varepsilon_{ij} - \delta_{ij}) E_j(\underline{r}, t) = 4\pi P_i(\underline{r}, t) \tag{7}$$

between the tensor of electric permittivity, the components of the electric field, and those of the electric polarization induced in the medium. This relation is valid for anisotropic bodies, as well as for isotropic ones subjected to the action of a strong electric field (3) inducing, in general, nonlinear polarization (4), so that induced electric anisotropy is quite generally described by a permittivity tensor ε_{ij}. In Eq. (7) we have applied (and shall henceforth apply) Einstein's summation convention, stating that an index appearing twice (here j) indicates summation over $j = x, y, z$.

C. Linear Effects

If the electric field acting on the medium is of the form given by Eq. (3), the electric polarization of first order occurring in the expansion (4) can be written as follows:

$$\underline{P}^{(1)}(\underline{r}, t) = \sum_a \underline{P}^{(1)}(\omega_a, \underline{k}_a) \exp\{i(\underline{k}_a \cdot \underline{r} - \omega_a t)\} \tag{8}$$

where the polarization $\underline{P}(\omega_a, \underline{k}_a)$ induced at frequency ω_a is a linear function of the electric field intensity. The ith polarization component on the axes x, y, z is of the general form:

$$P_i^{(1)}(\omega_a, \underline{k}_a) = \chi_{ij}(\omega_a, \underline{k}_a) E_j(\omega_a, \underline{k}_a) \tag{9}$$

Above, the second-rank tensor $\chi_{ij}(\omega_a, \underline{k}_a)$ defines the linear electric susceptibility of the medium. Its dependence on the oscillation frequency ω_a is referred to as time dispersion or frequency dispersion, whereas its dependence on the wavevector $\underline{k}_a$ is termed spatial dispersion [38].

By Eqs. (7) to (9), we have in a linear approximation:

$$\varepsilon_{ij}(\omega_a, \underline{k}_a) = \delta_{ij} + 4\pi\chi_{ij}(\omega_a, \underline{k}_a) \tag{10}$$

a relation between the electric permittivity and susceptibility tensors.

In Eq. (10) the permittivity tensor, and consequently the refractive index, do not depend on the electric field intensity of the light wave. This obviously results from the linear polarization (9), which causes the solutions of Maxwell's equations to obey, in a first approximation, the superposition principle [cf. Eq. (8)], and states that different waves propagate simultaneously in a medium independently of one another, without distortion. Under the above-stated conditions, all optical phenomena and laws partake of the nature of a first approximation and belong to linear optics [39].

D. Quadratic Effects

At sufficiently high intensities of the electric field (3) acting on the medium, we have to include higher-order polarizations into our considerations in the expansion (4), in particular, polarization of order 2, of the form:

$$\underline{P}^{(2)}(\underline{r},t) = \sum_{ab} \underline{P}^{(2)}(\omega_a + \omega_b) \exp\{i[(\underline{k}_a + \underline{k}_b) \cdot \underline{r} - (\omega_a + \omega_b)t]\} \tag{11}$$

The nonlinear polarization $\underline{P}^{(2)}(\omega_a + \omega_b)$ induced in the medium at frequency $\omega_a + \omega_b$ is a quadratic function of the field intensity. Its component, in the absence of spatial dispersion, is:

$$P_i^{(2)} (\omega_a + \omega_b) = \chi_{ijk}(\omega_a, \omega_b) E_j(\omega_a) E_k(\omega_b) \tag{12}$$

where the third-rank tensor $\chi_{ijk}(\omega_a, \omega_b)$ defines the nonlinear electric susceptibility of order 2.

In optically inactive media having a center of symmetry, all the tensor elements χ_{ijk} vanish. Hence, second-order polarization (12) can occur only in bodies not possessing a center of symmetry (e.g., piezoelectric and ferroelectric crystals), in regular crystals, and in isotropic optically active bodies [40]. For the case $\omega_a = \omega_b = \omega$, the nonlinear susceptibility tensor is symmetric with respect to the pair of indexes j,k, whence we have $\chi_{ijk} = \chi_{ikj}$, leaving but 18 of the 27 nonzero tensor elements mutually independent [41]. In optically transparent media (in regions remote from electron dispersion and absorption) one has symmetry in all the indexes, and the totally symmetric tensor χ_{ijk} then presents 10 mutually independent elements. A further reduction in number of the nonzero and independent elements of the symmetric tensor χ_{ijk} depends solely on the crystallographic symmetry class of the body [41].

On considering the shape of Eq. (11) one notes that the polarization of order 2 defines a first deviation from the superposition principle for waves. In particular, if but a single wave is propagating in the medium, we have:

$$\underline{P}^{(2)} (\underline{r}, t) = 2\underline{P}^{(2)} (0) + 2\underline{P}^{(2)} (2\omega) \cos 2(\omega t - \underline{k} \cdot \underline{r}) \tag{11a}$$

This is a remarkable result showing that a wave of fundamental frequency ω incident on the medium produces therein, in a second approximation, a polarization which is constant in time

$$P_i^{(2)} (0) = \chi_{ijk}(\omega, -\omega) E_j(\omega) E_k(-\omega) \tag{13}$$

and, moreover, a polarization varying in time at doubled vibration frequency:

$$P_i^{(2)}(2\omega) = \chi_{ijk}(\omega,\omega)E_j(\omega)E_k(\omega) \tag{14}$$

1. Second Harmonic Generation

Franken et al. [42], when observing the propagation in quartz crystal of red light of wavelength λ_ω = 6943 Å from a ruby laser, found a new, harmonic radiation component of a frequency double that of the incident wave, in the occurrence ultraviolet light of wavelength $\lambda_{2\omega}$ = 3472 Å (Fig. 1). Quartz is a crystal not having a center of symmetry, and owing to the nonlinear polarization (14), a wave of frequency ω generates therein a wave 2ω, referred to as the second optical harmonic. The discovery of second-harmonic generation has played an important role in the study of the nonlinear optical properties of numerous materials [16,24,43,44]. Investigation of second-harmonic generation in reflected light yields information regarding the structure of surface layers [16]. When studied in transmitted light, it permits the elucidation of volume structure in crystalline bodies [44] and of molecular structure in organic crystals [45].

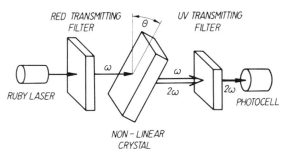

FIG. 1. Arrangement for second-harmonic generation in nonlinear crystals (e.g., quartz) [42]. The path traversed by the beam within the crystal plate is modified by varying the latter's inclination with respect to the propagation direction of the laser beam. The active thickness of the crystal is made to ensure proper phase-matching between the fundamental frequency ω and second-harmonic 2ω wave. The second-harmonic beam generated in the crystal is observed on transmission through a filter, totally absorbing the fundamental (red) beam, and its intensity is recorded by means of the monochromator and photomultiplier.

Second-harmonic generation requires large single crystals of high optical quality. However, when the growing of large single crystals is difficult or not feasible, the crystal powder method can be used [46]; it is simple and handy, and permits rapid assays of the nonlinear properties of new (polycrystalline, multidomain, etc.) materials. Interesting studies have been performed on second-harmonic generation due to angular fluctuations in NH_4Cl close to the critical point [47,47a] and domains in ferroelectric crystals [47b]. Of late, Bergman et al. [48] reported second-harmonic generation studies in crystalline polar polymers (polyvinylidene fluoride, mylar, etc.) and, recently, anomalously high nonlinear optical effects in m-nitroaniline [49], 5-nitroura-cil [49a] and benzene derivatives [49b]. Also, first results on optical second harmonic generation in biological systems have been reported [50].

2. Optical Rectification

The time-independent nonlinear polarization (13) accompanying second-harmonic generation has been observed by Bass et al. [51] in ferroelectric KDP (KH_2PO_4) crystal. Ward and coworkers [52] made direct absolute measurements of the optical-rectification tensor elements $\chi_{ijk}(\omega,-\omega)$ in KDP, ADP ($NH_4H_2PO_4$), and quartz. The effect is termed optical rectification by analogy to the rectification of ac current and radiofrequency electromagnetic fields so highly important to electro- and radiotechniques. By using a light beam of high intensity one can polarize electrically a body without center of symmetry.

3. Linear Electro-Optical Effect

On putting $\omega_a = \omega$, $\underline{k}_a = \underline{k}$, and $\omega_b = \underline{k}_b = 0$ in Eqs. (11) and (12), one obtains

$$P_i^{(2)}(\omega) = \chi_{ijk}(\omega,0)E_j(\omega)E_k(0) \tag{15}$$

This polarization defines the electro-optical processes referred
to as the Pockels effect [37].

On inserting into Eq. (7) the polarizations (9) and (15), one
obtains the following change in electric permittivity tensor due
to a dc electric field:

$$\Delta\varepsilon_{ij}^{\omega}(E) = 4\pi\chi_{ijk}(\omega,0)E_k(0) \tag{16}$$

Pockels' effect given by Eq. (16) is observable in crystals without a center of symmetry and in optically active bodies as linear
electro-optical rotation. In recent years, the study of the
electro-optical properties of materials has gained in significance
with regard to the production of coherent light harmonics, the
modulation of laser beams, parametric processes, and so forth
[53]. Lately, Di Domenico and Wemple [54] developed a microscopic
theory of the electro-optical properties of ferroelectrics of the
oxygen octahedron type. This has also been done for molecular
crystals and hexagonal semiconductors [55].

E.   Cubic Effects

In bodies possessing a center of symmetry, in the absence of
natural optical and magnetic activity, no second-order polarization (11) occurs in the electric dipole approximation (12).
Bodies of this kind exhibit in the expansion (4), besides the
linear polarization (8), an electric polarization of order 3
which we write as follows:

$$\underline{P}^{(3)}(\underline{r},t) = \sum_{abc} \underline{P}^{(3)}(\omega_a + \omega_b + \omega_c)$$
$$\exp\{i[(\underline{k}_a + \underline{k}_b + \underline{k}_c)\cdot\underline{r} - (\omega_a + \omega_b + \omega_c)t]\} \tag{17}$$

This nonlinear polarization $P^{(3)}(\omega_a + \omega_b + \omega_c)$ induced in an
isotropic medium at frequency $\omega_a + \omega_b + \omega_c$ is a cubic function
of the electric field intensity, and can be written explicitly as

follows in the absence of spatial dispersion:

$$P_i^{(3)}(\omega_a + \omega_b + \omega_c) = \chi_{ijkl}(\omega_a, \omega_b, \omega_c) E_j(\omega_a) E_k(\omega_b) E_l(\omega_c) \quad (18)$$

The fourth-rank tensor χ_{ijkl} defining nonlinear third-order susceptibility has nonzero elements in all bodies, including isotropic ones, thus gases and liquids. In the general case χ_{ijkl} has 81 nonzero tensor elements. Any accretion of symmetry conditions reduces their number for the various crystallographical classes [24,41]. In the regular system and in isotropic bodies, χ_{ijkl} has 21 nonzero elements, only 3 of which are mutually independent [56]:

$$\chi_{ijkl} = \chi_{xxyy}\delta_{ij}\delta_{kl} + \chi_{xyxy}\delta_{ik}\delta_{jl} + \chi_{yxxy}\delta_{il}\delta_{kj} \quad (19)$$

with the relation [41]:

$$\chi_{xxxx} = \chi_{yyyy} = \chi_{zzzz} = \chi_{xxyy} + \chi_{xyxy} + \chi_{yxxy} \quad (19a)$$

1. Tripling of Oscillation Frequency

Equation (17) defines a new deviation from the superposition principle, and Eq. (18) expresses the interaction of three waves in a cubic medium leading to the emergence of a fourth wave of frequency $\omega_a + \omega_b + \omega_c$. In particular, if a single wave $\omega_a = \omega_b = \omega_c = \omega$ is incident on such a medium, Eq. (18) yields:

$$P^{(3)}(\underline{r},t) = 3P^{(3)}(\omega) \cos(\omega t - \underline{k} \cdot \underline{r})$$
$$+ P^{(3)}(3\omega) \cos 3(\omega t - \underline{k} \cdot \underline{r}) \quad (20)$$

We thus note that, under the influence of a wave of fundamental frequency ω, two waves of nonlinear polarization arise in the cubic medium, one of them at the frequency:

$$P_i^{(3)}(\omega) = \chi_{ijkl}(\omega, \omega, -\omega) E_j(\omega) E_k(\omega) E_l(-\omega) \quad (20a)$$

and the other at the tripled frequency

$$P_i^{(3)}(3\omega) = \chi_{ijkl}(\omega,\omega,\omega)E_j(\omega)E_k(\omega)E_l(\omega) \qquad (20b)$$

Terhune and his coworkers [28,57] were the first to attempt the production of a ultraviolet third harmonic of wavelength $\lambda_{3\omega} = 2313$ Å using a ruby laser, in calcite as well as in dielectric liquids. Bey et al. [58] provided the earliest experimental evidence that phase matching can be achieved in harmonic processes by the introduction of anomalous dispersion into a normally unmatched medium; they obtained phase-matched third-harmonic generation at 3530 Å of the neodymium laser line at 10,600 Å by the addition of dye molecules to a liquid medium (fuchsin red dye in solution in hexafluoroacetone sesquihydrate). These experiments were extended by Chang and Galbraith [59] to other solvents of different index mismatch and to another dye, methylene blue.

Goldberg and Schnur [60] investigated harmonics generation in liquid crystals, in particular, cholesteryl nonanoate (CN), and other cholesteric and nematic systems. The solid phase of CN exhibited a high second-harmonic intensity (of the same order as in quartz crystal powder), which, on heating, fell steeply by 4 orders of magnitude near $79°C$ on transition from solid to cholesteric phase. In the inverse process of slow cooling from the isotropic liquid phase through the cholesteric and smectic mesophases, a clear second-harmonic signal was observed only at the point of recrystallization to the solid. Third-harmonic generation was apparent from all three phases both at heating and cooling, its intensity in the solid being 100 times stronger than in the liquid phase. Similar results have been obtained in other liquid crystals [61]; in no case did they yield a second-harmonic when in mesomorphic phases, which have to be regarded as centrosymmetric. Recently, Rentzepis and coworkers [62] reported an efficiency for third-harmonic conversion in pure organic liquids 100 times larger than in fused silica. The above-cited studies prove that some role in the process of harmonics generation is played by the molecular correlations so highly relevant to

multiharmonic light scattering and optical birefringence in isotropic media [11,63].

Ward and New [64] observed third-harmonic generation at
2314 A in simple gases (He, Ne, Ar, etc.), using a focused ruby
laser beam. A technique for the efficient production of ultraviolet radiation by third-harmonic generation in phase-matched metal
vapors was proposed by Harris et al. [65]. Also, the conditions
for frequency tripling in infrared by anharmonic molecular vibrations have been analyzed [66].

2. Direct-Current Field-Induced Second Optical Harmonic

With $\omega_a = \omega_b = \omega$ and $\omega_c = 0$ in Eq. (18), the third-order
polarization at frequency 2ω,

$$P_i^{(3)}(2\omega) = \chi_{ijkl}(\omega,\omega,0)E_j(\omega)E_k(\omega)E_l(0) \tag{21}$$

describes the generation of the second harmonic of light by a
centrosymmetric body immersed in a dc electric field $E(0)$.

The process of second-harmonic generation by electrically
polarized bodies was first observed by Terhune et al. [28] in calcite crystal, which has a center of symmetry and belongs to the
crystallographical class $3m(D_{3d})$. Their experiment, repeated by
Bjorkholm and Siegman [29] with accurate CW measurements (optimal-
ly focused beams from a He-Ne laser operating at 6328 Å), revealed, moreover, some slight second-harmonic generation in the
absence of a dc electric field due to induced electric-quadrupole
and magnetic-dipole polarization [67]. Since the second-harmonic
intensity $I(2\omega)$ is proportional to the square of the polarization (21), it should be expected to depend on the square of the
laser beam intensity I and the square of the dc field intensity
$E(0)$.

Mayer et al. [32] studied dc electric field-induced second-
harmonic generation in dipolar and quadrupolar gases, whereas Fin
and Ward [33] used inert gases.

In isotropic bodies, the tensor of third-order nonlinear susceptibility is of the form (19), whence the polarization (21) takes the form:

$$P_i^{(3)}(2\omega) = \chi_1^{2\omega}E_i(\omega)E_j(\omega)E_j(0) + \chi_2^{2\omega}E_j(\omega)E_j(\omega)E_i(0) \tag{21a}$$

where we have introduced the notation:

$$\chi_1^{2\omega} = \chi_{xxyy}(\omega,\omega,0) + \chi_{xyxy}(\omega,\omega,0)$$

$$\chi_2^{2\omega} = \chi_{yxxy}(\omega,\omega,0) \tag{21b}$$

and

$$\chi_3^{2\omega} = \chi_{yyyy}(\omega,\omega,0) = \chi_1^{2\omega} + \chi_2^{2\omega}$$

Now let us assume the laser beam to propagate along the laboratory z axis perpendicularly to the dc electric field $E(0)$, applied along the y axis. Equation (21a) now yields the following two mutually perpendicular polarization components at 2ω:

$$P_x^{(3)}(2\omega) = \chi_1^{2\omega}E_x(\omega)E_y(\omega)E_y(0)$$

$$P_y^{(3)}(2\omega) = \{\chi_2^{2\omega}E_x^2(\omega) + \chi_3^{2\omega}E_y^2(\omega)\}E_y(0) \tag{21c}$$

The above equations enabled Mayer [32] to determine experimentally the nonlinear susceptibility coefficients (21b). Namely, if the incident light beam is linearly polarized, the coefficient $\chi_3^{2\omega}$ can be determined for its oscillations parallel to the dc field direction, $E(\omega) \parallel E_y(0)$, whereas at oscillations perpendicular to the field, $E(\omega) \perp E_y(0)$, one determines $\chi_2^{2\omega}$. When applying circularly polarized light, measurement of the component $P_x^{(3)}(2\omega)$ permits the determination of the coefficient $\chi_1^{2\omega}$, whereas measurement of $P_y^{(3)}(2\omega)$ yields the value of $\left|\chi_2^{2\omega} + \chi_3^{2\omega}\right|$.

In the absence of optical dispersion, the tensor (19) can be dealt with as being totally symmetric, and one has [68]:

$$\chi^{2\omega}_{xxyy} = \chi^{2\omega}_{xyxy} = \chi^{2\omega}_{yxxy} = \frac{1}{3}\chi^{2\omega}_{yyyy} \tag{21d}$$

leading to Bloembergen's symmetry relations [31]:

$$\chi^{2\omega}_1 + \chi^{2\omega}_2 = \chi^{2\omega}_3$$

$$\chi^{2\omega}_1 : \chi^{2\omega}_2 : \chi^{2\omega}_3 = 2:1:3 \tag{22}$$

3. Light-Induced Birefringence Phenomena

We shall now consider the case when a wave of frequency $\omega_A = \omega_a$ plays the part of the analyzing agent and another wave of frequency $\omega_I = \omega_b = -\omega_c$, is the agent inducing in the medium an anisotropy of the form:

$$\Delta\varepsilon_{ij}(\omega_A) = 4\pi\chi_{ijkl}(\omega_A,\omega_I,-\omega_I)E_k(\omega_I)E_l(-\omega_I) \tag{23}$$

resulting by Eqs. (7) and (18).

The change in electric permittivity tensor (23), induced by the strong electric field $E(\omega_I)$ and analyzed by means of the field of frequency ω_A, can with regard to (19) be written as follows for the case of an isotropic body:

$$\Delta\varepsilon_{ij}(\omega_A) = 4\pi\{\chi_{xxyy}(\omega_A,\omega_I,-\omega_I)\delta_{ij}E_k(\omega_I)E_k(-\omega_I)$$

$$+ \chi_{xyxy}(\omega_A,\omega_I,-\omega_I)E_i(\omega_I)E_j(-\omega_I)$$

$$+ \chi_{yxxy}(\omega_A,\omega_I,-\omega_I)E_i(-\omega_I)E_j(\omega_I) \tag{24}$$

Let the analyzing light beam propagate along z. Eq. (24) now yields for the difference of diagonal elements of the permittivity tensor:

$$\Delta\varepsilon_{xx}(\omega_A) - \Delta\varepsilon_{yy}(\omega_A) = 4\pi\{\chi_{xyxy}(\omega_A,\omega_I,-\omega_I)$$

$$+ \chi_{yxxy}(\omega_A,\omega_I,-\omega_I)\}\{E_x(\omega_I)E_x(-\omega_I) - E_y(\omega_I)E_y(-\omega_I)\} \tag{24a}$$

This expression describes the birefringence induced optically in an isotropic medium, i.e., the optical Kerr effect discovered by Mayer and Gires [7] and at present studied in numerous molecular

liquids [8,9,69] (Fig. 2). Obviously, putting the inducing field
frequency equal to zero $(\omega_I = 0)$, Eq. (24a) defines the classical Kerr effect due to a dc electric field.

For molecular substances in the gas state, in the absence of
dispersion and absorption, the nonlinear optical birefringence
susceptibilities (24a) are given as follows [5,6]:

$$\chi_{xyxy}(\omega_A,\omega_I,-\omega_I) = \chi_{yxxy}(\omega_A,\omega_I,-\omega_I)$$

$$= \frac{\rho}{90}\left\{3c_{ijij}^{\omega_A,\omega_I} - c_{iijj}^{\omega_A,\omega_I}\right.$$

$$\left. + \frac{1}{kT}\left(3a_{ij}^{\omega_A}a_{ij}^{\omega_I} - a_{ii}^{\omega_A}a_{jj}^{\omega_I}\right)\right\} \tag{24b}$$

where $\rho = N/V$ is the number density of molecules.

The first, not directly temperature-dependent term of (24b)
accounts for Voigt's effect [1,39], which consists in nonlinear
distortion of the atom or molecule defined by the hyperpolarizabi-
lity tensor $c_{ijkl}^{\omega_A,\omega_I}$ at frequency ω_A and ω_I. The other, directly temperature-dependent term, results from Langevin's [2]
reorientation of molecules in the optical field [5]. The tensors
$a_{ij}^{\omega_A}$ and $a_{ij}^{\omega_I}$ describe linear polarizabilities at the frequency
ω_A and ω_I, respectively.

In liquid systems, optically induced birefringence depends on
the angular molecular correlations [11,70], spatial molecular
redistribution [71,72], and other molecular-statistical processes
[73,74]. Theories of optical Kerr effect dispersion exist in the
quantum mechanical [75], quantum field theoretical [76], and
classical [77] approaches. In recent years, the optical Kerr effect has found applications in the technique of ultrashort laser
pulses [78].

The general formula (24) moreover permits the calculation of
differences between the nondiagonal elements of the permittivity
tensor:

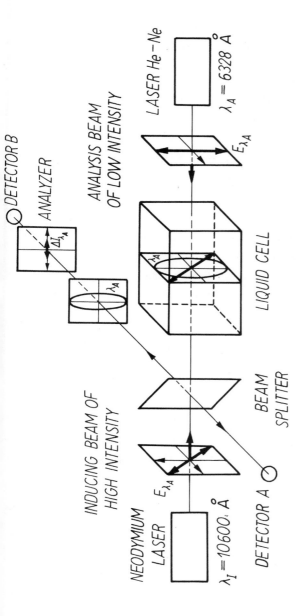

FIG. 2. Principle of optically induced birefringence experiment [8,9]. The liquid, irradiated with the intense plane-polarized wave λ_I, behaves like a uniaxial crystal with optical axis parallel to the electric vector $E(\lambda_I)$. The linearly and vertically polarized analyzing wave λ_A of low intensity becomes elliptical. The two laser beams, i.e., the one inducing the nonlinearity and the analyzing one, are polarized at $45°$ to one another and propagate in opposite directions. An analyzer whose plane of polarization is perpendicular to that of the initial polarizer (that is to say, horizontally), permits the measurement of this ellipticity. In the absence of an intense wave, no signal appears on the detector B because the analyzer and polarizer are at right angles. During propagation of the intense wave, the component $\Delta I(\lambda_A)$ of the analyzed wave is detected at B. The intensity of the strong inducing beam is measured with the detector A.

$$\Delta\varepsilon_{xy}(\omega_A) - \Delta\varepsilon_{yx}(\omega_A) = 4\pi\{\chi_{xyxy}(\omega_A,\omega_I,-\omega_I)$$
$$- \chi_{yxxy}(\omega_A,\omega_I,-\omega_I)\}\{E_x(\omega_I)E_y(-\omega_I)$$
$$- E_y(\omega_I)E_x(-\omega_I)\} \qquad (24c)$$

The above circular birefringence is induced in isotropic media by circularly and elliptically polarized light [79].

For a slowly varying analyzing electric field $\omega_A \ll \omega_I$, Eqs. (23) and (24) define the change in electric permittivity due to intense light [6], whereas Eq. (24a) defines laser light-induced electric anisotropy in liquids [27,34].

4. Self-Induced Optical Birefringence

Consider the case when a single, intense laser beam $(\omega_a = \omega_b = -\omega_c = \omega)$ is incident on an isotropic medium. Equations (18) and (19) now yield the following, mutually perpendicular polarization components (perpendicular to the propagation direction z):

$$P_x^{(3)}(\omega) = \chi_{xxxx}(\omega,\omega,-\omega)E_x(\omega)\{E_x(\omega)E_x(-\omega) + E_y(\omega)E_y(-\omega)\}$$
$$+ \chi_{yxxy}(\omega,\omega,-\omega)E_y(\omega)\{E_y(\omega)E_x(-\omega) + E_x(\omega)E_y(-\omega)\}$$
$$(25)$$
$$P_y^{(3)}(\omega) = \chi_{yyyy}(\omega,\omega,-\omega)E_y(\omega)\{E_x(\omega)E_x(-\omega) + E_y(\omega)E_y(-\omega)\}$$
$$+ \chi_{yxxy}(\omega,\omega,-\omega)E_x(\omega)\{E_x(\omega)E_y(-\omega) - E_y(\omega)E_x(-\omega)\}$$

whence we derive for the difference in diagonal elements of the permittivity tensor at frequency ω:

$$\Delta\varepsilon_{xx}(\omega) - \Delta\varepsilon_{yy}(\omega) = 4\pi\chi_{yxxy}(\omega,\omega,-\omega)\{E_x(\omega)E_x(-\omega) - E_y(\omega)E_y(-\omega)\}$$
$$(25a)$$

This expression defines the optical birefringence self-induced by a linearly polarized laser beam [5,8].

Quite similarly, one obtains from (25) the following difference between nondiagonal electric permittivity tensor elements:

$$\Delta\varepsilon_{xy}(\omega) - \Delta\varepsilon_{yx}(\omega) = 8\pi\chi_{yxxy}(\omega,\omega,-\omega)\{E_y(\omega)E_x(-\omega) - E_x(\omega)E_y(-\omega)\}$$

(25b)

describing the self-induced rotation of the polarization ellipse detected by Maker et al. [80] when using intense elliptically polarized laser light.

III. GENERATION OF SECOND-HARMONIC LASER LIGHT BY ELECTRICALLY POLARIZED ISOTROPIC MEDIA

When a naturally isotropic body is placed in a dc or ac electric field, its symmetry undergoes a deterioration and it becomes axially symmetric, with macroscopic symmetry of the type $C_{\infty V}$. The conditions existing in an isotropic medium with electrically destroyed center of symmetry favor second-harmonic generation of the electric dipole kind. The absence of a center of symmetry is a condition necessary, though not in all cases sufficient, for second-harmonic generation. Above, we have given a discussion of second-harmonic generation by electrically polarized bodies in a phenomenological approach. We now proceed to a detailed discussion of this new electro-optical effect in a statistical-molecular approach, with the aim of rendering apparent the microscopic mechanisms underlying it.

A. Statistical-Molecular Theory

Consider a volume V, containing N-like dipolar molecules whose optical and electric properties remain unaffected by short-range interaction. As long as the number density $\rho = N/V$ is not excessively large, the second-order nonlinear susceptibility of the system in the presence of a dc electric field E can be written in the statistical-molecular form [56]:

$$\chi_{ijk}^{2\omega}(E) = \frac{\rho}{4} \int \left(b_{ijk}^{2\omega} + c_{ijkl}^{2\omega} E_l + \ldots \right) \sqrt{f(\Omega,E)\,d\Omega} \qquad (26)$$

where the integral extends over all possible molecular orientation Ω in the elementary body angle $d\Omega$. The third-rank tensor
$b_{ijk}^{2\omega} = b_{ijk}(\omega,\omega)$ defines the molecular nonlinear polarizability
of order 2 at frequency 2ω, whereas the fourth-rank tensor
$c_{ijkl}^{2\omega} = c_{ijkl}(\omega,\omega,0)$ defines the molecular nonlinear polarizability of order 3.

The statistical distribution function $f(\Omega,E)$ describing the
orientation of the dipolar molecules with respect to the direction
of the field $E(0)$ is given by classical Maxwell-Boltzmann statistics, to a linear approximation in E, as:

$$f(\Omega,E) = f(\Omega,0)\left(1 + \frac{1}{kT}\, m_i E_i \right) \qquad (27)$$

where m_i is the ith component of the permanent dipole moment
of a molecule.

By inserting (27) into (26) and averaging over all possible
orientations of molecules with respect to the laboratory axes
x,y,z we obtain [56]:

$$\chi_{ijk}^{2\omega}(E) = \chi_{xxyy}^{2\omega}\delta_{ij}E_k + \chi_{xyxy}^{2\omega}\delta_{ik}E_j + \chi_{yxxy}^{2\omega}\delta_{jk}E_i \qquad (28)$$

with electric susceptibility coefficients $\chi_{xxyy}^{2\omega},\ldots$ given,
respectively, by the molecular parameters m_i, $b_{ijk}^{2\omega}$, and $c_{ijkl}^{2\omega}$.
In the absence of optical dispersion, these tensors are totally
symmetric and the relation (21d) holds, whence:

$$\chi_{yyyy}^{2\omega} = \frac{\rho}{20}\left(c_{iijj}^{2\omega} + \frac{1}{kT}\, b_{iij}^{2\omega} m_j \right) \qquad (29)$$

Thus, in the microscopic approach, the phenomenological
nonlinear susceptibilities of dipolar substances are found to consist of two terms: (a) one not directly dependent on temperature
due to a purely distortional process residing in nonlinear
polarization of the electron shell occurring for all molecular
symmetries, since $c_{ijkl}^{2\omega}$ has nonzero elements even in the case of

atoms (Voigt effect), and (b) a directly temperature-dependent
term in accordance with Langevin's theory related with the statistical process of reorientation of the molecular dipoles in the dc
field and Born's nonlinear distortional process [3] described
by tensor $b_{ijk}^{2\omega}$. Recently, Hameka [81] developed a quantum-
mechanical theory of field-induced SHG in gases.

B. Semimacroscopic Theory

Results valid for isotropic bodies of arbitrary structure and
degree of condensation are conveniently obtained by using
Kirkwood's [82] semimacroscopic dielectric theory extended to
nonlinear phenomena [83]. Semimacroscopically, we consider a
spherical sample of macroscopic size, volume V, and dielectric
permittivity ε, in a continuous isotropic medium of electric
permittivity ε_e. In general, the mean macroscopic electric field
E_M existing within the sample differs from the external field E;
if E is weak, E_M is given by the formula well-known from
electrostatics:

$$\underline{E}_M = \frac{3\varepsilon_e}{\varepsilon + 2\varepsilon_e} \underline{E} \tag{30}$$

Similar relations hold for fields oscillating at frequency ω or
2ω.

Semimacroscopic theory [31] in linear approximation leads
back to Eq. (28); however, in the absence of optical dispersion,
we have instead of (29):

$$\chi_{yyyy}^{2\omega} = \frac{L^{2\omega}}{20V} <C_{iijj}^{2\omega} + \frac{1}{kT} B_{iij}^{2\omega} M_j> \tag{31}$$

the symbol $< >$ denoting appropriate statistical averaging with
molecular correlations. The tensors of (31) written in capital
roman letters now refer to the whole volume V of the sample; $\underline{M}$
is its total electric dipole moment, and $B_{ijk}^{2\omega}$, $C_{ijkl}^{2\omega}$ are

tensors of its nonlinear polarizabilities at 2ω. The macroscopic parameter $L^{2\omega}$ is of the form:

$$L^{2\omega} = \left(\frac{\varepsilon_{2\omega} + 2\varepsilon_e}{3\varepsilon_e}\right)\left(\frac{\varepsilon_\omega + 2\varepsilon_e}{3\varepsilon_e}\right)^2\left(\frac{\varepsilon_0 + 2\varepsilon_e}{3\varepsilon_e}\right) \tag{32}$$

1. Dipolar Liquids

With N dipolar molecules per volume V of the liquid, one has:

$$M_i = \sum_{s=1}^{N} m_i^{(s)}, \quad B_{ijk}^{2\omega} = \sum_{s=1}^{N} b_{ijk}^{2\omega(s)}, \quad C_{ijkl}^{2\omega} = \sum_{s=1}^{N} c_{ijkl}^{2\omega(s)} \tag{33}$$

the tensors $m_i^{(s)}$, $b_{ijk}^{2\omega(s)}$, and $c_{ijkl}^{2\omega(s)}$ referring to the sth molecule, immersed in the medium.

By (32), the nonlinear susceptibility (31) can be written in molecular-statistical form as:

$$\chi_{yyyy}^{2\omega} = \frac{L^{2\omega}}{20V} < \sum_{s=1}^{N} c_{iijj}^{2\omega(s)} + \frac{1}{kT} \sum_{s=1}^{N} \sum_{t=1}^{N} b_{iij}^{2\omega(s)} m_j^{(t)} > \tag{34}$$

In particular, for molecules symmetric about their dipole moment, we thus obtain, as a satisfactory approximation:

$$\chi_{yyyy}^{2\omega} = \frac{\rho}{4} L^{2\omega}\left(c^{2\omega} + \frac{3mb^{2\omega}}{5kT} K\right) \tag{34a}$$

where by

$$c^{2\omega} = \frac{c_{iijj}^{2\omega}}{5} = \frac{c_{3333}^{2\omega} + 2c_{1122}^{2\omega} + 4c_{1133}^{2\omega} + 2c_{1111}^{2\omega}}{5} \tag{35}$$

we have denoted the mean nonlinear polarizability of order 3, and by

$$b^{2\omega} = \frac{b_{ii3}^{2\omega}}{3} = \frac{b_{333}^{2\omega} + 2b_{113}^{2\omega}}{3} \tag{36}$$

the mean nonlinear polarizability of order 2, whereas

$$K = 1 + \rho \int \cos \theta_{st} g(\tau_{st}) d\tau_{st} \qquad (37)$$

defines Kirkwood's angular correlation parameter [82], θ_{st} being the angle between the electric dipoles of molecules s and t at mutual configuration τ_{st} statistically interdependent by way of the pair correlation function $g(\tau_{st})$.

Comparing Eqs. (34) and (34a), on the one hand, and (29), on the other, we note that in dipolar liquids the temperature-dependent term undergoes a modification by the Kirkwood correlation parameter (37), which is larger or less than 1 according to the mutual configuration of the dipoles (e.g., nearly parallel, nearly antiparallel, of perpendicular configuration).

2. Nondipolar Liquids

Here, the temperature-dependent term of Eqs. (29) and (34) vanishes. A result like this, however, can hold only as long as one neglects the existence, in liquids, of intense molecular fields $\underline{F}$, which are in fact very strong in comparison with the currently applied external fields. In regions of short-range ordering, the molecular field $\underline{F}$ fluctuates in time and space, inducing dipole and higher moments in neighboring molecules, thus impairing the symmetry of the molecules themselves and of their neighborhood. Hence we can write, for centrosymmetric molecules, in place of (33):

$$M_i = \sum_{s=1}^{N} \left\{ a_{ij}^{(s)} F_j^{(s)} + \frac{1}{2} b_{ijk}^{(s)} F_j^{(s)} F_k^{(s)} + \dots \right\}$$

$$B_{ijk}^{2\omega} = \sum_{s=1}^{N} \left\{ c_{ijkl}^{2\omega(s)} F_l^{(s)} + \dots \right\} \qquad (38)$$

where $a_{ij}^{(s)}$ is the linear electric polarizability tensor of the molecule.

On inserting the expansion (38) into the general equation (31) and neglecting, for simplicity, the anisotropies in tensors

a_{ij} and c_{ijkl}, we have:

$$\chi_{yyyy}^{2\omega} = \frac{\rho}{4} c^{2\omega} L^{2\omega} \left(1 + \frac{a}{kT} <F^2>\right)$$ (39)

where

$$a = \frac{a_{ii}}{3} = \frac{a_{11} + a_{22} + a_{33}}{3}$$

stands for the mean linear polarizability and $<F^2>$ for the mean statistical value of the square of the electric molecular field which, generally, is nonzero for nondipolar liquids.

In the case of axially symmetric molecules having a permanent electric quadrupole moment Θ (e.g., the molecules CO_2 and C_6H_6) we have [63]:

$$<F^2> = \Theta^2 <r^{-8}>$$ (40)

where the mean statistical value $<r^{-8}>$ is given at $n = 8$ by the definition [84]:

$$<r^{-n}> = 4\pi\rho \int r_{st}^{-n+2} g(r_{st}) dr_{st}$$ (41)

with $g(r_{st})$ denoting the radial distribution function of a pair of molecules s,t mutually distant by r_{st}.

Since the molecular fields produced by various electric multipoles are well known [85], formula (39) can be applied to other molecular symmetries, too. In particular, for tetrahedrally symmetric molecules (CH_4, CCl_4, etc.) whose lowest nonzero moment is the octupole Ω_{123}, one has [63]:

$$<F^2> = \frac{16}{5} \Omega_{123}^2 <r^{-10}>$$ (40a)

Mean statistical values (41) can be computed numerically for real gases using the Lennard-Jones potential, as well as for liquids resorting to the simple Kirkwood model [86] of rigid spheres of diameter d and volume $V = \pi d^3/6$, leading to the simple result [84]

$$<r^{-n}> = \frac{4\pi\rho}{n - 3} \left(\frac{\pi}{6V}\right)^{(n-3)/3} ; \quad n \geq 4 \tag{41a}$$

One thus finds that in conformity with Eq. (39), nondipolar liquids too present a temperature-dependent term of the nonlinear
electric susceptibility as a result of reorientation of the electric dipoles which are induced in molecules by the fluctuating
electric fields of their nearest neighbors in the liquid.

C. Strong Electric Reorientation of Macromolecules

The theories presented in Secs. IIIA, B hold for molecular
substances, where the reorientation due to a dc electric field is
relatively weak, as given by the distribution function (27). In
substances consisting of macromolecules and colloid particles
strong reorientation is achieved quite easily. On the assumption
of axial symmetry, the distribution function takes the form [87,
88]:

$$f(\theta,E_y) = \frac{\exp (p \cos \theta \pm q \cos^2 \theta)}{\int \exp (p \cos \theta \pm q \cos^2 \theta)d\Omega} \tag{42}$$

where θ is the angle subtended by the symmetry axis of a
macromolecule and the field direction E_y, and $d\Omega = \sin \theta \, d \theta \, d\phi$.
In the Boltzmann factor,

$$p = \frac{mE_y}{kT} \tag{42a}$$

is the dimensionless Langevin-Debye reorientation parameter for
permanent electric dipoles, and

$$q = \frac{\left|a_{33} - a_{11}\right|}{2kT} E_y^2 \tag{42b}$$

the Langevin parameter of electric dipoles induced in macromolecules having the polarizabilities a_{33} and a_{11}, respectively,

along and perpendicular to their symmetry axis (see Chap. 3).

In the function (42), the sign "+" is for prolate macromolecules having positive anisotropy, $a_{33} - a_{11} < 0$, e.g., CS_2, whereas the lower sign "-" refers to disk-shaped ones, of negative anisotropy $a_{33} - a_{11} < 0$, e.g., C_6H_6.

Once the distribution function (42) is known explicitly, we are in a position to calculate the nonlinear susceptibility (26) for strong reorientation of macromolecules or of axially symmetric molecules of the point groups C_{3v}; e.g., NH_3, CH_3Cl, CH_3I, and C_{4v}; BrF_5 and $C_{\infty v}$; HCN, N_2O.

If the isotropic medium is placed in a strong electric field, the polarization induced by laser light at frequency 2ω is no longer a linear function of the field $E(0)$, as stated by Eq. (21), but takes the form of some nonlinear function; we express this generally by writing:

$$P_i^{2\omega}(E) = \chi_{ijk}^{2\omega}(E)E_j(\omega)E_k(\omega) \tag{43}$$

or, for light propagation along z and a dc field applied along y:

$$P_x^{2\omega}(E_y) = 2\chi_{xxy}^{2\omega}(E_y)E_x(\omega)E_y(\omega)$$

$$P_y^{2\omega}(E_y) = \chi_{yxx}^{2\omega}(E_y)E_x^2(\omega) + \chi_{yyy}^{2\omega}(E_y)E_y^2(\omega) \tag{44}$$

with, in the absence of dispersion, $\chi_{xxy}^{2\omega}(E_y) = \chi_{yxx}^{2\omega}(E_y)$.

Let us assume the dc field to cause only a reorientation of the microsystems given by the distribution function (42), with no contribution from the distortional effect defined by the second term $c_{ijkl}^{2\omega}E_l$ of the expansion (26); on the same assumptions as before, we get:

$$\chi_{xxy}^{2\omega}(E_y) = \frac{\rho}{8}\left\{\left(b_{333}^{2\omega} - b_{113}^{2\omega}\right)L_1\left(p, \pm q\right) - \left(b_{333}^{2\omega} - 3b_{113}^{2\omega}\right)L_3\left(p, \pm q\right)\right\}$$

$$\chi_{yyy}^{2\omega}(E_y) = \frac{\rho}{4}\left\{3b_{113}^{2\omega}L_1\left(p, \pm q\right) + \left(b_{333}^{2\omega} - 3b_{113}^{2\omega}\right)L_3\left(p, \pm q\right)\right\} \tag{45}$$

involving the generalized Langevin functions of odd order $L_1(p,\pm q)$ and $L_3(p,\pm q)$, defined analytically in Ref. [88] and plotted in Figs. 3 and 4 vs. p for some values of $q = p^2/n$, for $n = 1, 4, 9, 16, 25, 36, \ldots$.

In the case of prolate (positively-anisotropic) microsystems the permanent dipoles as well as the induced ones tend concordantly to orient themselves into the electric field direction $\underline{E}$, so that with growing field intensity and parameter q the functions $L_1(p,+q)$ and $L_3(p,+q)$ tend very steeply from 0 to the limiting value 1, which defines the state of complete electric saturation of orientation of the mocrosystems (dashed curves in Figs. 3 and 4).

For negatively anisotropic microsystems, the picture is entirely different. If the permanent dipole $\underline{m}$ lies along the symmetry axis, the polarizability a_{33} parallel to the latter is less than the polarizability a_{11} perpendicular thereto.

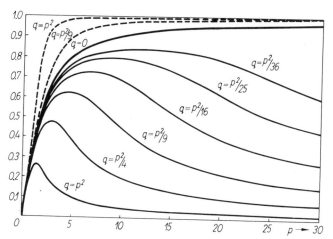

FIG. 3.   Langevin function $L_1(p,\pm q)$ vs. the reorientation parameters p and q.   Above the $L_1(p,0)$ curve for dipolar unpolarizable (q = 0) molecules lie the (dashed-line) curves $L_1(p,+q)$ for molecules having positive anisotropy.   Below it lie the curves of $L_1(p,-q)$ pertaining to molecules with negative anisotropy.

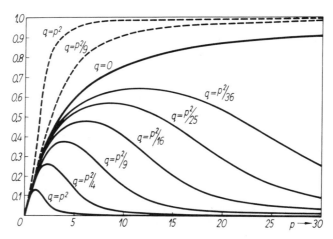

FIG. 4. Langevin functions $L_3(p, q)$ vs. the reorientation
parameters p and q. Above the curve $L_3(p,0)$ for dipolar
unpolarizable (q = 0) microsystems lie the (dashed-line) curves
$L_3(p,+q)$ for microsystems having positive anisotropy; below
$L_3(p,0)$ lie curves of the function $L_3(p,-q)$ relating to ones
having negative anisotropy.

Consequently, whereas the torque of the permanent electric dipole
tends to align the microsystem parallel to the applied field direction, the torque due to the induced dipole tends to orient it
perpendicular to the field $\underline{E}$ (see Ref. [89]). With growing
field intensity E, increasing the induced dipole parameter
(42b), the microsystem goes over from parallel to perpendicular
orientation with respect to $\underline{E}$. As a result of this reorientation, the Langevin functions $L_1(p,-q)$ and $L_3(p,-q)$ decrease
with growing E tending rapidly to 0 as q, the parameter of
induced dipole reorientation, grows (continuous-line curves in
Figs. 3 and 4 below the curve for q = 0).

D. Macroscopic and Microscopic Symmetry Relations

At moderate reorientation by the electric field E (at
p < 1, q < 1), the Langevin functions can be written to a
satisfactory approximation in the form of the expansions [88]:

$$L_1(p, \pm q) = \frac{p}{3} - \frac{p^3}{45} \pm \frac{4pq}{45} + \cdots$$

$$L_3(p, \pm q) = \frac{p}{5} - \frac{p^3}{105} \pm \frac{8pq}{105} + \cdots \qquad (46)$$

which, inserted into (45), lead to the fulfillment of the symmetry relations (22) in an approximation linear in the field E_y:

$$\chi^{2\omega}_{xxy}(E_y) = \frac{1}{3}\chi^{2\omega}_{yyy}(E_y) = \frac{\rho m}{60kT}\left(b^{2\omega}_{333} + 2b^{2\omega}_{113}\right)E_y \qquad (45a)$$

In the case of complete alignment of all the microsystems into the field direction $(p \rightarrow \infty, q \rightarrow \infty$, when the field is strong or the dipole moments and polarizabilities of the microsystems are large) the Langevin functions tend rapidly to 1 for positive anisotropy, and Eqs. (45) reduce to:

$$\chi^{2\omega}_{xxy}(E_y \rightarrow\infty) = \frac{\rho}{4}b^{2\omega}_{113}$$

$$\chi^{2\omega}_{yyy}(E_y \rightarrow\infty) = \frac{\rho}{4}b^{2\omega}_{333} \qquad (45b)$$

In that of complete alignment of negatively anisotropic microsystems, the Langevin functions tend to 0; in consequence, the components (45) vanish.

The fulfillment of the macroscopic symmetry relations (22) would necessarily have entailed that of the following microscopic relation for the individual microsystems:

$$b^{2\omega}_{333} = 3b^{2\omega}_{113} \qquad (47)$$

The Bloembergen symmetry relations in the form (22) are strictly fulfilled in the approximation linear in the field E, on neglecting electron dispersion. In very strong fields E the relations (22) are, in general, not fulfilled. For electric saturation, the expression (45b) suggests that the symmetry of the electro-optical properties of the individual microsystems become identical to the macroscopic symmetry of the isotropic body as a whole which, in the presence of an electric field, is endowed with symmetry of the type $C_{\infty V}$. The preceding conclusion seems

obvious, since at concordant alignment of all the microsystems in
the direction of the external field, the properties of the body
as a whole have to coincide with those of the microsystem.

E. Rarefied Molecular Substances

As yet, experimental data are too scarce to permit a full
comparative analysis of the above theory. Obviously, one cannot
hope to achieve complete electric or optical saturation [90] in
gases and molecular liquids for fear that the effect will be pre-
ceded by electric (optical) breakdown. In the dipolar substances
studied by Mayer, where the electric reorientation parameters were
at the best of order $p = 10^{-3}E_y$ and $q = 10^{-9}E_y^2$, it would be
necessary to apply a very strong electric field upward of
10^6 V/cm to approach saturation (see the graphs of Figs. 3 and 4).
In this situation, considerable optical ordering of the molecules,
or even some degree of optical saturation, appear much more feasible (in Mayer's measurements, the laser field E_ω amounted to
1.5×10^6 V/cm). In fact, it should be taken into consideration
that some increase in the numerical value of p and q can be
due to the circumstance that, generally, molecules in states excited by a strong electric field have dipole moments and
polarizabilities much in excess of those in their ground state.

Available measurements of $I^{2\omega}$ as yet do not permit a
quantitative decision as to which of the microscopic processes
discussed here, namely, nonlinear electron distortion or molecular
reorientation, plays the predominant role. The fact that an
anomalous increase in intensity $I^{2\omega}$ with growing field E_y was
observed by Mayer [32] (Fig. 5) to occur exclusively in the substances CH_3I, C_2H_5I and C_2H_5Br whose molecules have considerable dipole moments of order 2×10^{-18} esu cm and a positive
anisotropy of electric and optical polarizability speaks in favor
of reorientation. On the other hand, in chloroform and other

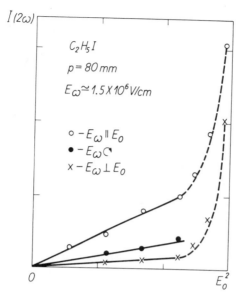

FIG. 5. Mayer's results [32] for the intensity $I^{2\omega}$, measured at frequency 2ω in ethyl iodide vs. the square E_0^2 of the applied dc electric field. Unit on abscissas axis: 2.7×10 (V/cm) ; unit of ordinate: 25 photoelectrons in the detector equivalent to 6.2×10 photons emitted into the body angle of observation. Stimulating peak power: 7 MW; breakdown threshold E_0^2 lies at $E_0^2 = 8.1 \times 10$ (V/cm) ; field intensity of the collimated ruby laser beam $E_\omega = 1.5 \times 10$ V/cm. Linearity (continuous straight lines in the graph), expressing quadratic dependence on E_0, maintains itself by nigh up to breakdown E_0^b; upward of $E_0 > 0.8 E_0^b$ anomalous effects are observed (dashed-line curves).

weakly dipolar and nondipolar substances, he failed to observe an anomalous increase in $I^{2\omega}$; in the light of our theory this can plausibly be explained by the fact that $CHCl_3$ molecules have a much smaller dipole moment as well as negative anisotropy of polarizability causing, in conformity with the graphs of Figs. 3 and 4, a decrease in value of the nonlinear susceptibilities and hence of $I^{2\omega}$. Numerical evaluations show that in dipolar substances the molecular reorientation process decidedly predominates over that of nonlinear electron distortion. The final decision can be expected from studies of second-harmonic generation as a

function of temperature, concentration and light frequency or the
frequency of a slowly varying electric field $E(\omega_0) = E(0)e^{i\omega_0 t}$
oscillating at ω_0 much below optical frequencies ω. With the
isotropic medium polarized by means of an ac field E, one has to
replace the constant moment $\underline{m}$ in Eq. (29) by a time-oscillating
dipole moment $\underline{m}(t)$. Relaxation theory [91,92] leads to:

$$m_j(t) = \frac{m_j e^{i\omega_0 t}}{1 + i\omega_0 \tau_j} \tag{48}$$

where τ_j is the relaxation time corresponding to the inertia
moment for the principal axis j of the molecule. By appropriately raising the applied field frequency one can reduce, or entirely
eliminate, the process of reorientation in various substances.

F.   The Role of Molecular Correlations

In condensed substances, nonlinear optical susceptibility is
affected by various molecular correlations, such as molecular
redistribution [71,72] and angular correlations between anisotropic molecules [83] as well as the local field and its anisotropy
[71].   These correlations can act in a manner to enhance or lower
electric reorientation according to the structure of the individual molecules and the degree of their condensation.   Molecular
redistribution is highly important to nonlinear processes, since
it leads to anisotropy even in the case of interacting nearly
spherical molecules, like CCl_4 [71,72].   In the latter case,
only correlations of the radial kind intervene, causing a spatial
regrouping of the molecules giving rise to pairwise, three-mole-
cule, etc. aggregates, apt to undergo a reorientation by the external electric field.   The process of nonlinear molecular
distortion as such is but insignificantly affected by spatial
redistribution.

In the absence of an external field, correlations between anisotropic molecules lead to a time-fluctuating short-range ordering, i.e., a cybotactic structure [93]. An externally applied electric field can cause a transition from a phase with partly ordered to one with with more highly ordered molecules similar to the nematic mesophase of liquid crystals [73]. Transition processes involving macromolecules are mentioned in Chap. 3. In the field-induced mesophase of nearly nematic substances, SHG can become enhanced as is the case in liquid crystals [60]. A rise in $I^{2\omega}$ can moreover be caused by strong fluctuations of molecular angular orientation close to the critical point [47].

In nondipolar substances, with regard to Eqs. (31) and (39), the presence of multipole interactions can cause a temperature-statistical process, consisting in a dc field-orientation of dipoles induced in the molecules by fields of the electric quadrupoles and octupoles of neighboring molecules. We have to deal with a similar mechanism in the cooperative second harmonic scattering in liquids with centrosymmetric molecules recently detected by Lalanne, Martin, and this author [94].

Studies of SHG in molecular solutions versus the concentration can also be expected to permit the quantitative determination of contributions from the various above-discussed processes. Namely, the process of third-order nonlinear molecular distortion depends but insignificantly on statistical correlations and should thus be nearly additive, meaning that the nonlinear susceptibilities of the components of the solutions will not depend on concentration. Inversely, molecular reorientation depends strongly on radial and angular correlation, whence the nonlinear susceptibility tensor will depend on higher powers of the concentrations thus not fulfilling the law of additivity similarly to the electric polarization and Kerr constant of liquid solutions [74].

G. Macromolecular Solutions

The theory expounded here is directly applicable to solutions
of macromolecules and colloid particles. The latter have to be
considered as being rigid, and the solution has to be so dilute as
to exclude solute-solvent or solute-solute interactions.

In the class of synthetic polypeptides, the most often studied are the macromolecules of the helical poly-γ-benzyl-L-gluta-
mate, roughly of the shape of long cylinders, presenting large
dipole moments (of order 10^3 D) and high anisotropy of their
electric polarizability (of order 10^{-18} cm^3) [87]. In this
case, the reorientation parameters amount to $p = 0.1$ E and
$q = 10^{-5}$ E^2, which means that reorientation of permanent dipoles
predominates over that of induced ones; electric saturation can
set in at an external field intensity of as little as
$E > 10^4$ V/cm. In bentonite suspensions, we deal with plate-shaped
microsystems; here [89], $p = 0.5E$ and $q = 0.7E^2$, whence
$q/p^2 > 1$ and, with growing E, orientation of induced dipoles
becomes rapidly predominant leading to electric saturation already
at $E > 10^3$ V/cm and a reorientation of the microsystems with
their short axis in the E-direction and consequently a steep decrease to 0 of the nonlinear susceptibilities.

If the laser pulse duration is more than 10^{-4} sec, thus
sufficiently long to permit optical reorientation of the macromolecules, some optical anisotropy will additionally arise in the
medium contributing to an increase or decrease of the nonlinear
susceptibilities. One is easily convinced that continuously
operating intermediate power gas lasers are quite sufficient for
the study of substances of the kind under consideration. From
what has been said, it is obvious that we are able to induce, both
electrically and optically, considerable nonlinearities in solutions of rod- or thread-shaped macromolecules and colloid particles. These cause a strong rise in $I^{2\omega}$ with growing dc
electric field or growing intensity of the light wave itself; this

capacity of the substances under consideration of increasing their optical nonlineaties and exhibiting saturation predestines them to play an important role in problems of optoelectronics. Experimental work in this direction should, moreover, prove of the highest value in providing us with new data concerning the nonlinear properties of biomacromolecules, viruses, and colloid particles in general [74].

IV. NONLINEAR ELECTRO-OPTICAL POLARIZABILITIES

Processes of second-harmonic generation are strongly sensitive to the structure exhibited by crystals, gases, liquids, and macromolecular solutions. In the wake of the new electro-optical phenomena discussed above, a search is continuing for new materials, having the ability to produce these phenomena to a high degree.

Much information concerning nonlinear electric and optical properties has been gleaned from the study of other effects, not considered by us above. We shall, however, adduce the most important results thus obtained with regard to the values of the nonlinear tensors b_{ijk} and c_{ijkl}.

A. Second-Order Nonlinear Polarizabilities

The dipole moment of order 2, induced in a microsystem at frequency $\omega_a + \omega_b$, is:

$$m_i^{(2)}(\omega_a + \omega_b) = \beta b_{ijk}(\omega_a, \omega_b) E_j(\omega_a) E_k(\omega_b) \tag{49}$$

where the ordering factor β is equal to $1/4$ for $\omega_a = \omega_b$ and to $1/2$ if the two frequencies are different. In the static case $(\omega_a = \omega_b = 0)$, β amounts to $1/2$.

In the absence of electron absorption and dispersion, the
second-order nonlinear polarizability tensor $b_{ijk}(\omega_a,\omega_b)$ can be
dealt with as totally symmetric and the numbers of its nonzero and
independent elements for the various point groups are listed in
Table 1. Static values of the nonlinear polarizability
$b_{ijk}^0 = b_{ijk}(0,0)$ have been computed theoretically for some molecules [95,96]. Tensor elements $b_{ijk}^{\omega} = b_{ijk}(\omega,0)$ are determined
from dc Kerr effect measurements [74,96] and molecular light
scattering in liquids [83]. Values of $b_{ijk}^{2\omega} = b_{ijk}(\omega,\omega)$ can be
determined from nonlinear light scattering at doubled frequency
by gases and liquids [97,98] or from SHG by a gas or liquid in
the presence of a dc electric field [32].

B. Third-Order Nonlinear Polarizabilities

The dipole moment of order 3 induced in a microsystem at
frequency $\omega_a + \omega_b + \omega_c$ is:

$$m_i^{(3)}(\omega_a + \omega_b + \omega_c) = \gamma c_{ijkl}(\omega_a,\omega_b,\omega_c)E_j(\omega_a)E_k(\omega_b)E_l(\omega_c) \quad (50)$$

where the ordering factor γ is $1/4$ if all the frequencies are
different, $1/8$ if two differ, $1/24$ for $\omega_a = \omega_b = \omega_c$, and
$1/6$ in the single-field static case.

On assuming total symmetry for the tensor of third-order
nonlinear polarizability $c_{ijkl}(\omega_a,\omega_b,\omega_c)$ we obtain its nonzero
and independent elements as listed in Table 2 (see also Ref.
[99]). Direct theoretical calculations of $c_{ijkl}^0 = c_{ijkl}(0,0,0)$
have been effected for the atoms of inert gases and simple molecules [100]. Tensor elements $c_{ijkl}^{\omega} = c_{ijkl}(\omega,0,0)$ have been
determined as to their value for numerous molecules from static
Kerr effect studies [96,101,102] and values $c_{ijkl}^{\omega} = c_{ijkl}(\omega,\omega_I,-\omega_I)$ from laser light-induced optical birefringence
[10]. Measurements of dc field-induced SHG by gases [32] yield
the tensor elements $c_{ijkl}^{2\omega} = c_{ijkl}(\omega,\omega,0)$, which can also be
obtained from cooperative second harmonic scattering by liquids

TABLE 1

Nonzero and Independent Tensor Elements
$b_{\alpha\beta\gamma}$ of the Second-Order Nonlinear Elements
on the Assumption of Total Symmetry in the Indexes α, β, γ

Point group	Number of nonzero elements	Number of independent elements	Elements of $b_{\alpha\beta\gamma}$ are denoted only by their subscripts α, β, γ = 1,2,3
C_1	27	10	111; 222; 333; 112 = 121 = 211; 113 = 131 = 311; 223 = 232 = 322; 122 = 212 = 221; 133 = 313 = 331; 233 = 323 = 233; 123 = 132 = 213 = 231 = 312 = 321
C_3	15	4	333; 111 = -122 = -212 = -211; 222 = -211 = -121 = -112; 311 = 131 = 113 = 322 = 232 = 233
C_s	14	6	111; 222; 122 = 212 = 221; 211 = 121 = 112; 133 = 313 = 331; 233 = 323 = 332
C_2	13	4	333; 311 = 131 = 113; 322 = 232 = 223; 123 = 132 = 213 = 231 = 312 = 321
S_4	12	2	311 = 131 = 113 = -322 = 232 = -223; 123 = 132 = 213 = 231 = 312 = 321
C_{3v}	11	3	333; 222 = -211 = -121 = -112 311 = 131 = 113 = 322 = 232 = 223
C_{3h}	8	2	111 = -122 = -212 = -221; 222 = -211 = -121 = -112
C_{2v}	7	3	333; 311 = 131 = 113; 322 = 232 = 223
C_4, C_{4v} C_6, C_{6v} $C_{\infty v}$	7	2	333; 311 = 131 = 113 = 322 = 232 = 223
D_2, D_{2d} T, T_d	6	1	123 = 132 = 213 = 231 = 312 = 321
D_3	4	1	111 = -122 = -212 = -221
D_{3h}	4	1	222 = -211 = -121 = -112

TABLE 2

Tensor Elements $c_{\alpha\beta\gamma\delta}$ of Third-Order
Nonlinear Polarizability on the Assumption
of Total Symmetricity in the Indexes α, β, γ, δ

Point group	Number of nonzero elements	Number of independent elements	Elements of $c_{\alpha\beta\gamma\delta}$ are denoted only by their subscripts α, β, γ, δ = 1,2,3
C_i	81	15	1111; 2222; 3333 1122 = 1212 = 1221 = 2211 = 2121 = 2112; 1133 = 1313 = 1331 = 3311 = 3131 = 3113; 2233 = 2323 = 2332 = 3322 = 3232 = 3223; 1112 = 1121 = 1211 = 2111; 2221 = 2212 = 2122 = 1222; 1333 = 3133 = 3313 = 3331; 2333 = 3233 = 3323 = 3332; 1113 = 1131 = 1311 = 3111; 2223 = 2232 = 2322 = 3222; 1123 = 1213 = 1231 = 1132 = 1312 = 1321 = 2113 = 2131 = 2311 = 3112 = 3121 = 3211; 2213 = 2123 = 2132 = 2231 = 2321 = 2312 = 1223 = 1232 = 1322 = 3221 = 3212 = 3122; 1332 = 1323 = 1233 = 2133 = 2313 = 2331 = 3123 = 3213 = 3231 = 3132 = 3321 = 3312
C_3 C_{3i}	53	5	3333; 1111 = 2222 = 1122 + 1212 + 1221; 1122 = 1212 = 1221 = 2211 = 2121 = 2112; 1133 = 1313 = 1331 = 3311 = 3131 = 3113 = 2233 = 2323 = 2332 = 3322 = 3232 = 3223; 1113 = 1131 = 1311 = 3111 = −2123 = −2312 −2231 = −2321 = −2213 = −2132 = −1223 = −1232 = −1322 = −3122 = −3212 = −3221; 2223 = 2232 = 2322 = 3222 = −1213 = −1132 = −1312 = −1123 = −1231 = −1321 = −2113 = −2131 = −2311 = −3211 = −3121 = −3112
C_2 C_5 C_{2h}	41	9	1111; 2222; 3333; 1122 = 1212 = 1221 = 2211 = 2121 = 2112; 1133 = 1313 = 1331 = 3311 = 3131 = 3113; 2233 = 2323 = 2332 = 3322 = 3232 = 3223; 1112 = 1121 = 1211 = 2111; 2221 = 2212 = 2122 = 1222; 1332 = 1323 = 1233 = 2133 = 2313 = 2331 = 3123 = 3213 = 3231 = 3132 = 3321 = 3312

TABLE 2 (Continued)

Point group	Number of nonzero elements	Number of independent elements	Elements of $c_{\alpha\beta\gamma\delta}$ are denoted only by their subscripts $\alpha, \beta, \gamma, \delta = 1,2,3$
C_{3v} D_{3d} D_3	37	4	3333; 1111 = 2222 = 1122 + 1212 + 1221; 1122 = 1212 = 1221 = 2211 = 2121 = 2112; 1133 = 1313 = 1331 = 3311 = 3131 = 3113 = 2233 = 2323 = 2332 = 3322 = 3232 = 3223; 2223 = 2232 = 2322 = 3222 = −1213 = −1132 = −1312 = −1123 = −1231 = −1321 = −2113 = −2131 = −2311 = −3211 = −3121 = −3112
C_4 S_4 C_{4h}	29	5	3333; 1111 = 2222; 1122 = 1212 = 1221 = 2211 = 2121 = 2112; 1133 = 1313 = 1331 = 3311 = 3131 = 3113 = 2233 = 2323 = 2332 = 3322 = 3232 = 3223; 1112 = 1121 = 1211 = 2111 = −2221 = −2212 = −2122 = −1222
D_2 C_{2v} D_{2h}	21	6	1111; 2222; 3333; 1122 = 1212 = 1221 = 2211 = 2121 = 2112; 1133 = 1313 = 1331 = 3311 = 3131 = 3113; 2233 = 2323 = 2332 = 3322 = 3232 = 3223
D_4 C_{4v} D_{2d} D_{4h}	21	4	3333; 1111 = 2222; 1122 = 1212 = 1221 = 2211 = 2121 = 2112; 1133 = 1313 = 1331 = 3311 = 3131 = 3113 = 2233 = 2323 = 2332 = 3322 = 3232 = 3223
C_6 C_{3h} C_∞ C_{6h} $C_{\infty v}$ D_6 $C_{\infty h}$ C_{6v} $D_{\infty h}$ D_{3h} D_{6h}	19	3	3333; 1111 = 2222 = 1122 + 1212 + 1221; 1122 = 1212 = 1221 = 2211 = 2121 = 2112; 1133 = 1313 = 1331 = 3311 = 3131 = 3113 = 2233 = 2323 = 2332 = 3322 = 3232 = 3223

TABLE 2 (Continued)

Point group	Number of nonzero elements	Number of independent elements	Elements of $c_{\alpha\beta\gamma\delta}$ are denoted only by their subscripts $\alpha,\ \beta,\ \gamma,\ \delta = 1,2,3$
T T_h O T_d O_h	21	2	1111 = 2222 = 3333; 1122 = 1212 = 1221 = 2211 = 2121 = 2112 = 3311 = 3131 = 3113 = 1133 = 1313 = 1331 = 2233 = 2323 = 3223 = 3322 = 3232 = 3223
K K_h Y Y_h	21	1	Nonzero elements as in the tetrahedral groups with the relation 1111 = 1122 + 1212 + 1221

with centrosymmetric molecules [94]. Finally, tensor elements $c_{ijkl}^{3\omega} = c_{ijkl}(\omega,\omega,\omega)$ are determined by measuring third-harmonic generation in gases and liquids [57,64], and have recently been calculated theoretically for perfect gases [103].

Numerical values of the nonlinear polarizabilities of atoms and molecules determined by experiment or calculated from the theory are listed in Table 3. None are as yet available for macromolecules, with regard to which, however, the theory points to highly interesting and promising possibilities both in the experimental and theoretical domains [19a,104].

V.   CONCLUDING REMARKS AND PROSPECTS

The study of optical harmonics in transmitted light provides information on the microscopic structure, phase transitions (e.g., in ferroelectrics), and crystallographical symmetry of the medium. That of harmonics in reflected light provides data on the electro-optical properties of surfaces, double electric layers, and surface phenomena at the boundary of two media [16]. The new optical and electro-optical phenomena concisely discussed by us here constitute

TABLE 3

Numerical Values of Linear and Nonlinear Optical Polarizabilities of Atoms and Molecules Calculated Theoretically and Determined from Experimental Data

Atom or molecule	10^{-24} cm³	10^{-30} cm⁵ esu⁻¹		10^{-36} cm⁷ esu⁻²		
Units	a^ω	b^ω	$b^{2\omega}$	c^ω	$c^{2\omega}$	$c^{3\omega}$
He	0.205			0.027 [101] 0.022 [100]	0.023 [100]	0.024 [100] 0.021 [103]
Ne	0.394			0.051 [24] 0.055 [96]	0.24 [33]	0.041 [103] 0.214 [64]
Ar	1.66			0.59 [24] 0.73 [96]	2.74 [33]	3.024 [64] 0.672 [103]
Kr	2.52			1.4 [24] 1.6 [96]	6.69 [33]	9.264 [64] 1.704 [103]
Xe	4.11			3.9 [24] 4.0 [96]	18.52 [33]	23.496 [64] 5.328 [103]
H₂	0.79			0.28 [101]	0.36 [32]	1.920 [64]
N₂	1.76			0.7 [101]		2.568 [64]
O₂	1.60				0.64 [32]	
CO	1.95				1.81 [32]	
CO₂	2.65			4.5 [101]		2.744 [64]
CH₄	2.6		0.01 [98]	1.45 [96]	1.48 [32]	
CCl₄	10.5		0.03 [98]	3.5 [10] 6.51 [83]	3.6 [32] 2.1 [32]	

TABLE 3 (continued)

| Atom or molecule | 10^{-24} cm^3 | 10^{-30} cm^5 esu^{-1} | | | 10^{-36} cm^7 esu^{-2} | |
	a^ω	b^ω	$b^{2\omega}$	c^ω	$c^{2\omega}$	$c^{3\omega}$
CH_3F	3.55	-0.28 [96]		1.28 [96]		
CHF_3	2.81	+0.41 [96]	-0.06 [32]	0.92 [96]	0.27 [32]	
$CHCl_3$	8.23	14.4 [96]	0.33 [32]	4.8 [10]	4.4 [32]	
CS_2	8.74			56.5 [83]		
				50.1 [10]		
				54.4 [96]		
				57.4 [101]		
C_2H_6	4.47			1.9 [101]	4.8 [32]	
C_6H_6	10.32			6.5 [101]		
C_6H_{12}	10.87			1.2 [10]		
$C_6H_5CH_3$	12.29			27.2 [10]		
$C_6H_5NO_2$	12.17			85.0 [10]		
CH_3I	7.28		0.84 [32]		8.0 [32]	
SF_6	4.47			1.2 [101]	0.26 [32]	
C_2H_5I			1.1 [32]		13.6 [32]	

aOnly mean values are given: $a = a_{11}/3$, $b = b_{113}/3 = b_{333}$ (for the axial symmetry), $b = b_{123}$ (for tetrahedral symmetry), and $c = c_{iijj}/5 = c_{3333}$.

a highly effective method for the elucidation of the electronic, atomic, molecular and macromolecular structures of matter in a wide range of the electromagnetic spectrum.

Especially promising is the new technique of producing ultrashort light pulses [12], of duration 10^{-8} to 10^{-12} s and consequently comparable to the relaxation times of molecules [92], making it possible for us to "seize," as it were, the momentary structure of molecular and macromolecular bodies and to perform a separation between the various nonlinear processes, differing as to the times necessary for attaining the steady state. Thus, in certain conditions, one can separate the processes of nonlinear electron polarizability from those of molecular reorientation and redistribution. This will, moreover, depend on the electro-optical structure of the constituent molecules or macromolecules, on their size and interactions, and on whether they are more highly susceptible to nonlinear electron distortion or to orientation by an external electric field. The future of these studies lies with the now rapidly developing nonlinear molecular spectroscopy [20,105].

Quite recently, numerous papers have appeared reporting determinations of various nonlinear optical and electro-optical properties of atomic and molecular systems [106-145].

ACKNOWLEDGMENTS

The author wishes to thank Professor Dr. C. T. O'Konski for his valuable discussion at our Institute of Physics in September 1972. Thanks are also due to Professors A. D. Buckingham, R. Y. Chiao, H. Strauss and Tae-Kyu Ha for their friendly comments. The author is, moreover, indebted to K. Flatau, of this Institute, for the English translation.

REFERENCES

1. W. Voigt, Magneto- und Elektro-Optik, Teubner, Leipzig
    (1908).

2. P. Langevin, Ann. Chim. Phys., 5, 70 (1905); Radium 7, 249
    (1910).

3. M. Born, Optik, Springer, Berlin (1933).

4. K. A. Levine (ed.), Lasers: A. Series of Advances, Vols. I
    and II, Dekker, New York (1968); A. F. Haught, Ann. Rev.
    Phys. Chem., 19, 343 (1968); W. J. Jones, Quart. Rev., 23, 73
    (1969); R. C. Smith, Solid State Biophysics (S. J. Wyard,
    ed.), McGraw-Hill, New York (1970), p. 371.

5. A. D. Buckingham, Proc. Phys. Soc. (London), B69, 344 (1956).

6. A. Piekara and S. Kielich, J. Chem. Phys., 29, 1297 (1958);
    S. Kielich and A. Piekara, Acta Phys. Polon., 18, 439 (1959).

7. G. Mayer and F. Gires, Compt. Rend. Acad. Sci. Paris, 258,
    2039 (1964); F. Gires, Ann. Radioelec., 23, 281 (1968).

8. M. Paillette, Ann. Phys. (Paris), 4, 671 (1969).

9. F. B. Martin and J. R. Lalanne, Phys. Rev., A4, 1275 (1971);
    J. R. Lalanne, F. B. Martin, and P. Botherel, J. Colloid
    Interface Sci., 39, 601 (1972); S. Kielich, J. R. Lalanne,
    and F. B. Martin, J. Phys. 33, C1-191 (1972).

10. S. Kielich, Proc. Phys. Soc., 90, 847 (1967); S. Kielich,
    J. R. Lalanne, and F. B. Martin, Compt. Rend. Acad. Sci.
    Paris, B272, 731 (1971).

11. S. Kielich, J. Phys., 29, 619 (1968); IEEE J. Quantum
    Electron., QE-4, 744 (1968).

12. I. L. Fabielinsky, Usp. Fiz. Nauk, 104, 77 (1971);
    B. Ya. Zeldovich and T. I. Kuznietzova, Usp. Fiz. Nauk, 106,
    47 (1972).

13. P. D. Maker and R. W. Terhune, Phys. Rev., 137A, 801 (1965);
    C. C. Wang, Phys. Rev., 152, 149 (1966); R. W. Minck,
    R. W. Terhune, and C. C. Wang, Appl. Opt., 5, 1595 (1966).

14. R. Y. Chiao, E. Garmire, and C. H. Townes, Phys. Rev.
    Letters, 13, 479 (1964); S. A. Akhmanov, A. P. Sukhorukov,
    and R. V. Khokhlov, Usp. Fiz. Nauk, 93, 19 (1967);
    S. A. Akhmanov and R. V. Khokhlov, Usp. Fiz. Nauk, 95, 231
    (1968); Y. R. Shen and Y. J. Shaham, Phys. Rev., 163, 224
    (1967).

15. G. N. Steinberg, Phys. Rev., A4, 1182 (1971); E. L. Kerr,
    Phys. Rev., A4, 1195 (1971); S. Kielich, Acta Phys. Polon.,
    34, 1093 (1968); R. R. Alfano, L. L. Hope, and S. L. Shapiro,
    Phys. Rev., 6A, 433 (1972); T. K. Gustafson and C. H. Townes,

Phys. Rev., $\underline{6A}$, 1659 (1972).

16. P. A. Franken and J. W. Ward, Rev. Mod. Phys., $\underline{35}$, 23 (1963); J. F. Ward, Rev. Mod. Phys., $\underline{37}$, 1 (1965); N. Bloembergen, Opt. Acta $\underline{13}$, 311 (1966).

17. N. Bloembergen, Am. J. Phys., $\underline{35}$, 989 (1967); W. L. Peticolas, Ann. Rev. Phys. Chem., $\underline{18}$, 233 (1967); V. S. Starunov and I. L. Fabielinskiy, Usp. Fiz. Nauk, $\underline{98}$, 441 (1969).

18. R. L. Carman, R. Y. Chiao, and Kelley, Phys. Rev. Letters, $\underline{17}$, 1281 (1966); R. Y. Chiao and J. Godine, Phys. Rev., $\underline{185}$, 430 (1969).

19. S. Kielich, Proc. Phys. Soc., $\underline{86}$, 709 (1965); J. Phys., $\underline{28}$, 519 (1967); J. Colloid Interface Sci. 27, 432 (1968).

20. D. A. Long, Chem. Brit., $\underline{7}$, 108 (1971); Essays in Structural Chemistry (D. A. Long, A. J. Downs, and L. A. K. Stavley, eds.), Macmillan, London (1972), p. 18.

21. P. W. Atkins and R. G. Woolley, J. Chem. Soc. A, $\underline{1969}$, 515; P. W. Atkins, Chem. Brit., $\underline{7}$, 244 (1971).

22. S. Kielich, Opt. Commun., $\underline{1}$, 129 (1969); Opto-Electronics, $\underline{1}$, 75 (1969).

23. D. V. Vlasov and V. P. Zaitsev, JETP Letters, $\underline{14}$, 112 (1971).

24. S. Kielich, Opto-Electronics, $\underline{2}$, 125 (1970); Ferroelectrics $\underline{4}$, 257 (1972).

25. P. W. Atkins and A. D. Wilson, Mol. Phys., $\underline{24}$, 33 (1972).

26. S. A. Akhmanov and A. S. Chirkin, Statistical Effects in Nonlinear Optics (in Russian), Izd. Moskovskogo Universiteta, USSR (1971).

27. S. Kielich, Physica, $\underline{34}$, 365 (1967); Acta Phys. Polon., $\underline{32}$, 405 (1967).

28. R. W. Terhune, P. D. Maker, and C. M. Savage, Phys. Rev. Letters, $\underline{8}$, 404 (1962).

29. J. E. Bjorkholm and A. E. Siegman, Phys. Rev., $\underline{154}$, 851 (1967).

30. V. S. Suvorov and A. S. Sonin, Zh. Eksperim. i Teor. Fiz., $\underline{54}$, 1044 (1968); I. A. Pleshakov, V. S. Suvorov, and A. A. Filimonov, Izv. Akad. Nauk SSSR, $\underline{35}$, 1856 (1971).

31. S. Kielich, Chem. Phys. Letters, $\underline{2}$, 596 (1968); Acta Phys. Polon., $\underline{37A}$, 205 (1970).

32. G. Mayer, Compt. Rend. Acad. Sci. Paris, $\underline{267B}$, 54 (1968); G. Hauchecorne, F. Kerhervé, and G. Mayer, J. Phys., $\underline{32}$, 47 (1971).

33. R. S. Finn and J. F. Ward, Phys. Rev. Letters, $\underline{26}$, 285 (1971).

34. S. Kielich, J. Colloid Interface Sci., $\underline{33}$, 142 (1970); 34, 228 (1970).

35. J. F. Ready, Effects of High-Power Laser Radiation, Academic, New York (1971).

36. H. A. Lorentz, The Theory of Electrons, Teubner, Leipzig (1909); L. Rosenfeld, Theory of Electrons, North-Holland, Amsterdam (1951).

37. F. Pockels, Lehrbuch der Kristallooptik, Teubner, Leipzig (1906); J. F. Nye, Physical Properties of Crystals, Clarendon, Oxford (1957); S. Bhagavantam, Crystal Symmetry and Physical Properties, Academic, London (1966).

38. V. M. Agranovich and V. L. Ginzburg, Spatial Dispersion in Crystal Optics and the Theory of Excitons, Wiley, New York (1965).

39. M. V. Volkenshteyn, Molekularnaya Optika (in Russian), Gostekhizdat, Moscow (1951).

40. J. A. Giordmaine, Phys. Rev., $\underline{A138}$, 1599 (1965); C. L. Tang and H. Rabin, Phys. Rev., $\underline{3B}$, 4025 (1971).

41. P. N. Butcher, Nonlinear Optical Phenomena, Ohio State University, Columbus (1965).

42. P. A. Franken, A. E. Hill, C. W. Peters, and G. Weinreich, Phys. Rev. Letters, $\underline{7}$, 118 (1961).

43. J. G. Bergman, Jr. and S. K. Kurtz, Mater. Sci. Eng., $\underline{5}$, 235 (1969/1970).

44. J. Jerphagnon and S. K. Kurtz, Phys. Rev., $\underline{1B}$, 1739 (1970); J. Appl. Phys., $\underline{41}$, 1667 (1970); R. C. Miller and W. A. Nordland, Phys. Rev., $\underline{2B}$, 4896 (1970); $\underline{5B}$, 4931 (1972); Appl. Phys. Letters, $\underline{16}$, 174 (1970); J. E. Pearson, G. A. Evans, and A. Yariv, Opt. Commun., $\underline{4}$, 365 (1972).

45. H. Rabin, International Conference on Science and Technology of Nonmetallic Crystals, New Delhi, India, January 13-17 (1969), p. 167; J. R. Gott, J. Phys. B. Atom, Mol. Phys., $\underline{4}$, 116 (1971).

46. A. Graja, Phys. Status Solidi, $\underline{27}$, K93 (1968); Acta Phys. Polon., $\underline{36}$, 283 (1969); $\underline{A37}$, 539 (1970); S. K. Kurtz and T. T. Perry, J. Appl. Phys., $\underline{39}$, 3798 (1968); A. A. Filimonov, V. S. Suvorov, and I. S. Rez, Zh. Eksperim. i Teor. Fiz., $\underline{56}$, 1519 (1969); M. Bass, D. Bua, R. Mozzi, and R. R. Monchamp, Appl. Phys. Letters, $\underline{15}$, 393 (1969).

47. I. Freund, Phys. Rev. Letters, $\underline{19}$, 1288 (1967); $\underline{21}$, 1404 (1968); Chem. Phys. Letters, $\underline{1}$, 551 (1968).

47a. I. Freund and L. Kopf, Phys. Rev. Letters, $\underline{24}$, 1017 (1970).

47b. G. Dolino, J. Lojzerowicz, and M. Valade, Phys. Rev., $\underline{B2}$, 2194 (1970); G. Dolino, Phys. Rev., $\underline{B6}$, 4025 (1972).

48. J. G. Bergman, Jr., J. H. McFee, and G. R. Crane, Appl. Phys. Letters, 18, 203 (1971).

49. P. D. Southgate and D. S. Hall, Appl. Phys. Letters, 18, 456 (1971).

49a. J. G. Bergman, C. R. Crane, B. F. Levine, and C. G. Bethea, Appl. Phys. Letters, 20, 21 (1972).

49b. P. D. Southgate and D. S. Hall, J. Appl. Phys. 42, 4480 (1971); 43, 2765 (1972).

50. S. H. Fine and W. P. Hansen, Appl. Optics, 10, 2350 (1971).

51. M. Bass, P. A. Franken, J. F. Ward, and G. Weinreich, Phys. Rev. Letters, 9, 446 (1962).

52. J. F. Ward and P. A. Franken, Phys. Rev., 133A, 183 (1964); J. F. Ward, Phys. Rev., 143, 569 (1966); J. F. Ward and G. H. C. New, Proc. Roy. Soc. (London), A299, 238 (1967).

53. I. S. Rez, Usp. Fiz. Nauk, 93, 653 (1967); H. Fay, Rev. Sci. Instr., 38, 197 (1967); C. G. B. Garret, IEEE J. Quantum Electron., QE-4, 70 (1968).

54. M. Di Domenico, Jr. and S. H. Wemple, J. Appl. Phys., 40, 720, 735 (1969); S. H. Wemple and M. Di Domenico, Phys. Rev., 1B, 193 (1970).

55. F. N. H. Robinson, J. Phys., C1, 286 (1968)1 J. L. Stevenson and A. C. Skopski, J. Phys., C5, L233 (1972); E. S. Kohn, J. Appl. Phys., 40, 2688 (1969); A. S. Barker, Jr., and R. Loudon, Rev. Mod. Phys., 44, 18 (1972); J. L. Stevenson, J. Phys., D6, L13 (1973).

56. S. Kielich, IEEE J. Quantum Electron., QE-5, 562 (1969); Acta Phys. Polon., 36, 621 (1969).

57. R. W. Terhune, P. D. Maker, and C. M. Savage, Appl. Phys. Letters, 2, 54 (1963); Third Conference on Quantum Electronics, Paris, 1963, Columbia University Press, New York (1964) p. 1559-1579.

58. P. P. Bey, J. F. Giuliani, and H. Rabin, Phys. Rev. Letters, 19, 819 (1967); IEEE J. Quantum Electron., QE-4, 932 (1968).

59. R. K. Chang and L. K. Galbraith, Phys. Rev., 171, 993 (1968).

60. L. S. Goldberg and J. M. Schnur, Appl. Phys. Letters, 14, 306 (1969); Radio Electron. Enr., 39, 279 (1970).

61. J. W. Shelton and Y. R. Shen, Phys. Rev. Letters, 25, 23 (1970); 26, 538 (1971); Phys. Rev., 5A, 1867 (1972).

62. M. R. Topp, R. P. Jones, and P. M. Rentzepis, Opt. Commun., 4, 264 (1971).

63. S. Kielich, Acta Phys. Polon., 33, 89, 141 (1968); S. Kielich and M. Kozierowski, Opt. Commun., 4, 395 (1972).

64. J. F. Ward and G. H. C. New, Phys. Rev., 185, 57 (1969).

65. S. E. Harris and R. B. Miles, Appl. Phys. Letters, 19, 385
    (1971); J. F. Young, G. C. Bjorklund, A. H. Kung,
    R. B. Miles, and S. E. Harris, Phys. Rev. Letters, 27, 1551
    (1971).

66. Y. Ueda and K. Shimoda, J. Phys. Soc. Japan, 28, 198 (1970).

67. N. Bloembergen and P. S. Pershan, Phys. Rev., 128, 606
    (1962); P. S. Pershan, Phys. Rev., 130, 919 (1963).

68. S. S. Jha and N. Bloembergen, Phys. Rev. 171, 891 (1968).

69. N. George, R. W. Hellwarth, and C. R. Cooke, Electron
    Technol. Warsaw, 2, 229 (1969); R. Hellwarth and N. George,
    Opto-Electron., 1, 213 (1969); R. R. Alfano and
    S. L. Shapiro, Phys. Rev. Letters, 24, 1217 (1970);
    F. B. Martin and J. R. Lalanne, Opt. Commun., 2, 219 (1970);
    G. Delfino, Opt. Commun., 4, 60 (1971); R. W. Hellwarth,
    A. Owyoung, and N. George, Phys. Rev., 4A, 2342 (1971).

70. A. D. Buckingham, Discussions Faraday Soc., 43, 205 (1967);
    M. Takatsuji, Phys. Rev., 165, 171 (1968).

71. S. Kielich, Acta Phys. Polon., 19, 149 (1960); 22, 299
    (1962); 30, 683 (1966); Chem. Phys. Letters, 2, 112 (1968);
    10, 516 (1971); Opt. Commun., 4, 135 (1971); S. Kielich and
    S. Wozniak, Acta Phys. Polon., A39, 233 (1971).

72. R. W. Hellwarth, Phys. Rev., 152, 156 (1966); 163, 205
    (1967); J. Chem. Phys., 52, 2128 (1970).

73. J. Hanus, Phys. Rev., 178, 420 (1969); R. Cubeddu, R. Poloni,
    C. A. Sacchi, and O. Svelto, Phys. Rev., 2A, 1955 (1970).

74. S. Kielich, Dielectric and Related Molecular Processes, Vol.
    1 (M. Davies, ed.), Chemical Scoiety, London (1972), Chap. 7.

75. M. Takatsuji, Phys. Rev., 155, 980 (1967); R. Locqueneux,
    Int. J. Quantum Chem. 6, 1 (1972).

76. P. W. Atkins and L. D. Barron, Proc. Roy. Soc. (London),
    A306, 119 (1968).

77. R. Pecora, J. Chem. Phys., 50, 2650 (1969).

78. J. P. Laussade and A. Yariv, IEEE J. Quantum Electron.,
    QE-5, 435 (1969); J. C. Comly, A. Yariv, and E. M. Garmire,
    Appl. Phys. Letters, 15, 148 (1969); L. Dahlström, Opt.
    Commun., 3, 399 (1971).

79. F. Shimizu, J. Phys. Soc. Japan, 22, 1070 (1967).

80. P. D. Maker, R. W. Terhune, and C. M. Savage, Phys. Rev.
    Letters, 12, 507 (1964).; P. D. McWane and D. A. Sealer,
    Appl. Phys. Letters, 8, 2786 (1966).

81. H. F. Hameka, Can. J. Chem., 49, 1823 (1971).

82. J. G. Kirkwood, J. Chem. Phys., $\underline{7}$, 911 (1939).

83. S. Kielich, Acta Phys. Polon., $\underline{17}$, 239 (1958); $\underline{22}$, 299 (1962); Mol. Phys., $\underline{6}$, 49 (1963).

84. S. Kielich, J. Chem. Phys., $\underline{46}$, 4090 (1967); Chem. Phys. Letters, $\underline{2}$, 112 (1968); $\underline{7}$, 347 (1970).

85. S. Kielich and R. Zawodny, Chem. Phys. Letters, $\underline{12}$, 20 (1971).

86. J. G. Kirkwood, J. Chem. Phys., $\underline{4}$, 592 (1936).

87. C. T. O'Konski, Encyclopedia of Polymer Science and Technology, Wiley, (1968), p. 551; C. T. O'Konski, K. Yoshioka, and W. H. Orttung, J. Phys. Chem., $\underline{63}$, 1558 (1959).

88. S. Kielich, IEEE J. Quantum Electron., $\underline{QE-5}$, 562 (1969); Opto-Electronics, $\underline{2}$, 5 (1970).

89. M. J. Shah, J. Phys. Chem., $\underline{67}$, 2215 (1963); D. N. Holcomb and I. Tinoco, Jr., J. Phys. Chem., $\underline{67}$, 2691 (1963).

90. R. G. Brewer, J. R. Lifsitz, E. Garmire, R. Y. Chiao, and C. H. Townes, Phys. Rev., $\underline{166}$, 326 (1968).

91. P. Debye, Polar Molecules, Dover, New York (1945).

92. N. E. Hill, W. E. Vaughan, A. H. Price, and M. Davies, Dielectric Properties and Molecular Behaviour, Van Nostrand-Reinhold, London (1969).

93. H. A. Stuart, Molekülstruktur, Vol. 1, Springer, Berlin (1967).

94. S. Kielich, J. R. Lalanne, and F. B. Martin, Phys. Rev. Letters, $\underline{26}$, 1295 (1971); Acta Phys. Polon., $\underline{A41}$, 479 (1972).

95. J. M. O'Hare and R. P. Hurst, J. Chem. Phys., $\underline{46}$, 2356 (1967); A. D. McLean and M. Yoshimine, J. Chem. Phys., $\underline{46}$, 3682 (1967); S. P. Liebman and J. W. Moscowitz, J. Chem. Phys., $\underline{54}$, 3622 (1971); G. P. Arrighini, M. Maestro, and R. Moccia, Symp. Faraday Soc., $\underline{2}$, 48 (1968); R. G. Brewer and A. D. McLean, Phys. Rev. Letters, $\underline{21}$, 271 (1968); G. P. Arrighini, J. Tomasi, and C. Petrongolo, Theoret. Chim. Acta (Berlin) $\underline{18}$, 341 (1970).

96. A. D. Buckingham and B. J. Orr, Quart. Rev., $\underline{21}$, 195 (1967); Trans. Faraday Soc., $\underline{65}$, 673 (1969).

97. S. Kielich, Acta Phys. Polon., $\underline{26}$, 135 (1964); S. J. Cyvin, J. E. Rauch, and J. C. Decius, J. Chem. Phys., $\underline{43}$, 4083 (1965).

98. R. W. Terhune, P. D. Maker, and C. M. Savage, Phys. Rev. Letters, $\underline{14}$, 681 (1965); D. L. Weinberg, J. Chem. Phys., $\underline{47}$, 1307 (1967); P. D. Maker, Phys. Rev. $\underline{1A}$, 923 (1970).

99. A. D. Buckingham, Adv. Chem. Phys., $\underline{12}$, 107 (1967); L. L. Boyle, Int. J. Quantum Chem. $\underline{3}$, 231 (1969); $\underline{4}$, 38

(1970); Z. Ozgo, Acta Phys. Polon., $\underline{34}$, 1087 (1968);
R. Zawodny and Z. Ozgo, Bull. Soc. Amis Sci. Lettres,
Poznan, $\underline{22B}$, 83 (1970/71).

100.   M. N. Grasso, K. T. Chung, and R. P. Hurst, Phys. Rev., $\underline{167}$,
       1 (1968); P. Sitz and R. Yaris, J. Chem. Phys., $\underline{49}$, 3546
       (1968); R. E. Sitter, Jr. and R. P. Hurst, Phys. Rev., $\underline{5A}$, 5
       (1972).

101.   A. D. Buckingham and B. J. Orr, Proc. Roy Soc. (London),
       $\underline{A305}$, 259 (1968); A. D. Buckingham and D. A. Dunmur, Trans.
       Faraday Soc., $\underline{64}$, 1775 (1968); A. D. Buckingham,
       M. P. Bogaard, D. A. Dunmur, C. P. Hobbs, and B. J. Orr,
       Trans. Faraday Soc., $\underline{66}$, 1548 (1970); M. P. Bogaard,
       A. D. Buckingham, and G. L. D. Ritchie, Mol. Phys., $\underline{18}$, 575
       (1970).

102.   R. Locqueneux and P. Smet, Compt. Rend., $\underline{267B}$, 353, 433
       (1968); R. Locqueneux, Compt. Rend., $\underline{268B}$, 1157 (1969);
       Tai Yup Chang, J. Chem. Phys., $\underline{56}$, 1745 (1972).

103.   E. L. Daves, Phys. Rev., $\underline{169}$, 47 (1968); E. Devin and
       R. Locqueneux, Compt. Rend., $\underline{270B}$, 981 (1970); R. Klingbeil,
       V. G. Kaveeshwar, and R. P. Hurst, Phys. Rev., $\underline{4A}$, 1760
       (1971); B. P. Tripathi, R. K. Laroraya, and S. L. Srivastava,
       Phys. Rev., $\underline{4A}$, 2076 (1971); $\underline{5A}$, 1565 (1972).

104.   R. Bersohn, J. Am. Chem. Soc., $\underline{86}$, 3550 (1964); R. Bersohn,
       Y. H. Pao, and H. L. Prisch, J. Chem. Phys., $\underline{45}$, 3184 (1966);
       B. Fanconi and W. L. Peticolas, J. Chem. Phys., $\underline{50}$, 2244
       (1969); B. Fanconi, B. Tomlinson, L. A. Nafie, W. Small, and
       W. L. Peticolas, J. Chem. Phys., $\underline{51}$, 3993 (1969); S. Kielich
       and M. Kozierowski, Bull. Soc. Amis Sci. Lettres, Poznan,
       $\underline{22B}$, 15 (1970/71).

105.   D. A. Long and L. Stanton, Proc. Roy. Soc. (London), $\underline{A318}$,
       441 (1970); J. F. Verdieck, S. H. Peterson, C. M. Savage, and
       P. D. Maker, Chem. Phys. Letters, $\underline{7}$, 219 (1970); B. J. Orr
       and J. F. Ward, Mol. Phys., $\underline{20}$, 513 (1971); D. H. Auston,
       Opt. Commun., $\underline{3}$, 272 (1971); R. R. Alfano, A. Lempicki, and
       S. L. Shapiro, IEEE J. Quantum Electron., $\underline{QE-7}$, 416 (1971);
       S. Kielich, Conference on Interaction of Electrons with
       Strong Electromagnetic Fields, Balatonfured, 11-16 Sept. 1972
       (1973).

106.   S. Kielich, Foundations of Nonlinear Molecular Optics,
       A. Mickiewicz University Press, Poznan, Part I (1972), Part
       II (1973).

107.   N. S. Hush and M. L. Williams, Theoret. Chim. Acta, $\underline{25}$, 346
       (1972).

108.   S. Kielich, J. R. Lalanne, and F. B. Martin, IEEE J. Quantum
       Electron., $\underline{QE-9}$, 601 (1973).

109.   J. P. Herman, D. Ricard, and J. Ducuing, Appl. Phys. Letters,

$\underline{23}$, 178 (1973).

110. P. A. Fleury and J. P. Boon, Adv. Chem. Phys., $\underline{24}$, 1 (1973);
P. W. Atkins and A. D. Wilson, Solid State Surface Sci., $\underline{2}$,
153 (1973).

111. S. Kielich, J. R. Lalanne, and F. B. Martin, J. Raman
Spectr., $\underline{1}$, 119 (1973).

112. T. Bancewicz, Z. Ozgo, and S. Kielich, J. Raman Spectr., $\underline{1}$,
177 (1973); Phys. Letters, $\underline{44A}$, 407 (1973).

113. S. Kielich and Z. Ozgo, Opt. Commun., $\underline{8}$, 417 (1973).

114. D. Bedaux and N. Bloembergen, Physica, $\underline{69}$, 57 (1973).

115. S. E. Harris, Phys. Rev. Letters, $\underline{31}$, 341 (1973).

116. R. Klingbeil, Phys. Rev. $\underline{A7}$, 48, 376 (1973).

117. G. P. Bolognesi, S. Mezzetti, and F. Pandarese, Opt. Commun.,
$\underline{8}$, 267 (1973); D. C. Haueisen, Am. J. Phys., $\underline{41}$, 1251 (1973).

118. E. Leuliette-Devin and R. Locqueneux, Chem. Phys. Letters,
$\underline{19}$, 497 (1973).

119. J. P. Herman, Opt. Commun., $\underline{9}$, 74 (1973).

120. A. Feldman, D. Horowitz, and R. M. Waxler, IEEE J. Quantum
Electron., $\underline{QE-9}$, 1054 (1973).

121. A. Owyoung, IEEE J. Quantum Electron., $\underline{QE-9}$, 1064 (1973).

122. S. Kielich and M. Kozierowski, Acta Phys. Polon., $\underline{A45}$, 231
(1974).

123. W. M. Gelbart, Chem. Phys. Letters, $\underline{23}$, 53 (1973);
R. Samson and R. A. Pasmanter, Chem. Phys. Letters, $\underline{25}$, 405
(1974).

124. Z. Ozgo and S. Kielich, Physica, $\underline{72}$, 191 (1974).

125. I. J. Bigio and J. F. Ward, Phys. Rev., $\underline{A9}$, 35 (1974).

126. R. S. Finn and J. F. Ward, J. Chem. Phys., $\underline{60}$, 454 (1974).

127. M. D. Levenson and N. Bloembergen, J. Chem. Phys., $\underline{60}$, 1323
(1974).

128. B. F. Levine and C. G. Bethea, Appl. Phys. Letters, $\underline{24}$, 445
(1974).

129. B. Kasprowicz-Kielich, S. Kielich, and J. R. Lalanne,
Molecular Motions in Liquids (J. Lascombe, ed.), Reidel,
Dordrecht-Holland (1974), p. 563.

130. T. Bischofberger and E. Courtens, Phys. Rev. Letters, $\underline{32}$, 163
(1974).

131. K. C. Rustagi and J. Ducuing, Opt. Commun., $\underline{10}$, 258 (1974).

132. M. P. Bogaard and B. J. Orr, MTP Rev. Mol. Struct.
      Properties, Ser. II (1975).

133. R. Tanas, Optik, 40, 109 (1974).

134. R. L. Byer, Ann. Rev. Materials Sct, 4, 147 (1974).

135. H. Vogt, Appl. Phys., 5, 85 (1974).

136. Z. Blaszczak, A. Dobek, and A. Patkowski, Acta Phys. Polon.,
      A45, 269 (1974).

137. B. R. Jennings and H. J. Coles, Nature, 252, 33 (1974); Proc.
      Roy. Soc., 348A, 525 (1976).

138. H. J. Coles and B. R. Jennings, Phil. Mag., 32, 1051 (1975).

139. K. Sala and M. C. Richardson, Phys. Rev., 12A, 1036 (1975).

140. B. Kasprowicz-Kielich, S. Kielich, Adv. Mol. Relaxation
      Processes, 7, 275 (1975).

141. S. Kielich, R. Zawodny, Optics Communications, 15, 267
      (1975).

142. S. Wozniak and S. Kielich, J. Physique, 36, 1305 (1975).

143. J. F. Ward and I. J. Bigio, Phys. Rev., 11A, 60 (1975).

144. B. F. Levine and C. G. Bethea, J. Chem. Phys., 63, 2666
      (1975).

145. A. Dobek, A. Patkowski, and D. Labuda, J. Phys. D. Appl.
      Phys., 9, L21 (1976).

Chapter 13

MAGNETOELECTRO-OPTICS

Stanislaw Kielich

Nonlinear Optics Department
Institute of Physics
A. Mickiewicz University at Poznań
Poznań, Poland

I. INTRODUCTION

Michael Faraday, in 1846 [1], discovered to the surprise of his contemporaries that a dc magnetic field affects directly the optical properties of matter. He showed that, when polarized

445

light traverses matter in a direction parallel to that of an
externally applied magnetic field, the polarization plane of the
light wave is rotated by an angle θ proportional to the path l
traversed in the layer of the substance and to the magnetic field
strength H:

$$\theta = V(\lambda) lH \tag{1}$$

$V(\lambda)$ denoting the magneto-optical rotation constant or Verdet
constant [2], dependent on the nature of the medium, the temperature, and wavelength λ of the incident light.

By Fresnel's theory [3], plane-polarized (linearly polarized)
light has to be regarded as a superposition of two circularly
polarized waves of the same wavelength and amplitude. The rotation of the plane of polarization is then due to a difference between the refractive indexes n_+ and n_- of the optically
activated medium for right and left circularly polarized waves, at
the circular frequency $\omega = 2\pi c/\lambda$. In Faraday's effect the optically inactive medium becomes active under the influence of the dc
magnetic field, and we can write the following relation for the
amount of magneto-optical rotation:

$$n_+ - n_- = \frac{2c}{\omega} V(\lambda)H = \frac{\lambda}{\pi} V(\lambda)H \tag{2}$$

Optical activity, induced in matter by an externally applied
magnetic field, is to be distinguished from natural optical activity. The latter is an intrinsic property of media as a result of
molecular, or crystalline, optical asymmetry.

If the light propagates at right angles to the magnetic field
(Voigt configuration [4]), the naturally isotropic medium -- as
was first shown by Majorana [5] in colloidal solutions and subsequently by Cotton and Mouton [6] in liquids -- becomes optically
birefringent with optical axis parallel to the direction of the
magnetic field. Similarly, to Kerr's effect, the amount of this

magnetically induced birefringence is given by the difference between the refractive indexes for component light vibrations
parallel and perpendicular to the direction of the applied field
H:

$$n_{\parallel} - n_{\perp} = \lambda C(\lambda) H^2 \tag{3}$$

where $C(\lambda)$ is the Cotton-Mouton constant specific to the medium,
its thermodynamical state, and the light wavelength.

One of the earliest theoretical explanations of magneto-
optical phenomena was proposed by Voigt [4] on the basis of
Lorentz's [7] electron theory of matter. Voigt interpreted
magnetically induced birefringence as due to the direct action of
the magnetic field on the electrons of atoms and molecules, which
undergo an anisotropic distortion proportional to the square of
the field strength H in accordance with the law (3). By
Voigt's theory, the absolute retardation ratio should be:

$$\frac{n_{\parallel} - n}{n_{\perp} - n} = +3 \tag{3a}$$

whereas the majority of experiments led to Havelock's relation
[8]:

$$\frac{n_{\parallel} - n}{n_{\perp} - n} = -2 \tag{3b}$$

which results from Langevin's theory [9] of statistical molecular
orientation.

Magneto-optical phenomena are unremittingly under study and
have by now become the subject matter of monographs [10-17] and
numerous review articles [18-23]. Work on the Faraday effect provides data on the influence of a linear (first power in H) magnetic field on atoms and molecules [24,25], macromolecules [26]
and biomacromolecules [27]. Magnetic anisotropies of molecules
and macromolecules, as well as their nonlinear magneto-optical

properties, can be determined from Cotton-Mouton studies in rare-
fied gases [28], real gases and liquids [29], liquid molecular
solutions [30], polymer and biopolymer solutions [31], and
colloids [32].

Rapid progress in laboratory techniques has recently provided
us with new sources of light in the form of gas and crystal lasers
[33] and, concomitantly, with new methods of producing intense
pulsed magnetic fields [34]. Thus, conditions now favor the
search for new magneto-optical effects, arising under the simultaneous action of strong electric and magnetic fields, both static
and rapidly alternating. Such nonlinear magneto-electro-optical
effects, making apparent the direct and simultaneous action of
strong magnetic and electric fields on matter, can be observed in
molecular, macromolecular, and biological substances [35], and in
particular in semiconducting, magnetic, and metamagnetic materials
[36], crystalline bodies [14,37], and liquid crystals [38]. Among
the new nonlinear effects, we shall choose for discussion here the
inverse Faraday effect [39], the dc electric field effect on laser
beam induced changes in magneto-optical birefringence and rotation
[40], generation of laser light harmonics by magnetized bodies
[41,42] as well as generation processes and frequency mixing processes in crossed electric and magnetic fields.

Strictly, the theory of magneto-optical phenomena would require a quantum-mechanical approach. However, since this article
is aimed essentially at a concise presentation, or enumeration, of
new magneto-electro-optical effects (besides the already known
ones) and at pointing out the experimental conditions and materials in which they can be detected, we shall be resorting to a simpler, phenomenological treatment. It will moreover be our aim to
draw the reader's attention to the information to be gleaned from
these new effects with regard to the electro-magnetic properties
of molecules, macromolecules, and colloidal particles.

II.   LINEAR MAGNETOELECTRIC PROCESSES

The light refractive index n of an isotropic medium is
related to the latter's electric permittivity $\varepsilon(\omega)$ and magnetic
permittivity $\mu(\omega)$ at a given circular frequency ω by the well-
known Maxwell relation:

$$n^2 = \varepsilon(\omega)\mu(\omega) \tag{4}$$

In naturally anisotropic bodies, the electric and magnetic proper-
ties are described by the second-rank tensors ε_{ij} and μ_{ij}, defined by the equations:

$$(\varepsilon_{ij}^{\omega} - \delta_{ij})E_j(\underline{r},t) = 4\pi P_i^e(\underline{r},t)$$

$$(\mu_{ij}^{\omega} - \delta_{ij})H_j(\underline{r},t) = 4\pi P_i^m(\underline{r},t) \tag{5}$$

where we have applied the summation convention over the recurring
index j (running through the values x,y,z of axes of the
laboratory Cartesian reference system). δ_{ij} is Kronecker's symmetric unit tensor.

The vectors of electric polarization $\underline{P}^e(\underline{r},t)$ and magnetic
polarization $\underline{P}^m(\underline{r},t)$ are in general functions of the electric
vector $\underline{E}(\underline{r},t)$ and magnetic vector $\underline{H}(\underline{r},t)$ of the electromagnetic field of the light wave existing at the moment of time t and
space point $\underline{r}$. In a first approximation, the relations between
these vectors are linear ones, and in a phenomenological treatment
we can write:

$$P_i^e(\underline{r},t) = \chi_{ij}^{ee}E_j(\underline{r},t) + \chi_{ij}^{em}H_j(\underline{r},t) \tag{6}$$

$$P_i^m(\underline{r},t) = \chi_{ij}^{mm}H_j(\underline{r},t) + \chi_{ij}^{me}E_j(\underline{r},t) \tag{7}$$

where χ_{ij}^{ee} is the tensor of electric susceptibility and χ_{ij}^{mm}
that of magnetic susceptibility of the medium. These tensors des-

cribe linear processes of light refraction and propagation in optically transparent media and media exhibiting frequency dispersion and spatial dispersion [37].

The axial tensor of cross susceptibility χ_{ij}^{em} describes the electric polarization (a polar vector) induced in the medium by a magnetic field (an axial vector). This linear magneto-electric effect has, as yet, not been observed successfully in molecular substances [10,43] but has been studied in piezoelectric paramagnetic crystals [44] and, more recently, in metamagnetic ones [45]. The inverse effect, consisting in the induction of magnetic polarization by an electric field, is described by the axial tensor χ_{ij}^{me} of cross susceptibility (in particular, $\chi_{ij}^{em} = \chi_{ji}^{me}$).

The fundamental equations (5) are also applicable to isotropic bodies, in which the respective anisotropies to be discussed in subsequent subsections are induced by an external electric, magnetic, or electromagnetic field.

III. QUADRATIC CROSS PROCESSES

In addition to the linear magneto-electric processes occurring in electric magnetic fields of low intensity, we have to consider nonlinear cross (or mixing) processes, caused by intense fields. In particular, in a second approximation, we have to deal with the electric polarization:

$$P_i^e(\underline{r},t) = \chi_{ijk}^{eem}E_j(\underline{r},t)H_k(\underline{r},t) \tag{8}$$

where χ_{ijk}^{eem} is a third-rank axial tensor describing nonlinear magneto-electric susceptibility of the second order. This tensor is of interest in that it possesses nonzero tensor elements in all media, including the isotropic medium (see Table 1).

TABLE 1

The Number of Nonzero (N) and Independent (I)
Elements of Nonlinear Susceptibility Axial Tensors
χ_{ijk}, χ_{ijkl}, and χ_{ijklm} for Crystallographical Classes

	eem $\chi_{(ij)k}$		eem $\chi_{[ij]k}$		eeem $\chi_{(ij)k,l}$		eeem $\chi_{[ij]k,l}$		eeeem $\chi_{(ij)k,l,m}$		eeeem $\chi_{[ij]k,l,n}$	
Class	N	I	N	I	N	I	N	I	N	I	N	I
1 (C_1)	27	18	18	9	81	54	54	27	243	162	162	81
$\bar{1}$ (C_i)	27	18	18	9	0	0	0	0	243	162	162	81
m (C_s)	13	8	10	5	40	26	28	14	121	80	82	41
2 (C_2)	13	8	10	5	41	28	26	13	121	80	82	41
2/m (C_{2h})	13	8	10	5	0	0	0	0	121	80	82	41
222 (D_2)	6	3	6	3	21	15	12	6	60	39	42	21
mm2 (C_{2v})	6	3	6	3	20	13	14	7	60	39	42	21
mmm (D_{2h})	6	3	6	3	0	0	0	0	60	39	42	21
4 (C_4)	11	4	10	3	39	14	26	7	119	40	82	21
$\bar{4}$ (S_4)	11	4	10	3	40	14	24	6	119	40	82	21
4/m (C_{4h})	11	4	10	3	0	0	0	0	119	40	82	21
422 (D_4)	4	1	6	2	21	8	12	3	58	19	42	11
4mm (C_{4v})	4	1	6	2	18	6	14	4	58	19	42	11
$\bar{4}$2m (D_{2d})	4	1	6	2	20	7	12	3	58	19	42	11
4/mmm (D_{4h})	4	1	6	2	0	0	0	0	58	19	42	11
3 (C_3)	19	6	10	3	71	18	42	9	231	54	146	27
$\bar{3}$ (S_6)	19	6	10	3	0	0	0	0	231	54	146	27
32 (D_3)	8	2	6	2	37	10	20	4	114	26	74	14
3m (C_{3v})	8	2	6	2	34	8	22	5	114	26	74	14
$\bar{3}$m (D_{3d})	8	2	6	2	0	0	0	0	114	26	74	14
6 (C_6)	11	4	10	3	39	12	26	7	119	32	82	19
$\bar{6}$ (C_{3h})	11	4	10	3	32	6	16	2	119	32	82	19
6/m (C_{6h})	11	4	10	3	0	0	0	0	119	32	82	19
622 (D_6)	4	1	6	2	21	7	12	3	58	15	42	10
6mm (C_{6v})	4	1	6	2	18	5	14	4	58	15	42	10
$\bar{6}$m2 (D_{3h})	4	1	6	2	16	3	8	1	58	15	42	10
6/mmm (D_{6h})	4	1	6	2	0	0	0	0	58	15	42	10

TABLE 1 (continued)

Class	$\chi^{eem}_{(ij)k}$		$\chi^{eem}_{[ij]k}$		$\chi^{eeem}_{(ij)k,l}$		$\chi^{eeem}_{[ij]k,l}$		$\chi^{eeeem}_{(ij)k,l,m}$		$\chi^{eeeem}_{[ij]k,l,n}$	
	N	I	N	I	N	I	N	I	N	I	N	I
23 (T)	6	1	6	1	21	5	12	2	60	13	42	7
m3 (T_h)	6	1	6	1	0	0	0	0	60	13	42	7
432 (O)	0	0	6	1	21	3	12	1	54	6	42	4
$\overline{4}$3m (T_d)	0	0	6	1	18	2	12	1	54	6	42	4
m3m (O_h)	0	0	6	1	0	0	0	0	54	6	42	4
Y	0	0	6	1	21	2	12	1	54	3	42	3
Y_h	0	0	6	1	0	0	0	0	54	3	42	3
K	0	0	6	1	21	2	12	1	54	3	42	3
K_h	0	0	6	1	0	0	0	0	54	3	42	3

A. Magneto-Optical Effects

Assuming the magnetic field in Eq. (8) to be static, one obtains with regard to (5) and (6) the following first-order variation in electric permittivity tensor measured at the frequency ω:

$$\Delta\varepsilon^{(1)}_{ij}(\omega) = 4\pi\chi^{eem}_{ijk}(\omega)H_k(0) \tag{9}$$

If the electromagnetic wave propagates in a direction parallel to the magnetic field, assumed as acting along the z axis (Fig. 1), the nondiagonal tensor elements of (9) defining Faraday's effect yield:

$$\Delta\varepsilon_{xy}(\omega) = -\Delta\varepsilon_{yx}(\omega) = 4\pi\chi^{eem}_{xyz}(\omega)H_z(0) \tag{9a}$$

Since, for diamagnetic media, we have in approximation [10]:

$$n_+ - n_- = \frac{i\Delta\varepsilon_{xy}(\omega)}{n} \tag{10}$$

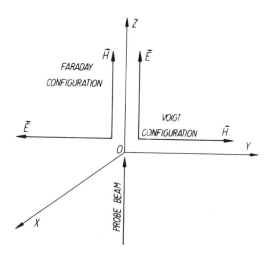

FIG. 1. Idealized experimental setup for the study of magneto-electro-optical phenomena. The probe light beam propagates along the Z axis and the molecular sample is located at the origin. The dc magnetic field H is applied either at Faraday configuration (along Z) or at Voigt configuration (along X or Y). The dc electric field E can be applied perpendicular to H (crossed fields), or parallel to H, or at any angle to H.

we find that in the case under consideration Verdet's constant given by Eq. (2):

$$V(\lambda) = \frac{4\pi^2}{n\lambda} \, i\chi_{xyz}^{eem}(\omega) \tag{2a}$$

is expressed by way of the magneto-electric susceptibility component $\chi_{xyz}^{eem}(\omega)$.

Recently, Ferguson and Romagnoli [46] proposed, on the basis of the magneto-electric polarization (8) and a supplementary polarization due to field gradients [47], a theory of transverse and longitudinal nonlinear Kerr magneto-optic effect in ferromagnetic metals.

B. Inverse Faraday Effect

Besides the linear magnetization by an electric field, Eq. (7), which can take place only in bodies without a center of symmetry [16], we have to deal with magnetic polarization caused by the square of an electric field [39]:

$$P_i^m(\underline{r},t) = \chi_{ijk}^{mee} E_j(\underline{r},t) E_k(\underline{r},t) \tag{11}$$

where, if the electromagnetic wave propagates in the direction of $\underline{k}$ (the wave vector), one has:

$$E(\underline{r},t) = E(\omega,\underline{k}) \exp\{i(\underline{k} \cdot \underline{r} - \omega t)\} + c.c. \tag{12}$$

We hence see that, by Eq. (11), an isotropic or indeed any other medium can undergo a magnetization not only by the square (second power) of a dc or ac electric field but, essentially, by light of intensity given by the tensor:

$$I_{jk} = \frac{E_j(\omega)E_k(-\omega)}{2} \tag{13}$$

We thus have for light beam-induced magnetization:

$$P_i^m(0) = 2\chi_{ijk}^{mee}(0) I_{jk} \tag{11a}$$

Now, considering that circularly polarized light propagating in the z direction has the amplitudes of right and left gyrating vibrations:

$$E_\pm = \frac{(E_x \pm iE_y)}{2} \tag{12a}$$

we obtain from Eq. (11a) the following magnetic polarization induced in a nonabsorbing material in the light propagation direction [39]:

$$P_z^m(0) = \frac{n}{2\pi^2} V(\lambda)(I_+ - I_-) \tag{14}$$

where the Verdet constant $V(\lambda)$ is defined by Eq. (2a), and I_+, I_- are intensities of right and left circularly polarized laser light. The above expression describes the inverse Faraday effect discovered by Pershan et al. [39] in diamagnetic and paramagnetic bodies illuminated with strong circularly polarized laser light.

The theory of optical magnetization of bodies was developed by Pershan et al. [39] in a phenomenological and quantum-mechanical approach, and by Kielich [48] in a molecular-statistical approach for diamagnetic and paramagnetic liquids. To Atkins and Miller [49] is due an extension of the quantum field theory of the inverse and optical Faraday effects.

C. Second Harmonic Generation

The nonlinear magneto-electric polarization (8) in conjunction with the field (12) involve, besides a constant component, a component which varies with the double frequence 2ω:

$$P_i^e(2\omega, 2\underline{k}) = \chi_{ijk}^{eem}(2\omega, 2\underline{k}) E_j(\omega, \underline{k}) H_k(\omega, \underline{k}) \tag{8a}$$

This polarization yields a magneto-electric contribution to second harmonic generation (SHG) in isotropic bodies [42], particularly metals [47], as well as to SHG at reflection from metals, semiconductors [50], and isotropic media [51].

IV. CUBIC MAGNETO-OPTICAL PROCESSES

When considering polarization processes of the third order induced by magnetic and electric fields, one comes upon a wide variety of components; for the sake of simplicity, we shall restrict ourselves here to the following one:

$$P_i^e(\underline{r},t) = \chi_{ijkl}^{eeem} E_j(\underline{r},t) E_k(\underline{r},t) H_l(\underline{r},t) \tag{15}$$

where the fourth-rank axial tensor χ_{ijkl}^{eeem} describes the nonlinear magneto-electric susceptibility of the third order. The nonzero and mutually independent elements of the tensor χ_{ijkl}^{eeem} have been calculated by Kielich and Zawodny [52] applying methods of group theory [16,53], assuming in general the presence of spatial dispersion.

A. Second-Order Magneto-Optical Birefringence and Rotation

Let us assume, in Eq. (15), one of the electric fields to be static likewise to the magnetic field. The second-order variations of the electric permittivity tensor analyzed at the frequency ω are now obtained in the form:

$$\Delta\varepsilon_{ij}^{(2)}(\omega) = 4\pi\chi_{ijkl}^{eeem}(\omega,\underline{k}) E_k(0) H_l(0) \tag{16}$$

For a further discussion of this expression, it is convenient to separate the axial tensor χ_{ijkl}^{eeem} into a symmetric part $\chi_{(ij)kl}^{eeem} = \chi_{ijkl}^{eeem} = \chi_{jikl}^{eeem}$ and a part $\chi_{[ij]kl}^{eeem} = \chi_{ijkl}^{eeem} = -\chi_{jikl}^{eeem}$ antisymmetric with respect to the indexes i and j (see Table 1). Thus, the difference between diagonal elements of the tensor (16) is:

$$\Delta\varepsilon_{xx}^{(2)}(\omega) - \Delta\varepsilon_{yy}^{(2)}(\omega) = 4\pi\{\chi_{xxkl}^{eeem}(\omega,\underline{k})$$

$$- \chi_{yykl}^{eeem}(\omega,\underline{k})\} E_k(0) H_l(0) \tag{16a}$$

whereas the difference between nondiagonal elements is:

$$\Delta\varepsilon_{xy}^{(2)}(\omega) - \Delta\varepsilon_{yx}^{(2)}(\omega) = 8\pi\chi_{xykl}^{eeem}(\omega,\underline{k}) E_k(0) H_l(0) \tag{16b}$$

Equations (16a) and (16b) describe, respectively, the optical birefringence and optical rotation induced in a medium by the

simultaneous action of a dc magnetic and electric field. When
proceeding to an experimental analysis of these equations, it is
necessary to take into consideration the mutual configuration of
the propagation direction of the analyzing light wave and the directions in which the fields $\underline{H}(0)$ and $\underline{E}(0)$ are applied in the
setup chosen by the experimenter. Two limiting configurations can
be distinguished, namely:

1. Faraday's configuration, in which the light wave propagates in
   a direction parallel to the magnetic field, and
2. Voigt's configuration, in which the light wave is directed at
   right angles to the magnetic field.

On taking further into consideration the various feasible spatial
configurations of the dc electric field, one gets the set of
experimental situations listed in Table 2 permitting the observation of new nonlinear magneto-electro-optical effects in the various crystallographical classes [54] as well as in isotropic
bodies composed of molecules without a center of symmetry, and
without planes of reflection symmetry [48,55]. Clearly, when
planning all the details of an experiment aimed at observing this
or that novel magneto-optical phenomenon, one has to calculate
from Fresnel's equation the relevant refractive indexes by resorting to Eqs. (4), (5), (9), and (16).

B. Magnetic Field-Induced Second Harmonic Generation

By (12) and (15), we obtain for the polarization component
induced at frequency 2ω by a dc magnetic field:

$$P_i^e(2\omega, 2\ \underline{k}) = \chi_{ijkl}^{eeem}(2\omega, 2\underline{k}) E_j(\omega, \underline{k}) E_k(\omega, \underline{k}) H_l(0) \tag{15a}$$

Cohan and Hameka [56] proposed a quantum-mechanical theory of
SHG by gases and liquids in the presence of a magnetic field.
This nonlinear process is allowed in systems of randomly oriented

TABLE 2

Predicted Experimental Situations for the
Observation of Second-Order Magneto-Optical Birefringence and Rotation

The analyzing light beam propagates along Z at:	Direction of dc electric field action	Magneto-Optical Birefringence [Eq. (16a)]		Magneto-Optical Rotation [Eq. (16b)]	
		Relevant tensor elements $\chi_{[ij]kl}$	Crystallographical classes admitting of the effect	Relevant tensor elements $\chi^{eem}_{[ij]kl}$	Crystallographical classes admitting of the effect
Voigt configuration (H_Y)	x	xxxy - yyxy	1, 2, mm2, 4, $\bar{4}$, 4mm, 3, 3m, 6, 6mm	xyxy = -yxxy	1, 2, 222, 4, $\bar{4}$, 422, $\bar{4}2m$, 3, 32, 6, 622, 23, 432, $\bar{4}3m$, Y, K
	y	xxyy - yyyy	1, 2, 222, 4, $\bar{4}$, 422, $\bar{4}2m$, 3, 32, 6, 622, 23, 432, $\bar{4}3m$, Y, K	xyyy = -yxyy	1, 2, mm2, 4, $\bar{4}$, 4mm, 3, 3m, 6, 6mm
	z	xxzy - yyzy	1, 3, 32, $\bar{6}$, $\bar{6}m2$	xyzy = -yxzy	1, m
Faraday configuration (H_Z)	x	xxxz - yyxz	1, m, 3, 3m, $\bar{6}$	xyxz = -yxxz	1, m
	y	xxyz - yyyz	1, m, 3, 32, $\bar{6}$, $\bar{6}m2$	xyyz = -yxyz	1, m
	z	xxzz - yyzz	1, 2, 222, 4, $\bar{4}2m$, $\bar{4}3m$, 23	xyzz = -yxzz	1, 2, mm2, 4, 4mm, 3, 3m, 6, 6mm

and noninteracting molecules that have neither a center of
symmetry nor a plane of reflection. Kielich and Zawodny [52], on
the basis of Eq. (15a), discussed SHG in all crystallographical
classes and showed that a dc magnetic field induces SHG in the
classes $422(D_4)$, $622(D_6)$, $432(0)$, as well as Y and K, in
which SHG is forbidden at $H(0) = 0$ and fulfillment of Kleinman's
symmetry conjecture [57]. Thus, experimental studies of dc magnetic field-induced SHG will surely become a powerful method of
deciding the crystallographical class of bodies, as well as molecules and macromolecules, the symmetry of which cannot be determined by other methods.

Lately, Hafele et al. [58], using a high-power Q-switched
CO_2 laser radiating at 10.6 μm, performed studies of the dc
magnetic field dependence of SHG in InSb near the energy gap.

V.   MAGNETO-OPTICAL PROCESSES OF HIGHER ORDER

Modern measuring techniques permit the investigation of
higher order magneto-optical processes, such as light-intensity-
dependent Faraday effect or Cotton-Mouton effect [59,60], doubling
and mixing of laser light frequencies in crossed electric and
magnetic fields [61], and third-harmonic generation (THG) in the
presence of a dc magnetic field [62,63]. Here, we shall consider
electric polarization of the fourth order, in the form:

$$P_i^e(\underline{r},t) = \chi_{ijkln}^{eeeem} E_j(\underline{r},t) E_k(\underline{r},t) E_l(\underline{r},t) H_n(r,t) \tag{17}$$

where the fifth-rank axial tensor χ_{ijkln}^{eeeem} describes nonlinear
magneto-optical susceptibility of the fourth order the nonzero
and independent tensor elements of which have been calculated by
group theoretical methods [64]. In Table 1, we give solely the

number of these elements for the partly symmetric tensor
$\chi^{eeeem}_{(ij)kln}$ and the antisymmetric tensor $\chi^{eeeem}_{[ij]kln} = \chi^{eeeem}_{ijkln} = \chi^{eeeem}_{jikln}$.

A. Third-Order Magneto-Optical Birefringence and Rotation

We shall now consider the experimental situation which arises
when, on a medium analyzed with light of frequency ω_A, another
light beam of frequency ω_I is incident inducing optical
nonlinearity. With regard to Eqs. (5) and (17), the third-order
variation of the tensor of electric permittivity takes the form:

$$\Delta\varepsilon^{(3)}_{ij}(\omega_A) = 4\pi\chi^{eeeem}_{ijkln}(\omega_A,\omega_I)E_k(\omega_I)E_l(-\omega_I)H_n(0) \tag{18}$$

whence we obtain the following expressions for the light intensity-
dependent magneto-optical birefringence and rotation:

$$\Delta\varepsilon^{(3)}_{xx}(\omega_A) - \Delta\varepsilon^{(3)}_{yy}(\omega_A) = 8\pi\{\chi^{eeeem}_{xxkln}(\omega_A,\omega_I)$$

$$- \chi^{eeeem}_{yykln}(\omega_A,\omega_I)\}I_{kl}H_n(0) \tag{18a}$$

$$\Delta\varepsilon^{(3)}_{xy}(\omega_A) - \Delta\varepsilon^{(3)}_{yx}(\omega_A) = 16\pi\chi^{eeeem}_{xykln}(\omega_A,\omega_I)I_{kl}H_n(0) \tag{18b}$$

In Table 3 are listed the experimental setups permitting the
observation of the new magneto-optical processes (18a) and (18b)
at Faraday's and Voigt's configurations, for various well-defined
propagation directions of the strong laser beam of intensity I.
    Recently, Kubota [60] first succeeded in revealing experimen-
tally light-intensity-dependent magneto-optical rotation in
Faraday configuration in crystalline semiconductors (CdS, ZnS).
In Kubota's experiment, a probe beam is obtained from a Xe flash
lamp, whereas a Q-switched Nd-glass or ruby laser is used as
intense light source of the frequency ω_I. Although the statistical-molecular theory of the light-intensity-dependent variations
in Faraday effect in gases and liquids was proposed by Kielich

[59,65] some years ago, reports of successful experiments are
still lacking.

On putting $\omega_I = 0$ in Eq. (18), one obtains an expression
for the influence of the square of a dc electric field on magneto-
optical effects. Obviously, when studying these effects in isotropic bodies, one has to take into account the unavoidable
presence of a strong Kerr or, respectively, Cotton-Mouton effect,
according to the configuration applied. In fact, we have the
additional variation in permittivity tensor:

$$\Delta\varepsilon_{ij}(\omega) = 4\pi\{\chi_{ijkl}^{eeee}E_k E_l$$

$$+ \chi_{ijkl}^{eemm}H_k H_l + \ldots + \chi_{ijklm}^{eeemm}E_k H_l H_m + \ldots$$

$$+ \chi_{ijklmn}^{eeeemm}E_k E_l H_m H_n + \ldots\} \tag{19}$$

Above, the tensor of nonlinear electric susceptibility χ_{ijkl}^{eeee} defines the quadratic Kerr effect, that of magneto-electric
susceptibility χ_{ijkl}^{eemm} -- a magnetically induced anisotropy [48]
or the quadratic Cotton-Mouton effect, and that of magneto-
electric susceptibility of order 5 χ_{ijklmn}^{eeeemm} -- the cross effect
due to the direct concomitant action of electric and magnetic
fields on the isotropic medium. This latter cross effect is
accessible to observation in solutions of macromolecules and colloid particles, where the effects of nonlinear electron Voigt distortion are accompanied by a strong Langevin reorientation of the
microsystems [40,66].

B. Doubling and Mixing in Crossed Electric and Magnetic Fields

The general formula (17) can be particularized to the form:

$$P_i^e(2\omega) = \chi_{ijkln}^{eeeem}(2\omega)E_j(\omega)E_k(\omega)E_l(0)H_n(0) \tag{17a}$$

describing SHG in the presence of dc electric and magnetic fields.
The process, a general discussion of which is due to Kielich [61],

TABLE 3

Predicted Experimental Situations for the
Observation of Third-Order Magneto-Optical Birefringence and Rotation

The analyzing light beam propagates along Z at:	Propagation direction of inducing laser light	Magneto-Optical Birefringence [Eq. (18a)]		Magneto-Optical Rotation [Eq. (18b)]	
		Relevant tensor elements $\chi_{(ij)kln}^{eeeem}$	Classes admitting the effect	Relevant tensor elements $\chi_{[ij]kln}^{eeeem}$	Classes admitting of the effect
Voigt configuration (H_y)	x	xxyyy - yyyyy, xxyzy - yyyzy, xxzyy - yyzyy, xxzzy - yyzzy	1, $\bar{1}$, m, 2, 2/m, 4, $\bar{4}$, 4/m, 3, $\bar{3}$, 6, $\bar{6}$, 6/m	xyyyy, xyyzy, xyzyy, xyzzy	All classes
	y	xxzzy - yyzzy, xxzxy - yyzxy, xxxzy - yyxzy, xxxxy - yyxxy	All classes	xyzzy, xyzxy, xyxzy, xyxxy	1, $\bar{1}$, m, 2, 2/m, 4, $\bar{4}$, 4/m, 3, $\bar{3}$, 32, 3m, $\bar{3}$m, 6, $\bar{6}$, 6/m
	z	xxxxy - yyxxy, xxxyy - yyxyy, xxyxy - yyyxy, xxyyy - yyyyy	1, $\bar{1}$, 3, $\bar{3}$, 32, 3m, $\bar{3}$m	xyxxy, xyxyy, xyyxy, xyyyy	1, $\bar{1}$, 3, $\bar{3}$, 32, 3m, $\bar{3}$m

Faraday configuration (H_z)		Classes		
x	xxyyz – yyyyz xxyzz – yyyzz xxzyz – yyzyz xxzzz – yyzzz	1, 1̄, m, 2, 2/m, 4, 4̄, 4/m, 3, 3̄, 6, 6̄, 6/m	xyyyz, xyyzz xyzyz, xyzzz	All classes
y	xxzzz – yyzzz xxzxz – yyzxz xxxzz – yyxzz xxxxz – yyxxz	1, 1̄, m, 2, 2/m, 4, 4̄, 4/m, 3, 3̄, 3m, 3̄m, 6, 6̄, 6/m	xyzzz, xyzxz xyxzz, xyxxz	All classes
z	xxxxz – yyxxz xxxyz – yyxyz xxyxz – yyyxz xxyyz – yyyyz	All classes	xyxxz, xyxyz xyyxz, xyyyz	All classes

is of especial interest, since it can take place in isotropic bodies, where the axial tensor χ_{ijklm}^{eeeem} reduces its 243 elements to 60 nonzero ones, 6 of which are generally mutually independent. Besides frequency doubling, one has likewise the possibility of performing experimental studies of sum-frequency and difference-frequency processes at various configurations of the fields $\underline{E}(0)$ and $\underline{H}(0)$, in particular when the two fields are at right angles. As yet, there is a lack of reports on experimental attempts to deal with this new nonlinear magneto-optical process.

C. Third-Harmonic Generation

The theoretical work of Lax, Kolodziejczak, and others [62] has proved the feasibility of enhancement in THG due to interband transitions and resonance in the presence of a dc magnetic field. The relevant experiments have been performed by Patel et al. [63] in InSb single-crystal sample by using a Q-switched CO_2 laser and a magnetic field growing to 54 kOe. Phenomenologically, the effect is described in a first approximation by the following polarization at frequency 3ω resulting from Eq. (17):

$$P_i^e(3\omega) = \chi_{ijkln}^{eeeem}(3\omega)E_j(\omega)E_k(\omega)E_l(\omega)H_n(0) \qquad (17b)$$

The above formula, moreover, describes THG in isotropic bodies with induced magnetic gyrotropy.

VI. EXPERIMENTAL PROSPECTS AND CONCLUSIONS

The preceding phenomenological treatment has permitted a concise discussion of the nature and observation conditions of a wide variety of new nonlinear magneto-electric-optical effects. Certain of these effects, like laser-induced magneto-optical rotation, SHG, and THG, are well known to have been detected and studied in semiconductor crystals [58,60,63], and everything

points to their impending detection in molecular and
macromolecular substances as well. To corroborate this statement,
it suffices to invoke the inverse Faraday effect [39] detected in
diamagnetic liquids and, in general, the recent rapid developments
in laser technique. Ingenious methods of observation, supported
by a high sensitivity of the measuring equipment and a judicious
choice of substances and materials, will surely lead in the near
future to detection of the other nonlinear magneto-optical effects
discussed in this chapter.

The accretion of knowledge to be gained by investigations of
these magneto-electro-optical effects in conjunction with other
nonlinear processes, e.g., inverse Cotton-Mouton effect [48] and
nonlinear light scattering in the presence of an electric and magnetic field [67], is obvious. In addition to the magneto-electric
anisotropies of molecules and macromolecules, one will be in a
position to determine nonlinear electron distortion processes
which, hitherto, have been successfully calculated for some simple
atoms and molecules only [68,69]. Despite the fact that general
quantum mechanical theories of nonlinear magneto-electric
polarizabilities [70] as well as discussions of the permutational
properties of these tensors [71] are available, there is still a
complete lack of papers claiming the evolvement of experimental
methods and procedures. We know of but one report, concerning the
observation of the influence of the square of a magnetic field on
the electric permittivity of diamagnetic liquids [72].

The new magneto-electric-optical effects (particularly,
nonlinear magnetic circular dichroism) can be detected soon and
easily in macromolecular and colloidal solutions [40], especially
liquid crystals. This statement is justified by the results
achieved along the lines initiated by Jezewski [73] and Kast [74]
showing that liquid crystals are highly sensitive not only to an
electric field but to a magnetic field as well [75]. Quite recently, some more papers have appeared on the topics dealt with
above [76-86].

REFERENCES

1. M. Faraday, Phil. Trans. Roy. Soc., 1846, 1; Phil. Mag., 28, 294 (1846); 29, 153 (1846).

2. E. Verdet, Ann. Chim. Phys., 41, 570 (1854); 52, 151 (1858).

3. A. Fresnel, Ann. Chim. Phys., 28, 147 (1825).

4. W. Voigt, Magneto- und Electro-Optik, Teubner, Leipzig (1908).

5. Q. Majorana, Compt. Rend., 135, 159, 235 (1902).

6. A. Cotton and M. Mouton, Compt. Rend., 141, 317, 349 (1905); 145, 229, 870 (1907).

7. H. A. Lorentz, The Theory of Electrons, Teubner, Leipzig (1909); L. Rosenfeld, Theory of Electrons, North Holland, Amsterdam (1951).

8. T. H. Havelock, Proc. Roy. Soc., (London), A77, 170 (1906); A80, 28 (1908).

9. P. Langevin, Ann. Chim. Phys., 5, 70 (1905); Radium, 7, 249 (1910).

10. J. H. Van Vleck, The Theory of Electric and Magnetic Susceptibilities, Oxford University Press, London (1932).

11. M. Born, Optik, Springer, Berlin (1933).

12. M. V. Volkenshteyn, Molekularnaya Optika, Gostekhizdat, Moscow (1951); Struktur und Physikalische Eigenschaften der Molekule, Teübner, Leipzig (1960).

13. W. Schütz, Handbuch der Experimental Physik, Magnetooptik, Vol. 16, Akademisches, Leipzig (1936).

14. J. F. Nye, Physical Properties of Crystals, Clarendon, Oxford (1957).

15. L. D. Landau and E. M. Lifshitz, Electrodynamics of Continuous Media, Pergamon, New York (1960).

16. S. Bhagavantam, Crystal Symmetry and Physical Properties, Academic Press, London (1966).

17. D. W. Davies, The Theory of Electric and Magnetic Properties of Molecules, Wiley, London (1967).

18. J. W. Beams, Rev. Mod. Phys., 4, 133 (1932); W. Heller, Rev. Mod. Phys., 14, 390 (1942).

19. A. Piekara and S. Kielich, J. Phys. Radium, 18, 490 (1957); J. Chem. Phys., 29, 1292 (1958); S. Kielich and A. Piekara, Acta Phys. Polon., 18, 439 (1959).

20. I. Tinoco Jr. and C. A. Bush, Biopolymers Symp. 1, 235 (1964).

21. A. D. Buckingham and P. J. Stephens, Ann. Rev. Phys. Chem.,
     17, 399 (1966); P. N. Schatz and A. J. Caffery, Quart. Rev.,
     23, 552 (1968); D. Caldwell, J. M. Thorne, and H. Eyring,
     Ann. Rev. Phys. Chem., 22, 259 (1971).

22. P. S. Pershan, J. Appl. Phys., 38, 1482 (1967); Y. R. Shen,
     Phys. Rev., 133A, 511 (1964).

23. S. Kielich, Dielectric and Related Molecular Processes, Vol.
     1 (M. Davies, ed.), Wright, London (1972), Chap. 7.

24. E. U. Condon, Rev. Mod. Phys., 9, 432 (1937);
     M. P. Groenewege, Mol. Phys., 5, 541 (1962); S. H. Lin and
     R. J. Bershon, J. Chem. Phys., 44, 3768 (1966).

25. R. Daudel, F. Gallais, and P. Smet, Int. J. Quantum Chem., 1,
     873 (1967); P. J. Stephens, J. Chem. Phys., 52, 3489 (1970).

26. R. A. Harris, J. Chem. Phys., 46, 3398 (1967); 47, 4481
     (1967).

27. M. V. Volkenshteyn, Fizyka Enzymow (in Polish), PWN,
     Warszawa (1970).

28. A. D. Buckingham, W. H. Prichard, and D. H. Whiffen, Trans.
     Faraday Soc., 63, 1057 (1967); R. Locqueneux, P. Smet, and
     J. Tillieu, J. Physique, 29, 631 (1968); P. W. Atkins and
     M. H. Miller, Mol. Phys., 15, 491 (1968).

29. S. Kielich, Acta Phys. Polon., 17, 209 (1958); 22, 65, 299
     (1962).

30. M. Surma, Acta Phys. Polon., 25, 485 (1964); R. J. W. Le Fevre
     and D. S. N. Murthy, Australian J. Chem., 19, 179, 1321
     (1966); 23, 193 (1970); R. J. W. Le Fevre, D. S. N. Murthy,
     and P. J. Stiles, Australian J. Chem., 21, 3059 (1968); 22,
     1421 (1969).

31. M. Y. Miekshenkov, Biofizika, 10, 747 (1965); 12, 30, 157
     (1967).

32. V. N. Tsvetkov and M. Sosinskii, Zh. Eksperim. i Teor. Fiz.,
     19, 543 (1949); Z. Blaszczak, A. Dobek, and A. Patkowski,
     Acta Phys. Polon., A40, 119 (1971).

33. J. F. Ready, Effects of High-Power Laser Radiation, Academic
     Press, New York, London (1971).

34. H. Kolm, B. Lax, F. Bitter, and R. Mills (eds.), High
     Magnetic Fields, Wiley, New York (1962); D. B. Montgomery,
     Solenoid Magnet Design, Wiley, New York (1969); H. Knoepfel,
     Pulsed High Magnetic Fields, North Holland, Amsterdam (1970).

35. M. F. Barnothy (ed.), Biological Effects of Magnetic Fields,
     Plenum, New York, Vol. 1 (1964), Vol. 2 (1969); Faraday Soc.,
     No. 3 (1969).

36. S. V. Vonsovskii, Magnetism (in Russian), Izd. Nauka, Moskva

(1971); H. Cofta, Metamagnetyki (in Polish), PWN, Warszawa (1971).

37. V. M. Agranovich and V. L. Ginzburg, Spatial Dispersion in Crystal Optics and the Theory of Excitons, Wiley, New York (1965).

38. G. W. Gray, Molecular Structure and the Properties of Liquid Crystals, Academic Press, London (1962); Liquid Cyrstals and Ordered Fluids, (J. F. Johnson and R. S. Porter, eds.), Plenum, New York (1970).

39. P. S. Pershan, J. P. Van der Ziel, and L. D. Malmstrom, Phys. Rev., 143, 574 (1966).

40. S. Kielich, J. Colloid Interface Sci., 30, 159 (1969).

41. S. Kielich, Proc. Phys. Soc., 86, 709 (1965); Acta Phys. Polon., 29, 875 (1966).

42. P. S. Pershan, Phys. Rev., 130, 919 (1963); E. Adler, Phys. Rev., 134, A728 (1964).

43. A. Huber, Z. Physik, 27, 619 (1926).

44. S. L. Hou and N. Bloembergen, Phys. Rev., 138, A1218 (1965).

45. L. M. Holmes and L. G. Van Uitert, Phys. Rev., B5, 147 (1972).

46. P. E. Ferguson and R. J. Romagnoli, Opt. Acta, 17, 667 (1970); R. J. Romagnoli and P. E. Ferguson, Opt. Acta, 18, 191 (1971); J. Appl. Phys., 42, 1712 (1971).

47. S. S. Jha, Phys. Rev., 140, A2020 (1965); 145, 500 (1966).

48. S. Kielich, Phys. Letters, 24A, 435 (1967); Acta Phys. Polon., 31, 929 (1967); 32, 405 (1967).

49. P. W. Atkins and M. H. Miller, Mol. Phys., 15, 503 (1968).

50. N. Bloembergen, Opt. Acta, 13, 311 (1966).

51. S. Kielich, Opto-Electronics, 2, 125 (1970); Ferroelectrics, 4, 257 (1972).

52. S. Kielich and R. Zawodny, Opt. Commun., 4, 132 (1971).

53. R. R. Birss, Symmetry and Magnetism, North-Holland, Amsterdam (1964).

54. S. Kielich and R. Zawodny, Acta Phys. Polon., A42, 337 (1972).

55. D. Mukherjee and M. Chowdhury, Physica, 58, 109 (1972).

56. N. V. Cohan and H. F. Hameka, Physica, 38, 320 (1967).

57. D. A. Kleinman, Phys. Rev., 126, 1977 (1962).

58. H. G. Hafele, R. Grosar, C. Islinger, H. Wachernig, S. D. Smith, R. B. Dennis, and B. S. Wherret, J. Phys. C. Solid St; Phys., 4, 2637 (1971).

59. S. Kielich, Phys. Letters, 25A, 517 (1967); Appl. Phys. Letters, 13, 152 (1968).

60. K. Kubota, J. Phys. Soc. Japan, 29, 986, 998 (1970).

61. S. Kielich, Opt. Commun., 2, 197 (1970).

62. M. Weiler, W. Zawadzki, and B. Lax, Phys. Rev. Letters, 18, 462 (1967); J. Kolodziejczak, Phys. Stat. Sol., 24, 323 (1967); 29, 645 (1968); Acta Phys. Polon., 33, 183 (1968).

63. C. K. N. Patel, R. E. Slusher, and P. A. Fleury, Phys. Rev. Letters, 17, 1011 (1966); N. Van Tran, J. H. McFee, and C. K. N. Patel, Phys. Rev. Letters, 21, 735 (1968).

64. S. Kielich and R. Zawodny, Acta Phys. Polon., A43, 579 (1973).

65. S. Kielich, Bull. Soc. Amis Sci. Lettres Poznan (Poland), Ser. B, 21, 47 (1968-1969).

66. S. Kielich, Bull. Soc. Amis Sci. Lettres Poznan, Ser. B, 21, 35 (1968/69).

67. S. Kielich, Acta Phys. Polon., 23, 321, 819 (1963); 37A, 447 (1970); Opt. Commun., 1, 345 (1970); S. Premilat and P. Horn, J. Chim. Phys., 62, 395 (1965); 63, 463 (1966).

68. J. Finkel, J. Mol. Spectr., 9, 421 (1962); J. Opt. Soc. Am., 58, 207 (1968); P. Smet and J. Tillieu, Compt. Rend. Acad. Sci. Paris, 260, 445 (1965); P. Smet, Int. J. Quantum Chem., 3, 417, 593 (1969); T. Y. Chang, J. Chem. Phys., 54, 1433 (1971); R. Locqueneux, Int. J. Quantum Chem., 6, 1 (1972).

69. A. D. Buckingham and J. A. Pople, Proc. Phys. Soc., B69, 1133 (1956); Proc. Cambridge Phys. Soc., 53, 262 (1957); J. D. Lyons and R. P. Hurst, Phys. Rev., 162, 698 (1970); T. Y. Chang, J. Chem. Phys., 56, 1752 (1972); R. Locqueneux, Int. J. Quant. Chem., 6, 1 (1972).

70. S. Kielich, Acta Phys. Polon., 30, 851 (1966); Physica, 32, 385 (1966).

71. L. L. Boyle, Int. J. Quantum Chem., 3, 231 (1969); 4, 413 (1970); L. L. Boyle and P. S. C. Matthews, Int. J. Quantum Chem., 381 (1971).

72. A. Piekara and A. Chelkowski, J. Phys. Radium, 18, 49 (1957).

73. M. Jezewski, J. Phys. Radium, 5, 59 (1924); Z. Physik, 40, 153 (1926); 159 (1928); M. Jezowski and M. Miesowicz, Phys. Z., 36, 107 (1935).

74. W. Kast, Ann. Physik, 73, 145 (1924); Z. Physik, 42, 81 (1927); 71, 39 (1931).

75. E. F. Carr, J. Chem. Phys., 42, 738 (1965); 43, 3905 (1965); R. B. Meyer, Appl. Phys. Letters, 14, 208 (1969); Groupe d'Etude des Cristaux Liquides (Orsay), J. Chem. Phys., 51,

816 (1969); Phys. Letters, 28A, 687 (1969); G. Durand,
L. Leger, F. Rondelez, and M. Veyssie, Phys. Rev. Letters,
22, 227 (1969); P. G. De Gennes, Mol. Cryst. Liquid Cryst.,
7, 325 (1969); 12, 193 (1971); J. D. Lee and A. C. Eringen,
J. Chem. Phys., 55, 4504 (1971); P. Martinaty, S. Candau, and
F. Debeauvais, Phys. Rev. Letters, 27, 1123 (1971); I. Raud
and P. Cladis, Mol. Cryst. Liquid Cryst., 15, 1 (1971);
T. J. Scheffer, Phys. Rev. Letters, 28, 593 (1972);
W. H. De Jeu, C. J. Gerritsma, and Th. W. Lathouwers, Chem.
Phys. Letters, 14, 503 (1972); F. Rondelez and J. P. Hulin,
Solid State Commun., 10, 1009 (1972); F. Brochars,
P. Pieranski, and E. Guyon, Phys. Rev. Letters, 28, 1681
(1972); J. Phys. Radium, 33, 681 (1972); 34, 35 (1973).

76.   H. Kelker, Mol. Crystals and Liquid Crystals, 21, 1 (1973);
A. Saupe, Ann. Rev. Phys. Chem., 24, 441 (1973); G. Elliott,
Chem. Brit., 9, 213 (1973).

77.   D. L. Partigal and E. Burstein, J. Phys. Chem. Solids, 32,
603 (1971).

78.   S. Bhagavantam, Proc. Indian Acad. Sci., A75, 1 (1972);
S. Bhagavantam and P. Chandrasekhar, Proc. Indian Acad. Sci.,
76, 13 (1972).

79.   R. M. Hornreich, IEEE Trans. Magn., 8, 584 (1972).

80.   B. A. Huberman, Solid State Commun., 12, 259 (1973).

81.   P. Chandrasekhar and T. P. Srinivasan, J. Phys., C6, 1085
(1973).

82.   S. Kielich and R. Zawodny, Opt. Acta, 20, 867 (1973).

83.   T. Hashimato, A. Sato, and Y. Fujiwara, J. Phys. Soc. Japan,
35, 81 (1973).

84.   S. G. Semanchinsky, Yu. V. Shaldin, and L. V. Soboleva,
Phys. Letters, 48A, 45 (1974).

85.   V. S. Zapasskij and P. P. Feofilov, Usp. Fiz. Nauk, 116, 41
(1975).

86.   S. Kielich and R. Zawodny, Optical Properties of Highly
Transparent Solids (S. S. Mitra and B. Bendow, eds.), Plenum
Press, New York (1975) p. 393.

Chapter 14

QUANTUM THEORY AND CALCULATION OF ELECTRIC POLARIZABILITY

Tae-Kyu Ha

Laboratory of Physical Chemistry
Swiss Federal Institute of Technology
Zurich, Switzerland

I. INTRODUCTION

Studies of the interactions between the static external electric field and the electrons of an atom, ion, or molecule can increase our understanding of its electronic structure. The application of an external field produces a distortion of the electronic charge distribution in an atomic or molecular system giving rise to induced moments. If the applied electric field is

uniform and the field strength is low, the distortion is described
by the induced dipole moment $\underline{\mu}$ which, to a first approximation,
is proportional to the first power of the field strength, $\underline{F}$ [1]

$$\underline{\mu} = \alpha \underline{F} \tag{1}$$

where $\underline{\mu}$ and $\underline{F}$ are vectors and the proportionality constant $\underline{\alpha}$
is well-known electric dipole polarizability, which is a second-
order tensor. Similar relationships involving higher-order
derivatives of the external electric field apply to the higher
multipole polarizabilities. However, if the applied electric
field strength is comparable to the average internal field
strength, one observes a deviation of $\underline{\mu}$ from a linear dependence
on $\underline{F}$ and Eq. (1) becomes [2]

$$\underline{\mu} = \alpha \underline{F} + \frac{1}{2} \underline{\beta} \underline{F}^2 + \frac{1}{6} \underline{\gamma} \underline{F}^3 + \dots \tag{2}$$

where the coefficients $\underline{\beta}$ and $\underline{\gamma}$ of the higher-order powers of
$\underline{F}$ are known as the first and second hyperpolarizabilities. They
are third and fourth rank tensors, respectively. In general, the
energy of a system depending on a uniform electric field, $\underline{F}$, of
arbitrary strength may be described by a generalized Taylor expansion about the origin $(F = 0)$,

$$E(F_i) = E^{(0)} - \mu_i^P F_i - \frac{1}{2} \alpha_{ij} F_i F_j$$
$$- \frac{1}{6} \beta_{ijk} F_i F_j F_k - \frac{1}{24} \gamma_{ijk\ell} F_i F_j F_k F_\ell - \dots \tag{3}$$

where $E^{(0)}$ is the field-free energy, μ_i^P is the permanent dipole moment component, and the indexes, i, j, k, and ℓ denote
Cartesian components, x, y, and z along the molecular axes.
The Einstein convention of summation over repeated indexes is
adopted here. We notice that the electric dipole polarizability
term, $\underline{\alpha}$, is second rank, and the hyperpolarizabilities $\underline{\beta}$ and
$\underline{\gamma}$ are third and fourth rank in the field strength, respectively.

In other words, the energy of interaction (ΔE), associated with the induction of the dipole moment, can be expressed as

$$\Delta E = -\frac{1}{2}\,\underline{\alpha F}^2 - \frac{1}{6}\,\underline{\beta F}^3 - \frac{1}{24}\,\underline{\gamma F}^4 \ldots \qquad (4)$$

The electric dipole polarizability is a second-order term, $\Delta E^{(2)} = -\frac{1}{2}\,\underline{\alpha F}^2$. According to the Rayleigh-Schrödinger perturbation theory,

$$\alpha_{ij} = -2 \sum_m{}' \frac{<\psi_0^{(0)}|\sum_\mu r_{i\mu}|\psi_m^{(0)}><\psi_m^{(0)}|\sum_\mu r_{j\mu}|\psi_0^{(0)}>}{E_0^{(0)} - E_m^{(0)}} \qquad (5)$$

$\psi_0^{(0)}$, $\psi_m^{(0)}$ and $E_0^{(0)}$, $E_m^{(0)}$ are eigenfunctions and eigenvalues of the unperturbed ground state and m^{th} excited states of the system, and $r_{i\mu}$ is the ith component of the dipole operator r_μ. We obtain similar expressions for the quadrupole and higher polarizabilities, if the corresponding operators are substituted [3] in Eq. (5). In an appropriate coordinate system, the dipole polarizability tensor, α_{ij}, can be characterized by the principal polarizabilities, α_{xx}, α_{yy}, and α_{zz} and it is convenient to define an average polarizability, $<\alpha>$, as

$$<\alpha> = \frac{1}{3}(\alpha_{xx} + \alpha_{yy} + \alpha_{zz}) \qquad (6)$$

and the polarizability anisotropy ($\Delta\alpha$) as

$$\Delta\alpha = \alpha_{zz} - \frac{1}{2}(\alpha_{xx} + \alpha_{yy}) \qquad (7)$$

The number of independent components of the polarizability can be further reduced by the symmetry of the system. If the system is spherically symmetric, the three principal polarizabilities are identical. If the system is axially symmetric, $\alpha_{\parallel} = \alpha_{zz}$, $\alpha_{\perp} = \alpha_{xx} = \alpha_{yy}$, the z axis being the symmetry axis. The hyperpolarizabilities $\underline{\beta}$ and $\underline{\gamma}$ can be expressed similarly using third- and fourth-order perturbation theory as follows [2]:

$$\beta_{ijk} = S \sum_{m,n \neq 0} \frac{\langle\Psi_0^{(0)}|\Sigma_\mu r_{i\mu}|\Psi_m^{(0)}\rangle\langle\Psi_m^{(0)}|\Sigma_\nu r_{j\nu}|\Psi_n^{(0)}\rangle\langle\Psi_n^{(0)}|\Sigma_\lambda r_{k\lambda}|\Psi_0^{(0)}\rangle}{(E_0^{(0)} - E_m^{(0)})(E_0^{(0)} - E_n^{(0)})}$$

$$- S \sum_{m \neq 0} \frac{\langle\Psi_0^{(0)}|\Sigma_\mu r_{i\mu}|\Psi_0^{(0)}\rangle\langle\Psi_0^{(0)}|\Sigma_\nu r_{j\nu}|\Psi_m^{(0)}\rangle\langle\Psi_m^{(0)}|\Sigma_\lambda r_{k\lambda}|\Psi_0^{(0)}\rangle}{(E^{(0)} - E_0^{(m)})^2} \tag{8}$$

and

$$\gamma_{ijk\ell} = S \sum_{m,n,p \neq 0} \frac{\langle\Psi_0^{(0)}|\Sigma_\mu r_{i\mu}|\Psi_m^{(0)}\rangle\langle\Psi_m^{(0)}|\Sigma_\nu r_{i\nu}|\Psi_n^{(0)}\rangle}{(E_m^{(0)} - E_0^{(0)})(E_n^{(0)} - E_0^{(0)})}$$

$$\times \frac{\langle\Psi_n^{(0)}|\Sigma_\lambda r_{k\lambda}|\Psi_p^{(0)}\rangle\langle\Psi_p^{(0)}|\Sigma_\sigma r_{\ell\sigma}|\Psi_0^{(0)}\rangle}{(E_p^{(0)} - E_0^{(0)})} \tag{9}$$

$$- S \sum_{m,n \neq 0} \frac{\langle\phi_0^{(0)}|\Sigma_\mu r_{i\mu}|\phi_m^{(0)}\rangle\langle\phi_m^{(0)}|\Sigma_\nu r_{j\nu}|\phi_n^{(0)}\rangle\langle\phi_n^{(0)}|\Sigma_\lambda r_{k\lambda}|\phi_0^{(0)}\rangle}{(E_m^{(0)} - E_0^{(0)})(E_n^{(0)} - E_0^{(0)})}$$

where the operator S denotes a sum of all permutations of the
subsequent vectors. The number of independent components of the
hyperpolarizabilities can also be reduced by the symmetry of the
system. The symmetry properties of $\underline{\beta}$ and $\underline{\gamma}$ will be discussed
in Sec. IV.

The electric polarizabilities and hyperpolarizabilities are
one-electron physical properties. In contrast to the usual one-
electron properties, such as permanent dipole and quadrupole moments, diamagnetic susceptibilities, electronic charge densities,
etc.; however, these are "transition" one-electron properties,
which are also known as second- and higher-order physical properties. In addition to the fact that the calculations of
polarizabilities and hyperpolarizabilities can aid in the interpretation of experimental measurements, they can also serve as
useful criteria for testing the accuracy of the wavefunctions of
the system, since the calculated values are known to be rather

sensitive to the wavefunctions employed. In Sec. II, we discuss
general methods of calculation, and in Sec. III, some numerical
results on atoms, ions, and molecules will be reviewed. The
hyperpolarizabilities will be treated separately in Sec. IV. In
this review, we restrict ourselves to static electric polarizabilities. For calculations of the frequency-dependent (dynamic)
polarizability, we refer to the several interesting studies [23,
171-175] which employ time-dependent perturbation theory.

II. METHODS OF CALCULATION

A. Nonempirical Calculations

1. Perturbation Methods

As Eq. (5) and Eqs. (8) and (9) show, the electric
polarizabilities and hyperpolarizabilities can be, in principle,
calculated if the wavefunction for the perturbed system is expanded in a complete set of unperturbed solutions using perturbation theory formulas. This strict perturbation approach requires,
of course, an infinite number of eigenfunctions of the electronically excited states of the unperturbed system, $\psi_m^{(0)}$. The summations should be performed over the continuum as well as the
excited bound states. This method is therefore known as the sum-
over-states method. The convergence of the perturbation expansion
is strongly dependent upon the accuracy of $\psi_0^{(0)}$ and $\psi_m^{(0)}$ in
this method. The usual method of obtaining the spectrum of
eigenfunctions for an unperturbed atomic or molecular system is
the Hartree-Fock method [4], as modified by Roothaan [5]. A
finite set of basis functions with energy determined linear coefficients (eigenvectors) is used to solve the Hartree-Fock equations in the Roothaan procedure, rather than numerical methods. A
finite set of bound excited states is thus obtained for the calculations of polarizabilities and hyperpolarizabilities. Unfortunately, however, the contributions of continuum states to the
polarizabilities may not be negligible [6,7] and, moreover, for

atomic and molecular systems, the excited states mostly lie in the
continuum if the Hartree-Fock model is adopted [8,9]. We have,
therefore, to sum over a small number of bound states and to inte-
grate over the continuum states. It is, in general, very difficult to obtain the continuum-wavefunction for atomic and molecular
systems, except for a few small atoms and molecules. Peng [10],
Kaneko [11], Allen [12], and Dalgarno [3,13] have developed general formalisms of this method for the calculation of electric
polarizabilities. More recently, Kelly [7] has demonstrated the
usefulness of this strict perturbational method by calculating the
electric dipole and quadrupole polarizabilities of beryllium and
oxygen atoms. He has used, however, the Bruekner-Goldstone
perturbation theory [14,15], instead of the Rayleigh-Schrödinger
perturbation theory, utilizing the diagram technique.

In the Hartree-Fock-Roothaan SCF procedure, the second-order
energy $E^{(2)}$, associated with the electric polarizability for the
closed-shell systems, can be expressed as

$$E^{(2)} = \sum_{\mu\nu} P_{\mu\nu}^{(0)} h_{\mu\nu}^{(2)} + \frac{1}{2} \sum_{\mu\nu} h_{\mu\nu}^{(1)} P_{\mu\nu}^{(1)} \tag{10}$$

if we employ Roothaan's equation [5],

$$\underline{FC} = \underline{SC}\varepsilon$$

where the ϕ_i are linear combinations of basis functions,

$$\phi_i = \sum_{\mu} C_{i\mu} \chi_{\mu} \tag{12}$$

the $S_{\mu\nu}$ are overlap integrals,

$$S_{\mu\nu} = \langle \chi_{\mu}(1) | \chi_{\nu}(1) \rangle \tag{13}$$

and $\underline{P}$ and h are the density matrix and core-Hamiltonian,
respectively,

$$P_{\mu\nu} = \frac{1}{2} \sum_{i} C_{i\mu} C_{i\nu} \tag{14}$$

$$h_{\mu\nu} = <\chi_\mu(1) \, | \, (- \frac{1}{2}\nabla_1^2 - \frac{Z}{r_1}) \, | \chi_\nu(1)> \tag{15}$$

The first-order wavefunction is determined by the equation,

$$\underline{F}^{(0)}\underline{C}^{(1)} + \underline{F}^{(1)}\underline{C}^{(0)} = \underline{SC}^{(0)}\varepsilon^{(1)} + \underline{SC}^{(1)}\varepsilon^{(0)} \tag{16}$$

$P_{\mu\nu}^{(0)}$ and $P_{\mu\nu}^{(1)}$ are the zeroth- and first-order perturbed density matrix elements, $h_{\mu\nu}^{(1)}$ and $h_{\mu\nu}^{(2)}$ are the first- and second-order perturbed core-Hamilton matrix elements, and $\underline{F}^{(0)}$, $\underline{F}^{(1)}$, $\underline{C}^{(0)}$, $\underline{C}^{(1)}$, $\varepsilon^{(0)}$, and $\varepsilon^{(1)}$ are the zeroth- and first-order Fock matrices and the corresponding eigenvectors and eigenvalues, respectively. Because of its slow convergence and difficulties in estimating the influence of continuum states, however, the sum-over-states method has not been widely used for the calculation of electric polarizabilities of atomic and molecular systems. Instead, variation-perturbation techniques, using various levels of approximation, have been employed.

2. The Coupled Hartree-Fock Scheme

The main difference between the sum-over-states method and the variation-perturbation techniques lies in the fact that in the variation-perturbation methods, the perturbed equations are solved in the presence of the external field, using the variational principle for the perturbed total energy of the atomic and molecular systems. The electric polarizabilities are then defined as the limiting values of the $\alpha_{ij}(\underline{F})$ when the electric field tends to zero. The sum-over-states method can be considered as a limiting case of the variation-perturbation method for a vanishing field [16].

As discussed in the introduction, the electric polarizabilities are related to the second-order energies, which are in turn determined by the first-order perturbed wavefunctions. If the Hamiltonian of the system in the presence of the electric field is defined as

$$H = H^{(0)} + \lambda W(1) \tag{17}$$

where $H^{(0)}$ corresponds to the field-free Hamiltonian, and

$$H\Psi_m = E_m \Psi_m \tag{18}$$

the Hartree-Fock equation can be written as

$$[F(1) + \lambda W(1) - \varepsilon_i]\phi_i^{(1)} = 0, \quad i = 1, 2, \ldots, N \tag{19}$$

for an N-electron system. The spin-orbitals, ϕ_i are mutually orthonormal, $W(1)$ is a one-electron perturbation operator such as $r_{i\mu}$, the electric dipole operator, ε_i is the molecular orbital energy and the Fock operator $F(1)$ has the following form:

$$F(1) = -\frac{1}{2}\nabla_1^2 - \frac{Z}{r_1} + \sum_{j=1}^{N} \langle \phi_j^{(2)} | r_{12}^{-1}(1 - P_{12})\phi_j(2) \rangle \tag{20}$$

where P_{12} is the permutation operator. Introducing perturbation expansions for ϕ_i and ε_i,

$$\phi_i = \phi_i^{(0)} + \lambda\phi_i^{(1)} + \ldots \tag{21}$$

$$\varepsilon_i = \varepsilon_i^{(0)} + \lambda\varepsilon_i^{(1)} + \ldots \tag{22}$$

we obtain to the first-order in λ

$$[F(1)^{(0)} - \varepsilon_i^{(0)}]\phi_i^{(1)}(1) + [F^{(1)}(1) - \varepsilon_i^{(1)}]\phi_i^{(0)}(1)$$

$$+ \sum_{j=1}^{N} [\langle \phi_j^{(1)}(2) | r_{12}^{-1}(1 - P_{12}) | \phi_j^{(0)}(2) \rangle + \langle \phi_j^{(0)}(2) | r_{12}^{-1}(1 - P_{12}) |$$

$$\phi_j^{(1)}(2) \rangle]\phi_i^{(0)}(1) = 0 \tag{23}$$

$F^{(0)}(1)$ is the field-free Fock operator,

$$F^{(0)}(1) = -\frac{1}{2}\nabla_1^2 - \frac{Z}{r_1} + \sum_{j=1}^{N} \langle \phi_j^{(0)}(2) | r_{12}^{-1}(1 - P_{12}) | \phi_j^{(0)}(2) \rangle$$

$$\tag{24}$$

which satisfies the well-known zeroth-order Hartree-Fock equation,

$$[F^{(0)}(1) - \varepsilon_i^{(0)}]\phi_i^{(0)}(1) = 0 \tag{25}$$

Equation (23) is known as the coupled Hartree-Fock equation of the first order [3,13,17]. To solve this coupled equation, an iterative procedure is followed until self-consistency is obtained. The orthonormality relationships

$$<\phi_i^{(0)}|\phi_j^{(0)}> = \delta_{ij} \tag{26}$$

and

$$<\phi_i^{(0)}|\phi_j^{(1)}> + <\phi_i^{(1)}|\phi_j^{(0)}> = 0 \tag{27}$$

provide additional constraints. The coupled Hartree-Fock method is a very accurate nonempirical method for calculating electric polarizabilities. This method is very time-consuming, since N coupled equations have to be solved for an N-electron system. Various uncoupling techniques have been proposed to reduce the time required to obtain a solution. In the following, we review some of the representative uncoupling schemes.

3. The Uncoupled Hartree-Fock Scheme

The uncoupling procedure proposed by Dalgarno is obtained from Eq. (23) by neglecting all terms in the sum over j, and can be written as [3,18]

$$[F^{(0)}(1) - \varepsilon_i^{(0)}]\phi_i^{(1)}(1) + [F^{(1)}(1) - \varepsilon_i^{(1)}]\phi_i^{(0)}(1) = 0 \tag{28}$$

where the more difficult two-electron integrals associated with the perturbed orbitals are completely neglected. Two alternative uncoupling procedures have been proposed by Langhoff et al. [17], and Karplus and Kolker [19]. By rearranging Eq. (23) as

$$[F_i^{(0)}(1) - \varepsilon_i^{(0)}]\phi_i^{(1)}(1) + [F^{(1)}(1) - \varepsilon_i^{(1)}]\phi_i^{(0)}(1)$$

$$+ \sum_{\substack{j=1 \\ (j \neq i)}}^{N} [<\phi_j^{(1)}(2)|r_{12}^{-1}(1 - P_{12})|\phi_j^{(0)}(2)>$$

$$+ <\phi_j^{(0)}(2)|r_{12}^{-1}(1 - P_{12})|\phi_j^{(1)}(2)>]\phi_i^{(0)}(1) = 0 \qquad (29)$$

with the modified Fock-operator $F_i^{(0)}(1)$,

$$F_i^{(0)}(1) = -\frac{1}{2}\nabla_1^2 - \frac{Z}{r_1} + \sum_{\substack{j=1 \\ (j \neq i)}}^{N} <\phi_j^{(0)}(2)|r_{12}^{-1}(1 - P_{12})|\phi_j^{(0)}(2)>$$

$$\qquad (30)$$

and neglecting the self-consistency term in the sum over j, in
the first-order equation (29), we obtain an uncoupled equation,

$$[F_i^{(0)}(1) - \varepsilon_i^{(0)}]\phi_i^{(1)}(1) + [F^{(1)}(1) - \varepsilon_i^{(1)}]\phi_i^{(0)}(1) = 0 \qquad (31)$$

In this scheme, all terms involving the first-order functions
$\phi_j^{(1)}$, other than the one being calculated, have been neglected.
Another simplified uncoupled Hartree-Fock equation can be obtained
by a modification of the exchange term by defining an alternative
Fock-operator, $\overline{F}_i^{(0)}$, of the following form:

$$\overline{F}_i^{(0)}(1) = -\frac{1}{2}\nabla_1^2 - \frac{Z}{r_1} + \sum_{j=1}^{N} [<\phi_j^{(0)}(2)|r_{12}^{-1}|\phi_j^{(0)}(2)>$$

$$-<\phi_j^{(0)}(2)|r_{12}^{-1}|\phi_j^{(0)}(2)>\phi_j^{(0)}(2)/\phi_i^{(0)}(1)] \qquad (32)$$

When this is introduced into the uncoupled equations (29) and
(31), we obtain a simplified uncoupled equation,

$$[\overline{F}_i^{(0)}(1) - \varepsilon_i^{(0)}]\phi_i^{(1)}(1) + [F^{(1)}(1) - \varepsilon_i^{(1)}]\phi_i^{(0)}(1) = 0 \qquad (33)$$

If $\phi_i^{(1)}$ is expressed in the product form $f_i\phi_i^{(0)}$, all of the
two-electron integrals in Eq. (33) can be eliminated, and we obtain the following very simple expression,

$$[-\frac{1}{2}\nabla_1^2 f_i(1)]\phi_i^{(0)}(1) - \nabla_1 f_i(1)\nabla_1\phi_i^{(0)}(1)$$

$$+ [F^{(1)}(1) - \varepsilon_i^{(1)}]\phi_i^{(0)}(1) = 0 \qquad (34)$$

This uncoupling scheme has proved to be computationally more convenient than any of the other uncoupling schemes. However, its simplicity is achieved only with some loss of accuracy with respect to the other schemes [17]. The uncoupled Hartree-Fock equations are, in general, inferior to the coupled Hartree-Fock equation in accuracy because the perturbed functions are not required to be solutions of the perturbed Hartree-Fock equation. However, the uncoupled equations can be used in computations for large many-electron systems. Thus, calculations of polarizabilities and hyperpolarizabilities have been performed for a wide variety of atoms and molecules using the uncoupled equations.

4. The Double-Perturbation Method

As mentioned in Sec. I, the electric polarizabilities and hyperpolarizabilities are one-electron, second-order, and higher-order physical properties, respectively. According to Brillouin's theorem [20], diagonal matrix elements of one-electron operators are given, correct to the first-order, by the Hartree-Fock approximation, and the errors due to neglect of electron correlation are second order. However, in constrast to first-order physical properties, the effect of electron-correlation may be appreciable for the electric polarizabilities. This is because transition matrix elements are involved, and these are, in general, not correct to first-order because of the contribution from virtual-orbital excitations. The double-perturbation method suggested by Dalgarno and coworkers [21,22] and Hirschfelder [23] provides a very convenient framework within which to analyze and improve the accuracy of the Hartree-Fock approximation.

The Rayleigh-Schrödinger perturbation formulas can be generalized to the double-perturbation problem if we use the

Hamiltonian,

$$H = H_0 + \lambda V + \tau W(1) \tag{35}$$

and expand the wavefunction and energy for the perturbed
nondegenerate states in a double power series in λ and τ,

$$\Psi = \sum_{n=0}^{\infty} \sum_{m=0}^{\infty} \lambda^n \tau^m \Psi^{(n,m)} \tag{36}$$

$$E = \sum_{n=0}^{\infty} \sum_{m=0}^{\infty} \lambda^n \tau^m E^{(n,m)} , \quad n, \; m = 1, \; 2, \; \ldots \tag{37}$$

H_0 is the approximate unperturbed Hamiltonian corresponding to
the approximate wavefunction $\Psi^{(0)}$, such as the Hartree-Fock
wavefunction. λV is the correlation for the "badness" of the
approximate wavefunction. This term corrects for the neglect of
electron-electron interactions in the unperturbed function, $\Psi^{(0)}$.
The second-order one-electron properties, $<W>$, which are associated with the operator $W(1)$ can be written, in general, as

$$<W> = <\chi, (W(1) - <W>) \Psi> \tag{38}$$

where χ is the solution of the equation,

$$(H - E)\chi + (W(1) - <W>)\Psi = 0 \tag{39}$$

On the other hand, we obtain from Eqs. (36) and (37) expressions
for the energy and wavefunction as a power series in λ,

$$E^{(m)} = \sum_{n=0}^{\infty} \lambda^n E^{(n,m)} \tag{40}$$

$$\Psi^{(m)} = \sum_{n=0}^{\infty} \lambda^n \Psi^{(n,m)} \tag{41}$$

From Eqs. (38), (40), and (41), and using the interchange theorem
suggested by Dalgarno, mixed perturbation energies and wavefunctions are expressed entirely by one-electron perturbed terms. We
obtain a rather convenient and easy-to-handle equation for the
second-order expectation values,

$$<W> = <\Psi^{(0,1)}, W\Psi^{(0)}> + \lambda[<\Psi^{(0,2)}, V\Psi^{(0)}>$$

$$+ <\Psi^{(0)}, V\Psi^{(0,2)}> + <\Psi^{(0,1)}, V\Psi^{(0,1)}>] + \ldots \qquad (42)$$

where the function $\Psi^{(0,2)}$ satisfies

$$(H_0 - E^{(0)})\Psi^{(0,2)} + W'\Psi^{(0,1)} - E^{(0,2)}\Psi^{(0)} = 0 \qquad (43)$$

The effect of electron-correlation on the electric polarizabilities can thus be calculated by the double-perturbation method, even though the exact wavefunction of the atomic and molecular systems is not known. Recently, Caves and Karplus [24] have compared the calculated electric dipole polarizability, obtained using the coupled Hartree-Fock equations, the uncoupled Hartree-Fock equations, and the double-perturbation treatment with the results obtained using Brueckner-Goldstone perturbation theory. They found the errors due to the electron-correlation effect to be about 10 to 20% smaller than the errors due to the uncoupling. On the other hand, Tuan et al. [25], and more recently Tuan and Davidz [26], proposed another uncoupled Hartree-Fock scheme, which differs from Dalgarno's uncoupled scheme [Eq. (28)] in that the zeroth-order effective Hamiltonian does not include any self-potential term or only includes half of the self-potential. They found that use of the double-perturbation method with their uncoupled scheme produced a first-order correction which gave results comparable in accuracy to the coupled Hartree-Fock scheme [170].

5. Other Nonempirical Methods

In addition to the methods discussed above, a variety of more or less accurate nonempirical methods have been employed in the past for the calculation of the electric polarizability. Earlier studies center around the determination of the perturbed wavefunction of the form,

$$\Psi = \Psi^{(0)}(1 + \Phi) \qquad (44)$$

within the framework of the variational method. Hassé [27],
Atanasoff [28], Steenhalt [29], and Bell and Long [30] have chosen
Φ equal to a constant term times the perturbing potential, while
Slater and Kirkwood [31], Hellmann [32], and later Buckingham
[33], Bravin [34], Pople and Schofield [35], and others [36-39]
introduced more flexibility into the perturbed wavefunction. They
considered perturbations of the individual orbitals in the form of
$\Phi = \sum_{i=1}^{N} W_i \Psi^{(0)}(r_i)$. As the zeroth-order wavefunction, $\Psi^{(0)}$,
they used either the Hartree-product or the Hartree-Fock wavefunc-
tion, depending upon their availability for the particular systems.

Extensive numerical calculations have been performed for
atoms and ions using the method known as Sternheimer's procedure
[40-45]. This method assumes the separability of the one-electron
perturbed function into various angular components leading to a
set of independent inhomogeneous differential equations of the
form,

$$(H^{(0)} - E^{(0)})\Psi_i^{(1)} = [-\nabla^2 + \frac{1}{\Psi_i^{(0)}} \nabla^2 \Psi_i^{(0)}]\Psi_i^{(1)} = -W\Psi_i^{(0)} \qquad (45)$$

These equations have been numerically integrated. In this equation, each perturbed function is associated uniquely with its
unperturbed orbital.

Recently, a geometric approximation of the form,

$$\alpha = \alpha_0 (1 - \frac{\alpha_1}{\alpha_0})^{-1} \qquad (46)$$

based upon the uncoupled Hartree-Fock schemes has been suggested
by various authors [26,46-48], where α_0 and α_1 correspond to
the zeroth-order and first-order values in the expansion series,

$$\alpha = \alpha_0 + \lambda\alpha_1 + \lambda^2\alpha_2 + \ldots \qquad (47)$$

for the Hartree-Fock perturbing potential with continuous parameter λ. While this approximation provides very accurate results
on small atoms and ions, its applicability to general atomic and

molecular systems is not yet fully investigated.

B. Semiempirical Calculations

1. The "Average Energy" Method

Since the convergence of the summation over m in the Eq.
(5) is usually very slow, various efforts have been directed at
the approximate evaluation of Eq. (5) in the framework of the
"average energy" (closure) approximation [50]. The energy difference in the denominator is replaced by an "average energy" $\overline{E}$.
Equation (5) is approximated by Eq. (48) using $\overline{E}$ and the
quantum-mechanical sum-rule:

$$\alpha \approx \frac{2}{\overline{E}} <\psi_0^{(0)}|\sum_\mu r_\mu^2|\psi_0^{(0)}> - <\psi_0^{(0)}|\sum_\mu r_\mu|\psi_0^{(0)}>^2 \tag{48}$$

A common choice for $\overline{E}$ is the ionization potential of the system.
This approximation is extremely crude and values are obtained
which are correct only within an order of magnitude. Further
refinements are possible if the average of the true excitation
energies weighted with respect to the matrix elements of the external perturbation, such as the electric field which polarizes
the atomic or molecular system, is used instead of the ionization
potential itself. This can be determined, for example, by [3],

$$\overline{E} - E_0^{(0)} = \frac{[<\psi_0^{(0)}|\sum_\mu r_\mu(H^{(0)} - E^{(0)})\sum_\mu r_\mu|\psi_0^{(0)}>]}{<\psi_0^{(0)}|\sum_\mu r_\mu|\psi_0^{(0)}>} \tag{49}$$

Recently, Kirtman and Benston [51] and Goodisman [52] have extended the "average energy" method and examined its applicability
to many-electron systems.

2. Calculations Using Semiempirical Molecular Orbital Methods

Since the generation of nonempirical wavefunctions for larger
molecules are still time-consuming, several attempts have been

made to develop a reliable method for evaluation of the electric
polarizabilities within the framework of semiempirical molecular
orbital theories. We may categorize them as the all-valence
electron theories and π-electron theories. The LCAO-MO method
and the free-electron MO method are the most representative methods for our purpose. Among various semiempirical all-valence
electron theories, the CNDO method developed by Pople et al [53]
has been extensively applied by several authors for the calculation of the electric polarizabilities. The main feature of the
CNDO method is the zero differential overlap assumption for electron repulsion integrals of the form,

$$<\chi_\mu(1)\chi_\nu(1)\,|r_{12}^{-1}|\,\chi_\lambda(2)\chi_\sigma(2)>$$

$$= \; <\chi_\mu(1)\chi_\mu(1)\,|r_{12}^{-1}|\,\chi_\lambda(2)\chi_\lambda(2)>\delta_{\mu\nu}\delta_{\lambda\alpha} \tag{50}$$

where the χ_μ are atomic orbitals. The resulting Fock matrix is
further simplified by employing empirical quantities such as the
ionization potential, the electron affinity, etc., for corresponding terms in the Fock matrix. The self-consistent field technique
is employed to obtain the eigenfunctions and eigenvalues for the
semiempirical Fock matrix. In calculating the electric
polarizability employing the CNDO method, we may use the sum-over-
states method or the variation-perturbation scheme. In the sum-
over-states method, the first-order equation (16) can be written
within the CNDO approximation as,

$$(F^{(0)} - \varepsilon_i^{(0)})c_i^{(1)} = (\varepsilon_i^{(1)} - F^{(1)})c_i^{(0)} \tag{51}$$

for the ith molecular orbital. The principal component of the
electric dipole polarizability α_{zz} can be expressed as

$$\alpha_{zz} = \frac{-\sum_{\mu,\nu} h_{\mu\nu}^{(1)} P_{\mu\nu}^{(1)}}{F_z^2} \tag{52}$$

since

$$h_{\mu\nu}^{(2)} = 0 \tag{53}$$

and

$$\sum_{\mu} P_{\mu\mu}^{(1)} = 0 \tag{54}$$

where F_z corresponds to the electric field in the z direction.
The matrix elements $h_{\mu\nu}$ and $P_{\mu\nu}$ have a very simple form in the
CNDO approximation. In the variation-perturbation method within
the CNDO approximation, the uncoupled Hartree-Fock schemes are
generally used. The perturbed orbital $\phi_i^{(1)}$ is expressed in the
product form,

$$\phi_i^{(1)} = f_i \phi_i^{(0)} \tag{55}$$

where $\phi_i^{(0)}$ denotes the unperturbed molecular orbital, which is
obtained from the CNDO method and f_i is determined by using the
variation principle for the second-order energy. The resulting
values of the electric polarizabilities are, of course, dependent
upon the empirical parameters employed.

For unsaturated hydrocarbons, the Pariser-Parr-Pople type
π-electron theories [54] have been especially successful in predicting molecular electronic spectra. The π-electron theory with
the assumption of zero differential overlap, in addition to the
σ-π separability assumption, has been used extensively for the
calculation of the electric polarizabilities of large π-electron
systems, in the framework of the variation-perturbation method.
On the other hand, the free-electron MO method [55] for large π-
electron systems has been employed in the framework of the sum-
over-states method for estimating the electric polarizabilities.
These calculations serve as rough estimates and are only of
qualitative importance.

3. Additivity of Bond Polarizabilities

Another approach to estimate the electric polarizabilities of
large molecules may be obtained by using an additivity relation.
The total polarizability is expressed as the sum of contributions
from various constituent groups in the molecule,

$$\alpha = \alpha_A + \alpha_B + \alpha_C + \ldots \tag{56}$$

The additivity of polarizability is a measure of the separability
of the molecule into electron groups, such as inner-shell electrons, electrons in the chemical bond region, nonbond region, etc.
The classical theory of this additivity relation has been studied
by Silberstein [56], Denbigh [57], Le Fevre [58], and Stuart [59].
Pitzer [60] has examined the problem quantum-mechanically and
found that the same results were obtained within the approximation
involved. Denbigh [57] has tabulated the bond polarizability for
a number of bonds, such as C-C, C=C, C≡C, C-H, C-Cl, C-Br, C=0,
C=S, C≡N, N-H, S-S, etc. It is found that the average polarizability can be estimated rather well with the bond-additivity rule,
but the anisotropy of the polarizability is very sensitive to the
interaction of various constituent groups and this scheme fails,
in general, for that purpose. In estimating the polarizabilities
of constituent atoms in various environments (e.g., an atom in a
bond of a diatomic molecule, etc.), for the application of the
bond-additivity rule, the semiempirical delta-function method [61,
62] has often been used. The delta-function strength of an atom
in a chemical bond is, in turn, a function of various empirical
parameters such as electronegativity, number of valence-shell
electrons, etc. Recently, the additivity of bond polarizabilities
was reviewed by Applequist et al. [62a].

III. NUMERICAL RESULTS AND DISCUSSIONS

The tables in the Appendix summarize the values of the electric polarizabilities and hyperpolarizabilities calculated using nonempirical methods. We have not included results from semiempirical calculations in the tables, since they are more or less qualitative estimates.

Exact calculations of the electric dipole polarizability for the one-electron atom have been carried out by Epstein [63], Schrödinger [64], Wentzel [65], and Doi [66]. For many-electron atomic and molecular systems, the approximate methods discussed above have been employed. In the following, we review some of the representative numerical results calculated from various approximate methods. They are compared to each other and with experimental values, if available. Experimental values of the electric polarizabilities can be obtained by deflection measurements [67], by the extrapolation of measurements of refractive indexes [68], by the extrapolation of the Rydberg-Ritz corrections for spectral series [69,70], from the indexes of refraction of salts in aqueous solutions [71], and from refraction data on crystals [72]. The shifts of optical spectra due to an electric field (electrochromism) also provides experimental data on the electric dipole polarizability in medium-sized and large molecules [73]. The experimental details of electrochromism have been discussed by Liptay in Chap. 5 of this volume. Recently, very accurate values of the polarizability anisotropy have been obtained from measurements of the light scattered by gases from a helium-neon gas laser beam [74].

A. Electric Dipole and
Quadrupole Polarizabilities in Atoms and Ions

Table 1 summarizes the calculated electric dipole polarizabilities for some neutral atoms and their positive and negative ions. We notice that for some systems with very few electrons, the perturbation method has been preferably utilized and it provides the most accurate results, such as those of Schwartz [75] for H^- and He, and those of Kelly [7] for Be. The values, calculated by the coupled Hartree-Fock scheme and by the perturbation method, are nearly the same in most cases. For heavier atoms and ions, the uncoupled Hartree-Fock schemes and Sternheimer's procedure have been preferred. In most cases, the uncoupled Hartree-Fock schemes provide slightly better results than those from Sternheimer's procedure, while those from the coupled-Hartree-Fock scheme, if available, are always better than both, as expected. For example, for Be, the values are: 9.5 $\overset{o}{A}{}^3$ from Sternheimer's procedure, 4.5 $\overset{o}{A}{}^3$ from the uncoupled Hartree-Fock equation, 6.75 $\overset{o}{A}{}^3$ from the coupled-Hartree-Fock equation. The most accurate value available is from the perturbation calculation by Kelly, which gives 6.93 $\overset{o}{A}{}^3$. Kaneko [11] has carried out numerical calculations on rare gas atoms using both the exact first-order perturbed Hartree-Fock equation and Sternheimer's equation [Eq. (45)], and found that Sternheimer's procedure gives values which are 10 to 20% too large. Yoshimine and Hurst [76] have carried out extensive calculations utilizing the uncoupled-Hartree-Fock scheme. They found a marked sensitivity in the results to the choice of the unperturbed Hartree-Fock function used in the calculation, as shown in Table 1. More recently, Weinhold [77] calculated the electric polarizabilities of the two-electron atoms H^-, He, and Li^+ by a lower-bound procedure, using well-correlated variational trial functions up to 104 terms in length. These are also included in Table 1. Although reasonable agreement with experiment was found for lighter atoms and ions, agreement became progressively worse for heavier atoms and ions in those

calculations. The discrepancy is found to be partially attributed
to electron correlation and relativistic effects, and partially
due to the approximations associated with the uncoupling procedure. A detailed comparison between the perturbation method
utilizing the Hartree-Fock model and Bruekner-Goldstone perturbation for Li has been made by Chang et al. [78]. In addition to
the nonempirical results, the electric dipole polarizabilities
have been reasonably well predicted using the semiempirical
method, by Dalgarno and Kingston [68] for alkali metals (Li, Na,
K, Rb, Cs), for which reliable oscillator strengths are available.
For atoms and ions of rare-gas structure (F^-, Ne, Na^+, Cl^-, Ar,
K^+, Br^-, Kr, Rb^+, I^-, Xe, Cs^+), the "average energy" method and
its extensions have been utilized by Buckingham [33], and more
recently by Heinrichs [79]. Except for the lightest atoms and
ions, the results are in good agreement with the experiment, although the approximations involved are rather crude.

Table 2 summarizes the electric quadrupole polarizabilities
for some atoms and ions calculated using methods similar to those
used to calculate electric dipole polarizabilities. Dalgarno [3],
Kaneko [11], and Lahiri and Mukherji [80] applied the coupled
Hartree-Fock scheme, while Khubchandani et al. [81], and Langhoff
and Hurst [82] and others employed uncoupled Hartree-Fock schemes.
Sternheimer's procedure has also been extensively applied by
various authors [41,45,83]. Langhoff and Hurst [82] have calculated the electric octupole polarizabilities for a variety of
atoms and ions.

B. Electric Dipole Polarizability
in Diatomic and Small Polyatomic Molecules

Table 3 and 4 summarize the electric dipole polarizabilities
for some diatomic and small polyatomic molecules, respectively.
The hydrogen molecule serves as a test example to examine the
quality of various methods and to verify the accuracy of the

unperturbed wavefunctions used for the calculations. Several
calculations have been done, most of them using variation-perturbation techniques. Early studies for the hydrogen molecule were
done by Hirschfelder [84], Bell and Long [30], and Ishiguro et al.
[38]. Recently, several other calculations have been reported,
the most accurate calculations being those of Kolos and Wolniewicz
[85]. Their zeroth-order wavefunctions include electron correlations. Wilkins and Taylor [86] have used the uncoupled Hartree-
Fock scheme of Karplus and Kolker applying the first-order
correction calculated from the double-perturbation method. Their
results are as accurate as those obtained from the coupled
Hartree-Fock method. Lim and Linder [87] have used both coupled
and uncoupled Hartree-Fock schemes, utilizing Gaussian functions
as the basis set. A comparison between the sum-over-states method
and the variation-perturbation technique has been made by
Ditchfield et al. [88] utilizing the Hartree-Fock-Roothaan SCF
procedure for H_2. Karplus and Kolker [19] determined the principal components of the electric polarizability for H_2, Li_2, N_2,
LiH, HF, LiF, and CO, by their uncoupled Hartree-Fock scheme.
Stevens and Lipscomb [89-91] used the perturbation method with
Hartree-Fock SCF wavefunctions in calculating polarizabilities for
LiH, HF, and F_2. Calculations made by Karplus and Kolker generally overestimated electric dipole polarizability, particularly for
the parallel component of the tensor, $\alpha_\parallel$. They have discussed
this point extensively and given numerical examples. The simplified version of the uncoupled Hartree-Fock scheme proposed by
Karplus and Kolker [Eq. (33)] has been applied by Harrison [92] to
Li_2, N_2, H_2O, and NH_3, using Gaussian lobe functions as basis functions for the zeroth-order wavefunctions. Moccia [93] used one-
center SCF wavefunctions for HF, H_2O, NH_3, and CH_4 to calculate the
electric dipole polarizability within the framework of a perturbation method similar to that of Lipscomb and Stevens. His results
are in accord with the best multicenter SCF wavefunctions.
Arrighini et al. [94] also studied H_2O, NH_3, CH_4, and H_2O_2 using

multicenter SCF wavefunctions with various size Slater-type orbital
basis sets. Drossbach and Schmittinger [95] used a very size-
limited basis set of Gaussian lobe functions to study the influence
of the electric field on the chemical bond in HF, CH_3Cl, and H_2O.
They utilized the variation-perturbation technique. Very extensive
calculations have been carried out by McLean and Yoshimine [96] for
a variety of linear molecules such as HF, HCl, LiF, BF, CO, NaBr,
KCl, RbF, AlF, SiO, PN, SrO, HCN, H_2O, $(OCN)^-$, FCN, SCO, $(SCN)^-$,
ClCN, FCCH, ClCCH, and NCCCH, from their accurate Hartree-Fock-
Roothaan SCF wavefunctions, which are very close to the Hartree-
Fock limit [97]. Their calculations can be considered to be
extensions of atomic calculations by Roothaan and Cohen [49] applied to linear molecules.

In addition to the nonempirical calculations, there are several semiempirical studies of small molecules. The main purpose
of these calculations is to develop reasonable semiquantitative
schemes for large molecules. Earlier semiempirical studies of N_2,
based upon a one-term variation expression in the Hartree-Fock
approximation were made by Abbott and Bolton [98], and Guy et al.
[36, 99]. Several attempts have recently been made to develop a
reliable method for the evaluation of the electric polarizability
tensor within the frame of the semiempirical all-valence electron
CNDO-method. Davies [100] employed a perturbation method within
the CNDO approximation for H_2, N_2, HF, C_2H_2, and C_2H_4. The resulting values are too low, particularly so for the perpendicular
tensor component, $\alpha_\perp$, compared with experiment, probably because
the restrictions imposed on the valence-orbital basis set does not
allow for an adequate representation of the perturbed molecular
orbital. Hush and Williams [101] employed a variation-perturbation technique within the CNDO approximation and made calculations
for H_2, N_2, O_2, F_2, CO, BF, LiF, NO, HF, CO_2, C_2H_2, FCCH, NH_3,
H_2O, CH_4, and HCN. Their results show that the average
polarizabilities are considerably underestimated, while the calculated values for the polarizability anisotropy are in good agreement with experiment. Using semiempirical corrections for

deficiencies in the atomic values, they later improved the results
for the rotational average values [102]. On the other hand,
Sadlej [103] attempted to simplify the scheme of Hush and Williams
by omitting the still time-consuming self-consistency process in
determining the perturbed orbitals. Nagarajan and Cotter [104]
applied the semiempirical delta-function method to predict the
polarizability tensor components for NCl_3 using the additivity
rule of the bond polarizabilities.

C. Electric Polarizability in Large Molecules

Among large organic molecules, π-electron systems have been
extensively studied with the Pariser-Parr-Pople type LCAO method
and also by the free-electron MO method. The π-electron SCF
perturbation theory has been formulated by Amos and Hall [105] and
used for polarizability calculations in some conjugated molecules
such as benzene, naphthalene, anthracene, azulene, butadiene, and
hexatriene. Similar formulations in terms of density matrices
have been given by Dierksen and McWeeny [106]. The separability
of $\underline{\alpha}$ into σ and π components has been assumed, and the
additivity rule for the σ-bond contribution usually been used.
Benerjee and Salem [107] estimated the polarizability tensor
components for benzene with Bolton's additivity scheme of the
bond polarizabilities for the σ component. Schweig [108] has
performed extensive calculations for large π-electron systems
such as benzenoid hydrocarbons, polyenes and various types of dye
molecules, utilizing the variation-perturbation technique in conjunction with the Pariser-Parr-Pople π-orbital method. Agree-
ments with experimental values were good. Seibold et al. [109]
used Schweig's method to interpret the change in polarizability
due to π-π^* transition for crocetindimethyl ester. The calculated change of polarizability was, however, about three times
smaller than the experimental one. Several authors [110-112] have

employed the free-electron MO method to estimate the electric
polarizability in large conjugated chain molecules (e.g., polyene),
utilizing the sum-over-states method. Labhart and coworkers [73],
and more recently Dietrich and Jaenicke [113] have estimated the
change of the polarizability due to electron excitation using the
free-electron method. The calculated results can serve as
qualitative estimates and may guide further experimental work for
large molecules.

IV. HYPERPOLARIZABILITIES

If the external electric field strength is extremely high,
for example of the order 10^7 V/cm for an atomic system, the
system will spontaneously ionize. However, prior to ionization we
expect departures from the linear relationship [Eq. (1)] for the
induced dipole moment. This nonlinearity [Eq. (2)] was first
considered by Coulson et al. [114] in connection with the benzene
molecule, by Matossi and Mayer [115] to account for certain effects in the polarizability theory associated with the Raman effect, and also by Senftleben and Gladisch [116]. Buckingham and
Pople [117] have shown that the electric hyperpolarizabilities for
S-state atoms and ions are directly related to the anisotropy in
index of refraction of the gas in an external field (Kerr effect).
The experimental values of the first hyperpolarizability (β) can
be obtained from experiments on the second harmonic scattering in
the fluids [118], from electric birefringence experiments [2],
from refractivity virial data [119], and from Rayleigh scattering
by polar liquids [2,120]. The experimental value of the second
hyperpolarizability (γ) can be obtained from an electric birefringence experiment [121]. The hyperpolarizabilities are of
special importance because of their close relationship to nonlinear optics [2,122].

As mentioned in the introduction, the electric hyperpolarizabilities $\underline{\beta}$ and $\underline{\gamma}$ are associated with the third- and fourth-order perturbations in the energy of the system, and can be expressed, similar to the linear polarizabilities, as shown in Eqs. (8) and (9), respectively, according to the third- and fourth-order Rayleigh-Schrödinger perturbation theory.

For static electric fields, the tensor, $\beta_{ij,k}$, is symmetric in the first two suffices (see Chap. 2),

$$\beta_{ij,k} = \beta \tag{57}$$

whereas the tensor $\gamma_{ij,k\ell}$ is symmetric in ij and $k\ell$ separately,

$$\gamma_{ij,k\ell} = \gamma_{ji,k\ell} = \gamma_{ji,\ell k} = \gamma_{ij,\ell k} \tag{58}$$

It is convenient to define the average hyperpolarizabilities as [2],

$$<\beta> = \frac{3}{5}(\beta_{xx,z} + \beta_{yy,z} + \beta_{zz,z}) \tag{59}$$

and

$$<\gamma> = \frac{1}{5}(\gamma_{xx,xx} + \gamma_{yy,yy} + \gamma_{zz,zz} + 2\gamma_{xx,yy} + 2\gamma_{yy,zz} + 2\gamma_{zz,xx}) \tag{60}$$

and the hyperpolarizability anisotropy as,

$$\Delta\beta = \beta_{zz,z} - \frac{3}{2}\beta_{xx,z} - \frac{3}{2}\beta_{yy,z} \tag{61}$$

in a Cartesian coordinate system. The number of independent components of the hyperpolarizabilities can be reduced according to the symmetry of the system. For example, while the second hyperpolarizability $\underline{\gamma}$ is nonzero for all atoms and molecules, all components of $\underline{\beta}$ are zero for systems possessing a center of inversion. For such systems, the average second hyperpolarizability, $<\gamma>$ is the only independent quantity. For systems with $C_{\infty v}$ symmetry, the values, $<\beta>$ and $\Delta\beta$ completely specify the tensor, $\underline{\beta}$. For systems with C_{3v} symmetry, an additional

parameter $\beta_{xx,x}$ is needed. With C_{2v} symmetry the further
parameter, $(\beta_{xx,z} - \beta_{yy,z})$ must be specified. For complete
specification of γ for systems with lower symmetries, additional
anisotropy parameters are required in addition to the average
hyperpolarizability, $<\gamma>$. Detailed and extensive studies have
been made by Buckingham and Stephen [123] and more recently by
Buckingham and Orr [2], and McLean and Yoshimine [124] on the
symmetry properties of β and γ. These authors have given the
number of independent components required to specify the tensors,
β and γ for various symmetry species. For calculations of the
electric hyperpolarizabilities in many electron-systems, either
the Eqs. (8) and (9) may be directly applied in the framework of
the perturbation method, or the variation-perturbation techniques
can be employed as in the case of the linear polarizabilities.
Table 5 in the Appendix summarizes the values calculated for some
atoms and ions, reported by several authors with various nomempi-
rical methods. Sewell [125] calculated the γ value of the
hydrogen atom. The helium atom serves as a test example to examine the quality of various methods and wavefunctions, since the
experimental value is available, 52.8 au, as cited by Buckingham
and Hibbard [126]. Drake and Cohen [127] used the Z-expansion
technique in the framework of both the sum-over-states method
and the uncoupled Hartree-Fock scheme, and obtained γ values
for helium of 28.5 au and 44.0 au, which are too low as compared
with the experimental value. Langhoff et al. [128], on the other
hand, employed a Hartree-Fock function as the zeroth-order wavefunction within the uncoupled Hartree-Fock scheme, and obtained a
value of 51.6 au. The result by Cohen [129] with the coupled
Hartree-Fock scheme, and Boyle et al. [121] with the modified Z-
expansion method are 24.0 au and 31.1 au, respectively. Sitz
and Yaris [130] have given a value of 42.6 au. Grosso et al.
[131] and Buckingham and Hibbard [126], utilizing Hylleraas type
wavefunctions as the zeroth-order functions, have given similar
values of 42.81 au and 43.10 au, respectively. We notice,

however, from reviewing these values, that there is significant
discrepancy between the calculated values and the experimental
value. Since the wavefunction used by Buckingham and Hibbard,
for example, is considered to be very accurate, the experimental
value of γ for helium may need further refinement. For other
atoms, the calculations made by Langhoff et al. [128] and by
Drake and Cohen [127] are the only systematic ones, and we present their results in Table 5. There are relatively few nonempirical calculations available on β and γ for small molecules.
O'Hare and Hurst [132] have used an uncoupled Hartree-Fock scheme
similar to that of Karplus and Kolker, utilizing extended basis
set SCF wavefunctions as zeroth-order functions, to calculate some
$\beta_{ij,k}$ values for polar diatomic molecules such as LiH, HF, CO,
and BF. They examined the influences of the zeroth-order wavefunctions and found that the hyperpolarizabilities are much more
sensitive to this choice than the linear polarizabilities.
Unfortunately, there are no experimental values for these molecules available as yet to be compared with the calculated ones.
Arrighini et al. [133] have employed the coupled Hartree-Fock
scheme to determine some $\beta_{ij,k}$ values for H_2O, NH_3, CH_4, and
CH_3F. Comparisons with the experimental values, if available,
show large discrepancies. However, the experimental values for
these molecules are not accurate enough to be used as reference
values.

     In addition to the nonempirical calculations, there are a few
semiempirical attempts to estimate the hyperpolarizabilities.
Heinrichs [79] has used the "average energy" method to determine
the γ values for some atoms and ions with a rare-gas structure,
and Buckingham and Orr [2] have listed some values of $\beta_{ij,k}$, for
substituted methanes and other small molecules, utilizing the
bond-additivity rule for the hyperpolarizabilities. Schweig [134]
has used the Pariser-Parr-Pople method for several large π systems, in the framework of the variation-perturbation technique,
to calculate β and γ tensor components. Molecules considered

in his study are alternant and nonalternant hydrocarbons, polyenes
and dye molecules, for which, however, no experimental results are
available.

ACKNOWLEDGMENTS

I thank J. W. Jost, W. Gelbart, A. D. Buckingham, and
H. Labhart for helpful suggestions leading to improvements in the
manuscript.

REFERENCES

1. P. Debye, Polar Molecules, Dover, New York (1945), p. 8.

2. A. D. Buckingham and B. J. Orr, Quart. Rev., 21, 195 (1967).

3. A. Dalgarno, Adv. Phys., 11, 281 (1962).

4. V. Fock, Z. Phys., 61, 126 (1930); 62, 795 (1930).

5. C. C. J. Roothaan, Rev. Mod. Phys., 23, 69, 89 (1951).

6. J. Tillieu and J. Guy, Compt. Rend. 238, 2498 (1954).

7. H. P. Kelly, Phys. Rev., B136, 896 (1964); 152, 62 (1966).

8. H. P. Kelly and H. S. Taylor, J. Chem. Phys., 40, 1478 (1964).

9. S. T. Epstein, Perturbation Theory and Its Applications in
   Quantum Mechanics, Wiley, New York (1966), p. 49.

10. H. Peng, Proc. Roy. Soc. (London), 178, 499 (1941).

11. S. Kaneko, J. Phys. Soc. (Japan), 14, 1600 (1959).

12. L. C. Allen, Phys. Rev., 118, 167 (1960).

13. A. Dalgarno, Proc. Roy. Soc. (London), A251, 282 (1959).

14. D. A. Bruekner, Phys. Rev., 97, 1353 (1955); 100, 36 (1955).

15. J. Goldstone, Proc. Roy. Soc. (London), A239, 267 (1957).

16. S. Fraga and G. Malli, Many-Electron Systems; Properties and
    Interaction, Saunders, Philadelphia, Pa. (1968).

17. P. W. Langhoff, M. Karplus, and R. P. Hurst, J. Chem. Phys.,
    44, 505 (1966).

18. A. Dalgarno and J. M. McNamee, J. Chem. Phys., 35, 1517
    (1961).

19. M. Karplus and H. J. Kolker, J. Chem. Phys., 38, 1263 (1963); 39, 2011 (1963).

20. L. Brillouin, Actualites Sci. Ind., 71, 159 (1933).

21. A. Dalgarno and J. T. Lewis, Proc. Roy. Soc. (London), A233, 70 (1955).

22. A. Dalgarno and A. L. Stewart, Proc. Roy. Soc. (London), A247, 245 (1958).

23. J. O. Hirschfelder, W. Byers-Brown, and S. T. Epstein, Adv. Quantum Chem., 1, 256 (1964).

24. T. C. Caves and M. Karplus, J. Chem. Phys., 50, 3649 (1969).

25. D. F. Tuan, S. T. Epstein, and J. O. Hirschfelder, J. Chem. Phys., 44, 431 (1966).

26. D. F. Tuan and A. Davidz, J. Chem. Phys., 55, 1286 (1971).

27. H. R. Hassé, Proc. Cambridge Phil. Soc., 26, 542 (1930); 27, 66 (1931).

28. J. V. Atanasoff, Phys. Rev., 36, 1232 (1930).

29. G. Steenhalt, Z. Phys., 93, 620 (1935).

30. R. P. Bell and D. A. Long, Proc. Roy. Soc. (London), A203, 364 (1950).

31. J. C. Slater and J. G. Kirkwood, Phys. Rev., 37, 696 (1931).

32. H. Hellmann, Acta Physicochim. (USSR), 2, 273 (1935).

33. R. A. Buckingham, Proc. Roy. Soc. (London), 160, 94, 113 (1937).

34. A. V. Bravin, J. Exp. Theoret. Phys. (USSR), 27, 384 (1954).

35. J. A. Pople and P. Schofield, Phil. Mag., 2, 591 (1957).

36. J. Guy and M. Harrand, Compt. Rend., 234, 616 (1952).

37. E. G. Wikner and T. P. Das, Phys. Rev., 107, 497 (1957).

38. E. Ishiguro, T. Arai, M. Mizushima, and M. Kotani, Proc. Phys. Soc. (London), 65, 178 (1952).

39. T. P. Das and R. Bersohn, Phys. Rev., 115, 897 (1959).

40. R. M. Sternheimer and H. M. Foley, Phys. Rev., 92, 1460 (1953).

41. R. M. Sternheimer, Phys. Rev., 96, 951 (1954).

42. R. M. Sternheimer, Phys. Rev., 107, 1565 (1957).

43. R. M. Sternheimer, Phys. Rev., 115, 1198 (1959).

44. R. M. Sternheimer, Phys. Rev., 127, 1220 (1962).

45. R. M. Sternheimer, Phys. Rev., 130, 1423 (1963).

46. H. P. Kelly, Phys. Letters, A25, 6 (1967).

47. J. M. Shulman and J. I. Musher, J. Chem. Phys., 49, 4845 (1968).

48. A. T. Amos and H. G. Robert, J. Chem. Phys., 50, 2376 (1969).

49. H. D. Cohen and C. C. J. Roothaan, J. Chem. Phys., 43, S34 (1965).

50. A. Unsold, Z. Phys., 43, 563 (1927).

51. B. Kirtman and M. L. Benston, J. Chem. Phys., 46, 472 (1967).

52. J. Goodisman, J. Chem. Phys., 47, 2707 (1967).

53. J. A. Pople, D. P. Santry, and G. A. Segal, J. Chem. Phys., 43, S130.(1965); J. A. Pople and G. A. Segal, J. Chem. Phys., 44, 3289 (1966).

54. R. Pariser and R. G. Parr, J. Chem. Phys., 21, 466, 767 (1953); J. A. Pople, Trans. Faraday Soc., 49, 1375 (1953).

55. J. R. Platt, Handbuch der Physik, Vol. 37/2, Springer, Berlin, Göttingen, and Heidelberg, p. 173 (1961).

56. L. Silberstein, Phil. Mag., 33, 92, 215, 521 (1927).

57. K. G. Denbigh, Trans. Faraday Soc., 36, 936 (1940).

58. C. G. Le Fevre and R. J. W. Le Fevre, Rev. Pure Appl. Chem., 5, 261 (1955).

59. H. A. Stuart, Molekülstruktur, Springer, Berlin (1967).

60. K. S. Pitzer, Adv. Chem. Phys., 2, 79 (1959).

61. K. Ruedenberg and R. G. Parr, J. Chem. Phys., 19, 1268 (1951); K. Ruedenberg and C. W. Scherr, J. Chem. Phys., 21, 1565 (1953).

62. A. A. Frost, J. Chem. Phys., 22, 1613 (1954); 23, 985 (1955); 25, 1150 (1956); A. A. Frost and F. A. Leland, J. Chem. Phys., 25, 1154 (1956).

62a. J. Applequist, J. R. Carl, and K. K. Fung, J. Am. Chem. Soc., 94, 2952 (1972).

63. P. S. Epstein, Phys. Rev., 28, 695 (1926).

64. E. Schrödinger, Ann. Phys., 80, 437 (1926).

65. G. Wentzel, Z. Phys., 38, 635 (1926).

66. S. Doi, Proc. Phys.-Math. Soc. (Japan), 10, 223 (1928).

67. A. Salop, E. Polak, and B. Biderson, Phys. Rev., 124, 1431 (1961).

68. A. Dalgarno and A. E. Kingston, Proc. Phys. Soc. (London), 73, 455 (1959).

69. M. Born and W. Heisenberg, Z. Phys., $\underline{23}$, 388 (1924).

70. J. E. Mayer and M. G. Mayer, Phys. Rev., $\underline{43}$, 605 (1933).

71. K. Fajans and G. Joos, Z. Phys., $\underline{23}$, 1 (1924).

72. J. R. Tessman, A. H. Kahn, and W. Shekley, Phys. Rev., $\underline{92}$, 890 (1953).

73. H. Labhart, Helv. Chim. Acta, $\underline{44}$, 447 (1961); H. Navangul and H. Labhart, Theoret. Chim. Acta, $\underline{17}$, 249 (1970); K. Jug and H. Labhart, Theoret. Chim. Acta, $\underline{24}$, 283 (1972).

74. N. J. Bridge and A. D. Buckingham, Proc. Roy. Soc. (London), $\underline{295A}$, 334 (1966).

75. C. Schwartz, Phys. Rev., $\underline{123}$, 1700 (1961).

76. M. Yoshimine and R. P. Hurst, Phys. Rev., $\underline{135}$, A612 (1964).

77. F. Weinhold, Proc. Roy. Soc. (London), $\underline{A327}$, 209 (1972).

78. E. S. Chang, R. T. Pu, and T. P. Das, Phys. Rev., $\underline{74}$, 16 (1968).

79. J. Heinrichs, J. Chem. Phys., $\underline{52}$, 6316 (1970).

80. J. Lahiri and A. Mukherji, Phys. Rev., $\underline{141}$, 428 (1966); $\underline{153}$, 386 (1967).

81. P. G. Khubchandani, R. R. Sharma, and T. P. Das, Phys. Rev., $\underline{126}$, 594 (1962).

82. P. W. Langhoff and R. P. Hurst, Phys. Rev., $\underline{139}$, A1415 (1965).

83. H. M. Foley, R. M. Sternheimer, and D. Tycko, Phys. Rev., $\underline{93}$, 734 (1954).

84. J. O. Hirschfelder, J. Chem. Phys., $\underline{3}$, 555 (1935).

85. W. Kolos and L. Wolniewicz, J. Chem. Phys., $\underline{46}$, 1426, 2356 (1967).

86. R. L. Wilkins and H. S. Taylor, J. Chem. Phys., $\underline{48}$, 1426, 4934 (1968).

87. T. K. Lim and B. Linder, Theoret. Chim. Acta, $\underline{19}$, 38 (1970).

88. R. Ditchfield, N. S. Ostlund, J. N. Murrell, and M. A. Turpin, Mol. Phys., $\underline{18}$, 433 (1970).

89. R. M. Stevens and W. N. Lipscomb, J. Chem. Phys., $\underline{40}$, 2238 (1964).

90. R. M. Stevens and W. N. Lipscomb, J. Chem. Phys., $\underline{41}$, 184 (1964).

91. R. M. Stevens and W. N. Lipscomb, J. Chem. Phys., $\underline{41}$, 3710 (1964).

92. J. F. Harrison, J. Chem. Phys., $\underline{49}$, 3321 (1968).

93. R. Moccia, Theoret. Chim. Acta, $\underline{8}$, 192 (1967).

94. G. P. Arrighini, M. Maestro, and R. Moccia, Chem. Phys. Letters, 1, 242 (1967).

95. P. Drossbach and P. Schmittinger, Z. Naturforsch., 25a, 823, 827, 834 (1970).

96. A. D. McLean and M. Yoshimine, J. Chem. Phys., 46, 3682 (1967); IBM J. Res. Develop., 12, 206 (1968).

97. A. D. McLean and M. Yoshimine, Tables of linear molecule wavefunctions, (Supplement to the paper, "Computation of molecular properties and structure"), IBM Res. Rep., Nov. (1967).

98. A. Abbott and H. C. Bolton, Proc. Roy. Soc. (London), A216, 477 (1953); A221, 135 (1954).

99. J. Guy, M. Harrand, and J. Tillieu, Ann. Phys., 9, 373 (1954).

100. D. W. Davies, Mol. Phys., 17, 473 (1969).

101. N. S. Hush and M. L. Williams, Chem. Phys. Letters, 5, 505 (1970).

102. N. S. Hush and M. L. Williams, Chem. Phys. Letters, 6, 163 (1970).

103. A. J. Sadlej, Theoret. Chim. Acta, 21, 159 (1971).

104. G. Nagarajan and S. B. Cotter, Z. Naturforsch., 26a, 1800 (1971).

105. A. T. Amos and G. G. Hall, Theoret. Chim. Acta, 5, 148 (1966); 6, 159 (1966).

106. G. Dierksen and R. McWeeny, J. Chem. Phys., 44, 3554 (1966).

107. K. Banerjee and L. Salem, Mol. Phys., 11, 405 (1966).

108. A. Schweig, Chem. Phys. Letters, 1, 163 (1967).

109. K. Seibold, H. Navangul, and H. Labhart, Chem. Phys. Letters, 3, 275 (1969).

110. M. N. Adamov and J. S. Milevskaya, Dokl. Acad. Nauk (USSR), 109, 57 (1956).

111. H. C. Bolton, Trans. Faraday Soc., 50, 1265 (1954).

112. Y. N. Zhivlink, Pot. i Spectroskopiya, 8, 220 (1960).

113. C. Dietrich and W. Jaenicke, Ber. Bunsen-Gesellschaft., 74, 47 (1970).

114. C. A. Coulson, A. McColl, and L. E. Sutton, Trans. Faraday Soc., 48, 106 (1952).

115. F. Matossi and R. Mayer, Phys. Rev., 74, 449 (1948).

116. H. Senftleben and H. Gladisch, Naturwissenschaften, 34, 187 (1947).

117. A. D. Buckingham and J. A. Pople, Phys. Soc. (London), A68, 905 (1955).

118. R. W. Terhune, P. D. Maker, and C. M. Savage, Phys. Rev. Letters, 14, 681 (1965).

119. A. R. Blythe, J. D. Lambert, P. J. Petter, and H. Spoel, Proc. Roy. Soc. (London), A255, 427 (1960).

120. A. D. Buckingham and M. J. Stephen, Trans. Faraday Soc., 53, 884 (1957).

121. L. L. Boyle, A. D. Buckingham, R. L. Disch, and D. A. Dunmur, J. Chem. Phys., 45, 1318 (1966).

122. N. Bloemenbergen, Nonlinear Optics, Benjamin, New York (1965).

123. A. D. Buckingham and M. J. Stephen, Trans. Faraday Soc., 53, 884 (1957).

124. A. D. McLean and M. Yoshimine, J. Chem. Phys., 47, 1927 (1967); A. D. Buckingham, Adv. Chem. Phys., 12, 107 (1967).

125. G. S. Sewell, Proc. Cambridge Phil. Soc., 45, 678 (1949).

126. A. D. Buckingham and P. G. Hibbard, Faraday Soc. Symp., 2, 41 (1968).

127. G. W. F. Drake and H. D. Cohen, J. Chem. Phys., 48, 1168 (1968).

128. P. W. Langhoff, J. D. Lyons, and R. P. Hurst, Phys. Rev., 148, 18 (1966).

129. H. D. Cohen, J. Chem. Phys., 43, 3558 (1965).

130. P. Sitz and R. Yaris, J. Chem. Phys., 49, 3546 (1968).

131. M. Grosso, K. T. Chung, and R. P. Hurst, Phys. Rev., 167, 1 (1968).

132. J. M. O'Hare and R. P. Hurst, J. Chem. Phys., 46, 2356 (1967).

133. G. P. Arrighini, M. Maestro, and R. Moccia, Faraday Soc. Symp., 2, 48 (1968).

134. A. Schweig, Chem. Phys. Letters, 1, 195 (1967).

135. A. L. Stewart, Proc. Phys. Soc. (London), 77, 447 (1961).

136. A. Dalgarno and J. M. McNamee, Proc. Phys. Soc. (London), 77, 673 (1961).

137. H. J. Kolker and H. H. Michels, J. Chem. Phys., 43, 1027 (1965).

138. S. Geltman, Astrophy. J., 136, 935 (1962).

139. D. R. Johnston, G. J. Oudemans, and R. H. Cole, J. Chem. Phys., 33, 1310 (1960).

140. R. E. Watson and A. J. Freeman, Phys. Rev., 123, 521 (1961).

141. D. Parkinson, Proc. Phys. Soc. (London), 75, 169 (1960).

142. M. Sundbom, Arkiv. Fysik, 13, 539 (1958).

143. A. Dalgarno and R. M. Pengally, Proc. Phys. Soc. (London), 89, 503 (1966).

144. G. M. Stacey, Proc. Phys. Soc. (London), 88, 896 (1966).

145. K. Murakawa and M. Yamamoto, J. Phys. Soc. (Japan), 21, 821 (1966).

146. G. E. Chamberlain and J. C. Zorn, Phys. Rev., 129, 677 (1963).

147. A. Dalgarno and H. A. J. McIntyre, Proc. Phys. Soc. (London), 85, 47 (1965).

148. J. Lahiri and A. Mukherji, J. Phys. Soc. (Japan), 21, 1178 (1966).

149. M. R. Flannery and A. L. Stewart, Proc. Phys. Soc. (London), 82, 188 (1963).

150. A. Dalgarno and D. Parkinson, Proc. Roy. Soc. (London), A250, 422 (1959).

151. R. A. Alpher and D. L. White, Phys. Fluids, 2, 1953 (1959).

152. M. M. Klein and K. A. Bruckner, Phys. Rev., 111, 1115 (1958).

153. W. E. Donath, J. Chem. Phys., 39, 2685 (1963).

154. H. D. Cohen, J. Chem. Phys., 45, 10 (1966).

155. H. H. Landolt and R. Bornstein, Zahlenwerte und Funktionen aus Physik, Chemie, Astronomie, Geophysik und Technik, Vol. 1, Part 3, Springer, Berlin (1950).

156. R. E. Watson and A. J. Freeman, Phys. Rev., 135, A1209 (1964).

157. V. V. Mitskyvichyuz, Lietuvos Fiz. Rinkinys, 4, 81 (1964).

158. V. M. Evodokimov, Opt. Spectr., 17, 165 (1964).

159. P. L. Altick, J. Chem. Phys., 40, 238 (1964).

160. A. Dalgarno and A. E. Kingston, Proc. Roy. Soc. (London), A259, 424 (1960).

161. T. P. Das and R. Bersohn, Phys. Rev., 102, 360 (1956).

162. G. Burns, J. Chem. Phys., 31, 1253 (1959).

163. G. Burns, Phys. Rev., 115, 357 (1959).

164. G. Burns and E. G. Wikner, Phys. Rev., 121, 155 (1961).

165. H. Hartmann and F. Becker, Z. Phys. Chem., 195, 186 (1950).

166. J. R. Hoyland, J. Chem. Phys., 41, 3153 (1964).

167. G. P. Arrighini, C. Guidotti, and O. Salvetti, J. Chem.
      Phys., 52, 1037 (1966).

168. R. Sänger, Z. Phys., 31, 306 (1930).

169. M. Krauss, "Compendium of ab initio calculations of molecular
      energies and properties", NBS Tech. Note, 438, 71 (1967).

170. J. I. Musher, J. Chem. Phys., 46, 369 (1967).

171. R. Yaris, J. Chem. Phys., 39, 2472 (1963); 40, 664 (1964).

172. M. Karplus and H. J. Kolker, J. Chem. Phys., 39, 1493 (1963).

173. C. Mavroyannis and M. J. Stephen, Mol. Phys., 5, 629 (1962).

174. A. Dalgarno, Perturbation Theory and Its Application in
      Quantum Mechanics (C. H. Wilcox, ed.), Wiley, New York (1966),
      p. 145.

175. M. H. Alexander and R. G. Gordon, J. Chem. Phys., 55, 4889
      (1971); 56, 3823 (1972).

176. W. D. Davison, Proc. Phys. Soc. (London), 87, 133 (1966).

177. N. C. Dutta, T. Ishihara, C. Matsubara, and T. P. Das, Phys.
      Rev. Letters, 22, 8 (1969).

178. M. P. Briggs, J. N. Murrell, and D. R. Salhub, Mol. Phys.,
      22, 907 (1971).

179. K. T. Chung and R. P. Hurst, Phys. Rev., 152, 35 (1966).

180. L. Essen, Proc. Phys. Soc. (London), B66, 189 (1953).

181. K. Bockasten, Arkiv. Fysik, 10, 567 (1956).

182. U. Öpik, J. Phys., B2, 1411 (1969).

APPENDIX

Table 1

Electric Dipole Polarizabilities for Atoms and Ions [a]

System		System		System	
H⁻	7.3(22)		23.97[144]	Be⁺	2.48[76,82]
	11.8[135,136]		23.9[145]		3.65[149]
	16.15(41)		22.0[146]*		2.53(17)
	13.34(42)	Li⁺	0.0315(41)	Be²⁺	0.007684(26)
	13.8[129]		0.307(42)		0.00769(17)
	14.88(37)		0.0305(37)		0.007684[148]
	16.6[76,82]		0.0304[82]		0.00765[129]
	39.28[137]		0.0304[76,136]		0.008153[82]
	31.41[75]		0.0299(22)		0.007[140,156]*
	29.93[138]		0.0292[137]	B	5.1[150]
	8.211[82]		0.0282[75]	B⁺	1.96[76,82]
	25.04[131]		0.02816[135]		1.69[147]
	29.90[77]		0.281[136]		1.682[129]
	30.530[77]		0.0280[129]		1.685(26)
	30.51(179)		0.02807(33)		1.40(17)
	30.2[138]*		0.0285168[77]		1.686[148]
He	0.236(41)		0.028502[77]		1.07(17)
	0.224(42)		0.0236(41)	B²⁺	0.654[76,82]
	0.2205(11)		0.025[140]*		1.16[149]
	0.220[136,76,82]		0.028262(181)*		0.660(17)
	0.218(37)		0.0285(182)*	B³⁺	0.002897[26,148]
	0.1959[135,69]	Li⁻	289.2[82]		0.00290(17)
	0.196[11,136 129]	Be	9.6[141]		0.00289[129]
	0.2165[137]		9.54[75]		0.00304[76,82]
	0.2050[75]		9.5[136]		0.0033[140,156]*
	0.206(22)		4.5[136]	C	2.1[150]
	0.204956[126]		4.58(8)	C²⁺	0.6669[148]
	0.205[131]		6.75[129]		0.496(17)
	0.20495(176)		6.76[147]		0.6680[148]
	0.2047[77]		6.93(7)		0.666[129]
	0.2085(177)		9.28(33)		0.67[147]
	0.2069[139]*		6.758[26]		0.653[76,82]
	0.2074(180)*		6.26(17)		0.497(17)
	0.2051[160]*		6.759[148]		0.398(17)
Li	25.0[141]		6.733(24)	C³⁺	0.245[76,82]
	24.9(44)		6.24(17)		0.509[149]
	20.0[142]		4.28(17)		0.153[148]
	21.0[76,82]		9.52[82]		
	21.7(17)		9.54[82]		
	24.84[78]		4.5(18)*		
	25.6[143]				

TABLE 1 (continued)

System	System	System

Column 1 — System

C^{4+} 0.001324[26,148]
0.00132[129,17]
0.00138[76,82]
0.0015[140,156]*

N 1.3[150]
1.13[151]*

N^{3+} 0.3310(26)
0.231(17)
0.3315[148]
0.329[129]
0.280[76,82]
0.221(17)
0.184(17)

N^{4+} 0.112[76,82]
0.267[149]
0.113[148]

N^{5+} 0.0006877(26)
0.000688(17)
0.0006882[148]
0.000685(17)
0.000712[76,82]

O 0.89[150]
0.783(7)
0.829[152]
0.77[151]*

O^{+} 0.49[141]

O^{-} 3.2[141]

O^{2-} 65.88[82]
134.3[82]

O^{4+} 0.1869(26)
0.114(17)
0.0883[148]
0.187[129]
0.140[76,82]
0.115(17)
0.0976(17)

O^{5+} 0.0586[76]
0.157[149]
0.0583[148]
0.0584[82]

F 0.6[150]

F^{-} 1.858(43)
1.81[76]

Column 2 — System

1.8[150]
1.206[153]
1.560[154]
1.902[82]
1.893[82]
0.99[69]*

F^{5+} 0.1168(26)
0.065(17)
0.117[148]
0.116[129]
0.078[76,82]
0.0653(17)
0.0566(17)
0.0781(19)

Ne 0.615(11)
0.3674[153]
0.38[150]
0.409[76]
0.398[155]*

Ne^{+} 0.21[141]

Ne^{2+} 0.15[141]

Na 27.1[76,82]
22.9(44)
24.6[160]
21.5 - 2[146]*

Na^{+} 0.145[141]
0.152(43)
0.1541[153]
0.163[76]
0.140[154]
0.171(34)
0.167[157]
0.0728[158]
0.17[70]*

Na^{-} 293.7[82]

Mg 19.4[76,82]
7.0 ± 1.8[159]*

Mg^{+} 5.51[76,82]

Mg^{2+} 0.0801[76]
0.0698[154]
0.08122[82]
0.10[140,156]*

Al^{+} 5.893[82,76]

Column 3 — System

Al^{2+} 2.05[76,82]

Al^{3+} 0.0446[76]
0.0398[154]
0.04525[82]
0.053[140,156]*

Si^{2+} 2.566[82]

Si^{3+} 0.994[76,82]

Si^{4+} 0.0271[76]
0.0275[82]
0.043[69]*

P^{3+} 1.347[76,82]

P^{4+} 0.556[76,82]

P^{5+} 0.0175[76,82]

S^{4+} 0.792[76,82]

S^{5+} 0.340[76,82]

Cl^{-} 7.19(43)
6.23[76]
7.772[82]
6.605[82]
3.05[69]*

Cl^{5+} 0.502[76,82]

Ar 2.40(11)
2.32[76]
2.03(35)
1.441[82]
2.141[82]
1.63[155]*

K 59.6[76]
44.4[141]
41.6[160]
38.0 ± 4[146]

K^{+} 1.24(43)
1.08[76]
0.80[70]*

Ca 48.9[76]
19.7 ± 0.6[159]

Ca^{+} 14.3(43)

Ca^{2+} 0.620[76]
0.54[140,156]*

TABLE 1 (continued)

System		System		System	
Sc	154.0[142]	Cu^+	0.982(43)	Cs	67.7(44)
Sc^{3+}	0.391[76]	Rb	49.1(44)	Cs^+	5.60(43)
Ti^{4+}	0.263[76]	Rb^+	2.92(43)	Hg^{2+}	2.78(43)
V^{5+}	0.185[76]	Xe	4.07[160]	U^{6+}	1.34(43)

(a) $\overset{\circ}{A}^3$ units; values in parentheses are references; and asterisks denote the experimental value.

Table 2

Electric Quadrupole Polarizabilities for Atoms and Ions[a]

System		System		System	
H^-	67.0[161]	C^{3+}	0.0671[82]	Al^{3+}	0.0091[82]
	66.5(42)	C^{4+}	0.0000431[82]		0.00915[162]
	47.5[82]		0.0000433[82]	Si^{2+}	1.657[82]
	94.76[82]	N^{4+}	0.0204[82]	Si^{3+}	0.5192[82]
He	0.0993(42)	N^{5+}	0.0000159[82]	Si^{4+}	0.00432[82]
	0.0979[136]	O^{4+}	0.0207[82]		0.00438[162]
	0.0967[82]		0.0213[82]	P^{3+}	0.5975[82]
	0.0981[82]	O^{5+}	0.00761[82]	P^{4+}	0.2085[82]
	0.1013[176]	O^{2-}	399.0[82]	P^{5+}	0.00225[82]
	0.101452[126]		1044.0[82]	S^{4+}	0.259[82]
Li	62.0(42)	F^{5+}	0.00847[82]	S^{5+}	0.0974[82]
	65.2[82]		0.00869[82]	Cl^-	13.77(42)
	61.5[82]	F^-	2.38[164]		13.1[164]
Li^+	0.00473(42)		3.46[82]		11.79[82]
	0.00472[136]		2.82[82]	Cl^{5+}	0.127[82]
	0.00471[161]		2.77[82]	Ar	2.19[162]
	0.00465[80]	Ne	0.37[162]		2.94[82]
	0.00466[82]		0.461[82]	K^+	0.733(42)
Li^-	4754.0[82]		0.258[82]		0.721[163]
Be	15.1[136,82]	Na	91.64[82]	Ca^{2+}	0.0309[162]
	14.21[80]		89.85[82]	Rb^+	2.99(42)
	14.2[147]	Na^+	0.067[161]		3.03[163]
	14.9[82]		0.0649[163]	Cs^+	7.80(42)
	15.27[82]		0.0562(42)		7.86[163]
Be^+	2.31[82]		0.064[82]	Cu^+	1.280(43)
Be^{2+}	0.000638[82]	Na^-	3371.0[82]		
	0.000642[82]	Mg	40.36[82]		
B^+	1.173[80]	Mg^{2+}	0.02188[82]		
	1.17[147]		0.0223[162]		
	1.20[82]	Al^+	6.208[82]		
	1.256[82]	Al^{2+}	1.625[82]		
B^{2+}	0.300[82]				
B^{3+}	0.00143[82]				
C^{2+}	0.217[82]				
	0.226[82]				

[a] $Å^5$ units; values in parentheses are references. The quadrupole polarizability is obtained if the dipole operator $\sum_\mu r_{i\mu}$ is replaced by the quadrupole operator $\sum_\mu r_{i\mu}^2 P_2(\cos \theta\mu)$, in Eq. (5).

Table 3

Electric Dipole Polarizabilities for Some Diatomic Molecules [a]

System	Values for $\alpha_\perp$, $\alpha_\parallel$, and $<\alpha>$
H_2	0.66,0.906,0.74(38); 0.65,0.89,0.73[165]; 1.45,2.46,1.78(39); 1.34,1.87,1.51(39); 0.80,1.29,0.97(19); 0.59,0.85,0.67(39); 0.65,0.75,0.68(39); 0.77, 1.11, 0.88[166]; 0.64,1.36,0.88(39); 0.72,1.19,0.88[165]; 0.70,0.76,0.72(39); 0.65,0.92,0.74(39); 0.66,0.91,0.74(38); 0.71,1.08,0.84[165]; 0.71,1.09,0.83[166]; 0.70,1.06,0.82[165]; 0.79,0.74,0.77(39); 0.69,0.95,0.77(39); 0.73,0.94,0.80[86]; 0.72,1.25,0.90(19); 0.72,1.29,0.91(19); 0.68,0.96,0.78[74]; 0.68,0.99,0.79[87]; 0.65,0.86,0.72(178); 0.7022,1.0021,0.8032[85]; 0.714,1.028,0.819[74]* [b]
Li_2	46.27,64.33,52.29(19); 40.65,59.85,47.05[132]; 40.31,62.67, 47.76[92]; 46.15,64.91,52.40[132]
N_2	1.27,4.67,2.40(19); 1.95,4.04,2.65[132]; 1.92,4.21,2.68[132]; 1.18,4.58,2.31(19); 1.25,4.70,2.40[132]; 1.53,2.23,1.43[74]* [b]
HF	0.53,0.78,0.61[93]; 0.62,0.86,0.70[90]; 0.31,0.74,0.45(19); 0.36,0.79,0.51(19); 0.36,0.91,0.54[132]; 0.91,1.21,1.01[132]; 0.79,1.00,0.86[132]; 0.72,0.96,0.80*[155]
F_2	0.40,1.55,0.78[91]; 0.77,2.15,1.23[91]; 0.43,2.14,1.00[132]; 0.95,2.25,1.38[132]
LiH	5.01,3.76,4.59(19); 5.08,3.35,4.50[132]; 4.20,3.76,4.05(19); 5.01,3.72,4.58[132]; 4.06,3.35,3.82[89]
CO	1.35,4.03,2.33(19); 1.46,4.23,2.38(19); 1.56,4.65,2.59[132]; 2.31,3.30,2.64[132]; 2.09,3.22,2.47[132]; 1.80, 2.33,1.98[74* [b]
BF	2.55,4.28,3.13[132]; 3.85,2.71,3.41[132]; 3.74,2.72,3.40[132]
LiF	3.40,5.38,4.06(19); 6.24,11.18,7.89(19)

[a] $Å^3$ units; values in parentheses are references; and asterisk denotes the experimental value.

[b] The experimental values are for a wavelength 6328 Å.

Table 4

Electric Dipole Polarizabilities
for Some Small Polyatomic Molecules

System	Values for α_{xx}, α_{yy}, α_{zz} and $<\alpha>$
H_2O	0.918,1.191,1.073,1.061[93]; 0.920,1.202,1.067,1.063[93]; 1.247,1.011,1.150,1.136[94]; 1.228,1.006,1.102,1.112[94]; 1.106,0.897,1.150,1.051[94]; 1.007,1.665,1.366,1.346[92]; 1.221,1.696,1.519,1.479[92]; 1.480,1.900,1.848,1.743[92]; 1.279,1.069,1.162,1.170[167]; $<\alpha>$ = 1.456*[168]
NH_3	1.742,1.624,1.703[93]; 1.450,1.30,1.40[93]; 1.772,1.627,1.724[93]; 1.130,0.529,0.93[94]; 1.517,1.124,1.386[94]; 1.741,2.240,2.074[92]; 2.153,2.305,2.254[92]; 2.663,2.553,2.589[92]; 2.18,2.42,2.26*[155]
CH_4	1.889[93]; 0.8177[94]; 1.4582[94]; 2.60*[155]
H_2O_2	1.449,0.644,0.449,0.847[94]

(a) $\mathring{A}^3$ units, values in the parentheses are references, and asterisks denote the experimental value. The $\alpha_{\parallel}$ component for several linear molecules have been reported by McLean and Yoshimine [96,169].

Table 5

Electric Dipole Hyperpolarizabilities ($\langle\gamma\rangle$) for Atoms and Ions[a]

System		System		System	
H	0.133[125]	B^{2+}	36.80[128]	Na^+	7.418[128]
H^-	$0.8157 \cdot 10^7$[128]		$-0.106 \cdot 10^4$[127]	Mg	$0.2190 \cdot 10^6$[128]
	$0.1736 \cdot 10^8$[131]		$-0.116 \cdot 10^4$[127]	Mg^+	$0.1057 \cdot 10^5$[128]
He	43.10[126]	B^{3+}	0.008102[128]	Mg^{2+}	0.9273[128]
	42.81[131]	C^{2+}	28.17[128]		
	34.1[128]		3.41[127]	Al^+	$0.5242 \cdot 10^4$[128]
	28.5[127]		5.06[127]	Al^{2+}	14.85[128]
	44.0[127]	C^{3+}	0.1974[128]	Al^{3+}	0.1754[128]
	42.6[127]		$-0.174 \cdot 10^3$[127]	Si^{2+}	135.3[128]
	51.5[128]		$-0.187 \cdot 10^3$[127]	Si^{3+}	-72.03[128]
	24.0[129]	N^{3+}	2.294[128]	Si^{4+}	0.04346[128]
	34.1[121]		0.878[127]		
	52.8[126,121]*		1.21[127]	P^{3+}	-35.42[128]
Li	$-74.57 \cdot 10^4$[128]	N^{4+}	0.1784[128]	P^{4+}	-26.64[128]
	$-67.5 \cdot 10^4$[121]		-41.5[127]	P^{5+}	0.01302[128]
	$-67.2 \cdot 10^4$[127]		-43.8[127]	S^{4+}	-19.70[128]
	$-54.4 \cdot 10^4$[127]	N^{5+}	$0.2004 \cdot 10^{-4}$[128]	S^{5+}	-9.634[128]
Li^+	0.2852[128]		$0.177 \cdot 10^{-4}$[127]		
	0.209[127]		$0.196 \cdot 10^{-4}$[127]	Cl^-	$0.6346 \cdot 10^5$[128]
	0.271[127]	O^{4+}	0.3001[128]	Cl^{5+}	-8.499[128]
	0.244[131]		0.282[127]	Ar	$0.2308 \cdot 10^4$[128]
Be	$0.7713 \ 10^5$[128]		0.370[127]	K	$0.3096 \cdot 10^7$[128]
	171.0[127]	O^{5+}	0.02198[128]	K^+	199.6[128]
	345.0[127]		-12.6[127]	K^-	$0.1252 \cdot 10^{11}$[128]
Be^+	$0.1759 \cdot 10^4$[128]		-13.2[127]		
	$-0.123 \cdot 10^5$[127]	F^-	$0.1168 \cdot 10^5$[128]	Ca	$0.1656 \cdot 10^7$[128]
	$-0.139 \cdot 10^5$[127]	F^{5+}	0.5338[128]	Ca^{2+}	36.60[128]
Be^{2+}	0.01019[128]		0.106[127]	Sc^{3+}	8.847[128]
	0.00783[127]		0.134[127]	Ti^{4+}	2.622[128]
	0.00946[127]	Ne	123.6[128]	V^{5+}	0.9073[128]
B^+	754.3[128]	Na	$0.1009 \cdot 10^7$[128]		
	18.4[127]				
	30.5[127]				

[a] Atomic units; values in parentheses are references; and asterisk denotes the experimental value. 1 a.u. = $5.04 \ 10^{-36}$ esu = $6.236 \ 10^{-61} \ C^4 \ m^4 \ J^{-3}$.

APPENDIX A

MEMENTOS OF JOHN KERR

Chester T. O'Konski

1. KERR'S ORIGINAL CELLS

The following, except for the parenthetical figure numbers,
is a quotation from "Kelvin Instruments and the Kelvin Museum" by
G. Green and J. T. Lloyd, Glasgow University Publications Office,
Glasgow, Scotland.

> The collection of Kerr cells in the Department
> include many large shaped blocks of glass which repre-
> sent a considerable technique in the drilling and shap-
> ing. This work was done for Kerr by a German glass
> worker who had settled in Glasgow.
> The original solid Kerr cell (Fig. A-1) is a
> simple slab of glass measuring 6 ins. by 3 ins. by 3/4
> ins. thick. Two holes are drilled along the long axis
> of the slab, leaving about 1/2 ins. of solid glass be-
> tween them. Kerr applied the output from an induction
> coil to two metal probes inserted in the holes, to ob-
> serve this effect.
> The original liquid Kerr cell (Fig. A-2) is a
> shaped slab with a cavity at its centre, enclosed at
> the sides by thin plates of glass. Various liquids
> were used in the cavity, the most effective being
> nitrobenzene.
> Dr. R. C. Gray, who is now retired, but will be
> remembered by many people for his outstanding lectures
> in medical physics, has the following anecdote of Kerr
> on the occasion, in 1898, when he was presented with
> the Royal Medal by Lord Lister. In the presentation
> Lister remarked "It has been a matter of admiration
> and wonder to subsequent investigators that Dr. Kerr
> should have been able to learn so much with the
> comparatively simple and ineffective apparatus at his

Fig. A-1. Kerr's cell for observing the electric birefringence of glass.

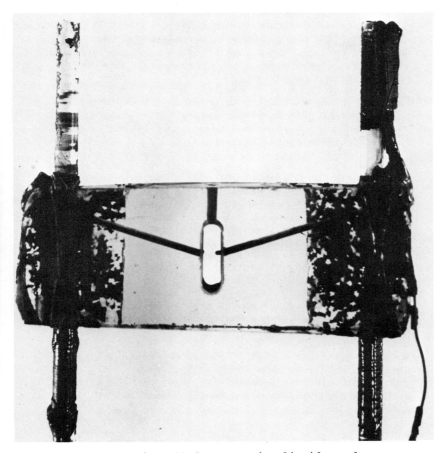

FIG. A-2. Kerr's cell for measuring liquid samples.

disposal." Apparently Kerr was much peeved at Lister's
remark: "Simple it may be," he protested, "but not
ineffective; rude, but not crude."

2. ROYAL MEDAL CITATION

This is an excerpt from the Yearbook of the Royal Society of
London, No. 3, pp. 156 - 157, Harrison and Sons, Ltd., St.
Martin's Lane, 1899.

Rev. John Kerr, L.L.D., F.R.S.

One of the Royal Medals is bestowed upon the
Rev. John Kerr, as the author of extremely important
experimental researches on the optical relations of
electricity and magnetism.

Dr. Kerr has made a name, which will always be remembered in the history of science, by his experiments
on the optical effects of electrical and mechanical
stress, and on the polarisation of light reflected from
the surface of a magnetised body.

His observations on electrical stress were recorded
in a series of papers published in the "Philosophical
Magazine" in 1875, 1879, and 1882, in which he demonstrated the fact that the velocity of polarised light in
a body subjected to electrostatic influence is different
according as the plane of polarisation is parallel or
perpendicular to the lines of electric force.

In these experiments, he was led to use mechanical-
ly strained glass as an auxiliary apparatus, and in
October, 1888, he published, in the "Philosophical
Magazine," an important paper on "The Birefringent Action
of Strained Glass."

In 1877, Dr. Kerr* showed that if plane polarised
light is reflected from a magnetised surface, the polarisation of the reflected ray is affected by the magnetic
state of the surface. This result was of the highest
theoretical interest, and it has been a matter of admiration and wonder to subsequent investigators that
Dr. Kerr should have been able to learn so much with the
comparatively simple and ineffective apparatus at his
disposal.

Both of Dr. Kerr's great researches have been the
starting points of numerous inquiries. His experiments
on electrical stress have been repeated and extended by
Gordon, Quincke, Rontgen, and others, while Fitzgerald,
Righi, Kundt, Lorentz, Sissingh, Zeeman, J. J. Thompson,
Du Bois, Goldhammer, Drude and Leathem, are among those
who have been occupied with the extension or theoretical
meaning of his work.

Dr. Kerr's researches rank among the most important
of those which have been made since the time of Faraday.

---

*Phil. Mag. (5), Vol. 3, p. 321.

3. SCIENTIFIC BIOGRAPHY OF
J. KERR, REPRINTED FROM PROC. ROY. SOC.
LONDON, 82A [OBITUARY NOTICES, PP. i-v] (1909)

JOHN KERR, 1824-1907

Dr. John Kerr was born on December 17, 1824, at
Ardrossan, in Ayrshire. In the Album of the University
of Glasgow for 1841, the year of his matriculation, he
is described as "Joannes Kerr filius natu secundus
Thomas piscarii in oppido Ardrossan." His father,
Thomas Kerr, seems to have removed to Skye when his son
was yet very young, for the boy received part of his
early education at a village school in that island.

His attendance at University classes extended over
the years from 1841 to 1849. He obtained prizes in
Natural Philosophy in 1845-6, when the class was
taught, during the illness of Prof. Meikleham, by
David Thomson, who afterwards became Professor of
Natural Philosophy in Aberdeen. In the following year
--the first of William Thomson's professorship--Kerr
appears to have devoted himself specially, no doubt under the influence of the new professor, to the study
of the mathematical theories of magnetism and electricity. In the four years from 1846 to 1849 he received
special prizes for examinations and essays in these subjects, and the "Earl of Eglinton's Prize of Twenty
Sovereigns" as the most distinguished student in mathematics and natural philosophy for the degree of M.A.
In 1849 he graduated with "Highest Distinction in
Physical Science," and his University career came to a
conclusion.

In 1848 and 1849 Prof. Thomson was converting the
old "professional wine-cellar" into a physical laboratory, the first in Great Britain, probably in the world,
and Kerr was one of the earliest members of the experimental corps who did pioneer work in the murky
seclusion of what was known among the students of the
time as the "coal-hole"! He was a divinity student,
and, like some others of his class, did excellent
scientific work; indeed, as Thomson declared in his
Bangor Address in 1885, among the divinity students of
the time were some of the best researchers, and they
became all the better clergymen for having seen something of scientific methods and handled scientific
instruments.

After some time spent in teaching, for he does not
seem to have assumed officially the clerical duties for
which he qualified, Kerr was appointed in 1857 to the
post of Lecturer in Mathematics in the Glasgow Free

Church Training College for Teachers. In this
institution he set up a small laboratory, modestly provided with apparatus, to a great extent at his own ex-
pense, and there he spent such time as he could spare
from his class work in carrying on physical researches,
aided, like Thomson himself, by a few devoted students.
He retained his post for forty-four years, and retired
in 1901 to spend the remaining six years of his life
among his instruments, working out special points connected with his optical researches.

His old students on that occasion showed their
gratitude to him for his teaching, and their respect
for his achievements and the fine qualities of mind and
heart which had endeared him to them as a friend as well
as a master by presenting him with some silver plate.

Kerr wrote one book only; but within its scope and
purpose it was a very good one. It was entitled 'An
Elementary Treatise on Rational Mechanics,' and no
doubt represented to a considerable extent such a view
of the subject as a mature student, who had passed
through the classes of Thomson, would naturally take.
Proofs attributed to Thomson appear in one or two
places and the book contains, like Thomson and Tait's
'Natural Philosophy,' which appeared a year later, a
chapter on "Cinematics." It adheres, however, to the
old division of the subject into statics and dynamics,
and therefore opens with a discussion of statics, in
which the parallelogram of forces is proved by the old
and little satisfying statical considerations. The
dynamical (kinetic) part is based on Newton's Laws of
Motion, a return to which was a feature of the
'Natural Philosophy'; but it does not include the extension of the third law to the modern theory of
energy, which, as Thomson and Tait first pointed out,
is contained in the Principia. It would appear probable, therefore, that this view of the third law had
not formed part of Thomson's teaching at the period
when Kerr was a student. On the while the 'Rational
Mechanics,' if revised and extended in some places
and abridged in others, would form a sound introduction to the study of dynamics for the students of the
present day.

Kerr's name will, however, be always associated
with his two great discoveries: the birefringence
developed in glass and other insulators when placed in
an intense electric field, and the change produced in
polarised light by its reflection from the polished
pole of a magnet. Led by the view already experimentally verified in various ways, that the Faraday lines
of induction in an insulating medium correspond to
some marked physical change in the medium, he set to

work to discover whether this change had any optical effect. Taking a plate of glass about two inches thick, he bored holes from its ends until their extremities were within about a quarter of an inch of meeting. In these he inserted the terminale of the secondary circuit of an induction coil. When the coil was in action, the intervening glass was subjected to intense strain. A beam of light, plane polarised by a Nicol's prism, was sent across the glass, at right angles to the line joining the terminals, then received on the other side by an analysing Nicol, arranged so as to give extinction of the beam when the glass was free from strain. Soon after the coil was excited, the light was seen to pass through the analysing Nicol, and to have become elliptically polarised. The maximum effect was produced when the wave-front of the incident light was arranged, as just described, to be parallel to the Faraday lines in the glass, and to have its plane of polarisation inclined to them at an angle of 45°. Midway between these two positions of the plane of polarisation, that is when that plane was parallel to the lines of force, or at right angles to them, no effect was produced.

Kerr measured the effect by means of a compensator of strained glass and found that the glass appeared to have received, by the action of the electric field, a crystalline structure such as would have been produced by compressing it along the lines of electric force. The effect was not instantaneously produced; it appeared about two seconds after the coil was started and increased for twenty or thirty seconds afterwards. In a piece of amber the opposite effect to that in glass was found to be produced, that is the material behaved as if it were extended along the lines of force. He devised glass cells, which he seems to have constructed to a great extent himself with great labour and much mechanical skill, for the extension of his experiment to liquids. He found that the effect was well marked in carbon disulphide and in paraffin oil, and was similar to that in glass. He obtained results for a large number of organic liquids, of which the mere enumeration would take considerable space.

His papers on this subject are contained in the "Philosophical Magazine" beginning with vol. 50 (1875), in which his first great paper appeared, and continued at intervals for many years.

The view was expressed by Quincke that, for the production of birefringence by electric strain, non-uniformity of the electric field was essential; but Kerr showed that when a uniform field was produced in a stratum of carbon disulphide by means of parallel

plate electrodes of considerable area, the effect was
still found to exist. He proved the same thing also
later for solid substances, and his results were confirmed by Quincke.

Kerr's second discovery was communicated first to
Section A of the British Association, at the Glasgow
Meeting in 1876; and the writer of this notice well re-
members the excitement which the exhibition of the
phenomenon caused among the distinguished physicists
there assembled. It was described in a paper published
in the "Philosophical Magazine" in the following year
(vol. 4, 1877).

The soft iron pole-piece of one core of a horse-
shoe electro-magnet was thrown upon it and reflected
into an analysing prism. Various effects were produced according to the position of the reflecting surface with reference to the direction of magnetisation
and to the plane of incidence of the light.

When the light falls normally on a pole of an
electro-magnet--that is when the polished surface is
perpendicular to the direction of magnetisation, and
the light is along that direction--the plane of
polarisation appears to be turned through a small
angle. In reality, however, the light has become
elliptically polarised in the act of reflection; but
each ellipse has a long axis and a very short one, and
a plane parallel to the long axis may be taken as the
new plane of polarisation of the light regarded as
still plane polarised. This new plane makes only a
small angle with the former position of the plane of
polarisation, and such that the direction of the
apparent turning is opposite to that in which the exciting currents are circulating round the pole. Thus
the apparent direction of turning depends on whether
the reflecting surface is a north or a south pole.

When the light is incident obliquely on the
polished surface (still at right angles to the direction of magnetisation), and has its plane of polarisation either in or at right angles to the plane of
incidence--so that, when the surface is unmagnetised,
there is no effect of reflection at the metallic surface on the position of the plane of polarisation--the
effect of exciting the magnet is to produce again a
small apparent rotation of the plane of polarisation.
In reality the light is elliptically polarised, as in
the former case. The direction of the apparent rotation is again opposite to that of the currents which
produce the magnetisation.

In other experiments, the reflecting surface was
magnetised tangentially. Kerr showed that when, in
this case, the plane of incidence is perpendicular to

the lines of magnetisation, and also when the incidence
is normal, no effect is produced. On the other hand,
when the incidence is oblique, elliptic polarisation,
similar to that already described, results. When the
light is polarised in the plane of incidence, the direction of the apparent turning is opposite to that of the
circulation of the exciting currents; on the other hand,
when the light is polarised at right angles to the plane
of incidence, the apparent rotation is in the same direction as these currents for angles of incidence between 0 and 75°, and changes sign for larger angles of
incidence.

These results have been verified and extended by
various experimenters, especially Righi and Kundt; and
mathematical theories of the effects have been worked
out, first by FitzGerald, and more recently by Larmor in
his general discussion of magneto-optic phenomena.
Kerr's papers are to be found in the 'Phil. Mag.,' loc.
cit. supra, and vol. 5 (1878).

In a paper published in the "Phil. Mag.," vol. 26
(1888), Kerr gave an account of experiments on the
double refraction of light in strained glass, and his
last paper ("Roy. Soc. Proc.," 1894) dealt again with
the subject of electro-optics.

Before Kerr had made his scientific discoveries his
University had recognised his educational work, and the
merit of his book on Rational Mechanics, by creating him
an honorary Doctor of Laws. In 1890, he was elected to
the Royal Society; and he received, in 1898, a Royal
Medal for his scientific researches. He died on August
18, 1907, at his residence in Glasgow.

Most of the apparatus and appliences--induction
coil, glass cells, electromagnet, etc.--with which Kerr
made his discoveries have been acquired by the University
of Glasgow, and are preserved there in the new Institute
of Natural Philosophy.

Dr. Kerr was a shining example of a man engaged in
engrossing routine work, and hampered by narrow resources,
yet devoting himself with splendid success to scientific
research of the highest order. His career shows what can
be achieved by patient, slow, unremitting work from day
to day, when guided by a true physical instinct and a
well trained and well balanced mind. The almost feverish
haste of many may well feel rebuked when it is remembered
that Kerr was 51 years of age ere he ventured to publish
his first patent, and that every detail of his work, when
it was made known stood the test of the scrutiny of the
best scientific investigators of the time. As has been
truly said by another, the name of this quiet and
unostentatious teacher and experimentalist will be linked
for all time with that of Faraday. He would not himself
have desired any better immortality.

INERTIAL EFFECTS IN ROTATIONAL DIFFUSION

Chester T. O'Konski and Lloyd S. Shepard

A particle in a viscous fluid encounters random collisions
with the surrounding molecules, and these will induce rotational
as well as translational diffusion; in addition, when such a
particle attains an angular momentum, the rotational energy will
be transmitted to the colliding molecules, and thus its motion
will be slowed.

Perrin's partial differential equation giving the time dependence of the orientation distribution function is given in
Chap. 3 of this volume (Eq. 19), together with equations of birefringence relaxation which are obtained from its solution. The
applicability of these equations depends upon whether or not the
inertial terms can be ignored for the particular system being
studied.

Einstein [1,2] developed the theory of Brownian motion and
considered the contributions of inertial effects. He indicated
that friction gives rise to a decaying transient in the angular
velocity, and this involves a term containing the inertial relaxation time, μB. In one-dimensional rotational diffusion, $\mu \equiv I$,
and $B \equiv 1/\zeta$, where I is the moment of inertia about the axis
being considered and ζ is the rotational frictional coefficient.
Einstein's conclusion was that the diffusion equation is accurate
when the time interval of observation of the particle under
consideration is large compared to μB. Ornstein [3], Furth [4],

and Uhlenbeck and Ornstein [5] developed formulas for the Brownian motion which explicitly include μB, and which reduce to Einstein's equation for sufficiently large time invervals compared to μB.

Here we will compare the inertial relaxation time (μB) for the rotational case with the diffusional relaxation time and show that for most systems, the time intervals for measurement of the electro-optic and dielectric relaxation times are sufficiently large compared to the inertial relaxation time that inertial effects can be neglected.

INERTIAL EFFECT FOR A SPHERE

It is easily shown that a sphere of radius a and density ρ set into rotation in a viscous fluid of viscosity η, will have its angular velocity $\dot{\theta}$ decrease exponentially with time, t, viz.:

$$\dot{\theta} = \dot{\theta}_0 \exp\left(\frac{-t}{\tau_I}\right) \tag{1}$$

Here τ_I is the inertial relaxation time given by

$$\tau_I = \mu B = \frac{I}{\zeta} = \frac{m}{20\pi\eta r} = \frac{r^2\rho}{15\eta} \tag{2}$$

Here I is the moment of inertia, m is the mass of the sphere, and ρ is its density.

The birefringence relaxation time is given in Chap. 3 as $\tau_n = 1/6\Theta$, where Θ is the rotational diffusion coefficient. Using the Einstein relation ($\Theta = kT/\zeta$) and Stokes law ($\zeta = 8\pi\eta r^3$), where ζ is the frictional coefficient for rotation of the sphere of radius r about an axis through its center, η is the viscosity of the medium, k is Boltzman's constant, and T the temperature. We obtain the following expression for the birefringence relaxation time:

$$\tau_n = \frac{1}{6\Theta} = \frac{\dot{\zeta}}{6kT} = \frac{4\pi\eta r^3}{3kT} \tag{3}$$

Inertial effects clearly are negligible when

$$\tau_n \gg \tau_I \tag{4}$$

For a sphere, this means

$$\frac{4\pi\eta r^3}{3kT} \gg \frac{r^2\rho}{15\eta} \tag{5}$$

Since τ_n increases as the cube of the radius, while τ_I increases as the square, inertial effects are negligible except for very small spheres. For water at 300 K, the above relation becomes

$$r \gg 2.02 \times 10^{-12} \rho \, (g^{-1} cm^4) \tag{6}$$

and indicates the importance of inertial effects only if r is approximately 10^{-12} cm or 10^{-4} Å! Of course the Stokes equation is not valid in that region, but the order of magnitude result demonstrates that inertial terms can be ignored for molecules in a liquid of ordinary viscosity.

INERTIAL EFFECT FOR A ROD

For a rod of length $2a$ and radius b, the birefringence relaxation time about the short axis through its center is:

$$\tau_n = \frac{1}{6\Theta} = \frac{4\pi\eta a^3}{9kT[\ln(2a/b) - \gamma(a/b)]} \tag{7}$$

where the expression for Θ is that given by Broersma [6], and γ is a tabulated quantity of the order of unity, and dependent upon the axial ratio.

The intertial relaxation time for such a rod is

$$\tau_I = \frac{I}{\zeta} = \frac{I\Theta}{kT} = \frac{b^2\rho[\ln(2a/b) - \gamma(a/b)]}{4\eta} \tag{8}$$

Equation (4) leads to

$$\frac{4\pi\eta a^3}{9kT[\ln (2a/b) - \gamma(a/b)]} >> \frac{b^2\rho[\ln (2a/b) - \gamma(a/b)]}{4\eta} \tag{9}$$

For a rod of axial ratio $a/b = 3$, $\gamma = 0.37$. In water at 300 K it follows that

$$a >> 1.66 \times 10^{-11}\rho \, (g^{-1}cm^4) \tag{10}$$

Thus, assuming a reasonable particle density, inertial effects are important only for very short rods $(a \approx 10^{-4} \, \text{Å})$.

We note that for both a sphere and a rod, τ_n is a linear function of η, while τ_I is an inverse function. Thus, the conclusions here would be different for solvents of anomalously low viscosity, such as superfluids.

These calculations assure us that in the measurement of birefringence relaxation times, inertial effects are negligible in all cases except, possibly, for a superfluid solvent.

The above discussion applies only for the case where the particles have no initial angular momentum. Another type of experiment could be devised whereby particles of micron size are given a large angular momentum impulse, followed by observations of their damped rotation. Then additional particle characteristics could be deduced from I/ζ ratios [7].

REFERENCES

1. A. Einstein, Ann. Physik, (4) __19__, 371 (1906).

2. A. Einstein, Investigations on the Theory of the Brownian Movement, Methuen, London (1926).

3. L. S. Ornstein, Proc. Amst., __21__, 96 (1918).

4. R. Furth, Z. Physik., __2__, 244 (1920).

5. G. E. Uhlenbeck and L. S. Ornstein, Phys. Rev., __36__, 823 (1930).

6. S. Broersma, J. Chem. Phys., __32__, 1626 (1960).

7. We thank Prof. S. Goren for a discussion of this possibility.